IMO Problems, Theorems, and Methods

Algebra

Mathematical Olympiad Series

ISSN: 1793-8570

Series Editors: Lee Peng Yee *(Nanyang Technological University, Singapore)*
Xiong Bin *(East China Normal University, China)*

Published

Vol. 25 *IMO Problems, Theorems, and Methods: Algebra*
by Jinhua Chen (East China Normal University, China) &
Bin Xiong (East China Normal University, China)

Vol. 24 *Leningrad Mathematical Olympiads (1961–1991)*
by Dmitri Fomin

Vol. 23 *Solving Problems in Point Geometry:*
Insights and Strategies for Mathematical Olympiad and Competitions
by Jingzhong Zhang (Guangzhou University, China &
Chinese Academy of Sciences, China) &
Xicheng Peng (Central China Normal University, China)

Vol. 22 *Mathematical Olympiad in China (2021–2022):*
Problems and Solutions
editor-in-chief Bin Xiong (East China Normal University, China)

Vol. 21 *Problem Solving Methods and Strategies in High School*
Mathematical Competitions
edited by Bin Xiong (East China Normal University, China) &
Yijie He (East China Normal University, China)

Vol. 20 *Hungarian Mathematical Olympiad (1964–1997):*
Problems and Solutions
by Fusheng Leng (Academia Sinica, China),
Xin Li (Babeltime Inc., USA) &
Huawei Zhu (Shenzhen Middle School, China)

Vol. 19 *Mathematical Olympiad in China (2019–2020):*
Problems and Solutions
edited by Bin Xiong (East China Normal University, China)

Vol. 18 *Mathematical Olympiad in China (2017–2018):*
Problems and Solutions
edited by Bin Xiong (East China Normal University, China)

The complete list of the published volumes in the series can be found at
http://www.worldscientific.com/series/mos

Vol. 25 | Mathematical Olympiad Series

IMO Problems, Theorems, and Methods

Algebra

Authors

Jinhua Chen
Bin Xiong
East China Normal University, China

Proofreader

Jiu Ding
School of Mathematics and Natural Sciences,
University of Southern Mississippi, USA

Copy Editors

Lingzhi Kong, Liyu Zhang, and Ming Ni
East China Normal University Press, China

East China Normal University Press

World Scientific

Published by

East China Normal University Press
3663 North Zhongshan Road
Shanghai 200062
China

and

World Scientific Publishing Co. Pte. Ltd.
5 Toh Tuck Link, Singapore 596224
USA office: 27 Warren Street, Suite 401-402, Hackensack, NJ 07601
UK office: 57 Shelton Street, Covent Garden, London WC2H 9HE

Library of Congress Control Number: 2025004976

British Library Cataloguing-in-Publication Data
A catalogue record for this book is available from the British Library.

Mathematical Olympiad Series — Vol. 25
IMO PROBLEMS, THEOREMS, AND METHODS
Algebra

Copyright © 2026 by East China Normal University Press and
World Scientific Publishing Co. Pte. Ltd.

ISBN 978-981-98-0327-9 (hardcover)
ISBN 978-981-98-0688-1 (paperback)
ISBN 978-981-98-0328-6 (ebook for institutions)
ISBN 978-981-98-0329-3 (ebook for individuals)

For any available supplementary material, please visit
https://www.worldscientific.com/worldscibooks/10.1142/14098#t=suppl

Desk Editors: Nambirajan Karuppiah/Angeline Husni

Typeset by Stallion Press
Email: enquiries@stallionpress.com

Preface

It is generally believed that formal mathematics competitions began with a contest held in Hungary in 1894, an event that gradually garnered attention worldwide. People aptly liken mathematics competitions to "Mental Gymnastics." In 1934, the Soviet Union straightforwardly termed it the "Mathematical Olympiad," a designation that reflects the Olympic spirit of pursuing excellence in intellect more vividly than the previous term, "mathematics competition."

By 1959, the internationalization of mathematics competitions had matured, leading to the inception of the "International Mathematical Olympiad" (IMO). The first IMO was held in Brasov, Romania in 1959. As of 2023, the IMO has successfully been held 64 times, except for 1980 when it was not conducted.

The IMO is typically held in July each year, and the format has become standardized: the official competition spans two days, with contestants tackling three problems in 4.5 hours each day, each problem worth a maximum of 7 points, totaling 42 points. Each participating team consists of six contestants, accompanied by a Leader and a Deputy Leader. Approximately half of the contestants receive medals, with about $1/12$ of the contestants earning gold medals, $2/12$ receiving silver medals, and $3/12$ obtaining bronze medals.

The IMO is currently one of the most influential secondary school mathematics competitions worldwide. In recent years, over 100 countries and regions have participated in this event, including all major nations globally.

Problems for the IMO are submitted by the participating teams and then reviewed and selected by a problem selection committee organized by the host country. This committee narrows down the submissions to approximately 30 Shortlist problems, covering algebra, geometry, number theory, and combinatorics, with about seven to eight problems on each topic. These are then presented to the Jury Meeting, composed of team leaders, who discuss and vote to decide on the six problems that will constitute the official competition paper. The host country does not provide any problems.

This event has played a significant role in promoting the exchange of mathematical education among nations, enhancing the level of mathematical education, facilitating mutual learning and understanding among young students worldwide, stimulating a broad interest in mathematics among secondary school students, and identifying and nurturing mathematically gifted students.

The development over more than 60 years is the result of the collective efforts of mathematicians, organizers, and contestants, and is worthy of reflection and study. Particularly deserving of study are the evolution of the competition problems, the mathematical ideas, and methods involved. Indeed, several colleagues from the International Mathematical Olympiad Research Center at East China Normal University had envisioned research and publication before the 60th IMO. For this purpose, we initiated several seminars involving over ten people. For special reasons, this work was delayed. Based on the mathematical domains covered by the IMO problems — algebra, geometry, number theory, and combinatorics — we planned to compile the work into four volumes, with the general title *IMO Problems, Theorems, and Methods*, to be included in the "IMO Study Series."

Each volume begins with an introduction that provides an overview of the IMO. Subsequent chapters introduce relevant foundational knowledge and methods, followed by a reclassification and organization of past IMO problems. For some problems, multiple solutions are provided, along with a difficulty analysis. It is worth noting that some problems do not fit neatly into a single topic, as they may involve both algebra and number theory, or algebra and combinatorics. We primarily categorize them based on the topic under which they were placed on the Shortlist.

The four volumes titled *IMO Problems, Theorems, and Methods* were conceived with an overall writing plan proposed by myself, with the authors collectively discussing and refining the plan. The majority of the initial drafts were completed by Jinhua Chen (Algebra), Tianqi Lin (Geometry), Gengyu Zhang (Number Theory), and Guangyu Xu (Combinatorics). The first three volumes were supplemented, consolidated, and finalized by myself, while the combinatorics volume was supplemented, consolidated, and finalized by Zhenhua Qu.

We extend our gratitude to the leaders and contestants of the Chinese IMO teams over the years, as some elegant solutions included in the book were contributed by them. During the compilation of this book, we consulted various domestic and international sources, which are too numerous to acknowledge individually here.

While the authors have diligently studied the IMO problems and provided thoughtful strategies and solutions, errors and inaccuracies may occur due to our limitations. We sincerely invite readers to offer corrections and feedback.

The translation of the algebra volume in this series was done by Jinhua Chen and Bin Xiong; the geometry volume was translated by Xinyuan Yang; the number theory volume was translated by Gengyu Zhang; and the combinatorics volume was translated by Zhenhua Qu and Jinhua Chen. Jiu Ding revised the translations of all four books.

Bin Xiong
June 2024

About the Authors

Jinhua Chen is a doctoral candidate in Mathematical Education at East China Normal University, with research interests in mathematics competitions and gifted education. He serves as an external mentor for mathematics competition courses at several high schools. Chen attended the Affiliated High School of South China Normal University and Guangdong Olympic School during his secondary education. He has excelled in various competitions, including the Chinese High School Mathematics League, the American Mathematics Competition 12, the Chinese College Mathematics Competition, and the COMAP Mathematical/Interdisciplinary Contest in Modeling.

Bin Xiong is a professor and doctoral supervisor at the School of Mathematical Sciences, East China Normal University. He also serves as the director of the Shanghai Key Laboratory of Core Mathematics and Practice, and the International Mathematical Olympiad Research Center. Professor Xiong is an expert with the State Council Special Allowance, and has been honored with the Shanghai May 1 Labor Medal as well as the Shanghai Model of Teaching and Educating. He has published over 100 scholarly papers in renowned national and international journals and has authored or co-authored more than 150 books. Additionally, Professor Xiong has served as the leader and head coach of the Chinese IMO team more than 10 times, and received the prestigious Paul Erdös Award in 2018 for his contribution to the development of mathematics competitions at the national and international level.

Contents

Introduction to the IMO

The International Mathematical Olympiad (IMO), established in the year 1959, represents one of the foremost intellectual endeavors at the highest tier for youth on a global scale. Prior to 1959, numerous countries around the world had already initiated the organization of mathematics competitions, thereby laying the groundwork for the inception of the IMO.

In 1891, the renowned physicist and President of the Hungarian Academy of Sciences, Loránd Eötvös (also known as Roland Eötvös), founded the Hungarian Mathematical and Physical Society. In 1894, he assumed the position of Minister of Education, and under his enthusiastic support, the Hungarian Mathematical and Physical Society initiated secondary school mathematics competitions. This competition, also known as the Eötvös Competition, offered winners the Eötvös Prize and the opportunity to pursue higher education. Subsequently, the Eötvös Competition was not held during 1919–1921 and 1944–1946 due to the world political events. In 1947, under the leadership of János Surányi, the Eötvös Competition was reinstated and renamed the Kürschák Competition (named after József Kürschák). This competition has played a significant role in Hungary in nurturing numerous mathematicians and scientists, including Győző Zemplén, Lipót Fejér, Theodore von Kármán, Alfréd Haar, Dénes Kőnig, Marcel Riesz, Gábor Szegő, Tibor Radó, Edward Teller, and Tibor Szele. Interestingly, George Pólya also participated in the competition, but did not hand in his paper.

With the aim of identifying and nurturing mathematical talents, prominent mathematicians such as Boris Delaunay (also known as Delone), Grigorii Fikhtengol'ts, Dmitry Faddeev, and others organized the inaugural Leningrad Mathematical Olympiad (LMO) in 1934, under the initiative of

1

Boris Delaunay. Winners of this competition were granted the privilege of direct admission to the Mathematics Department of Leningrad State University without the need for entrance examinations. Following the example set by the LMO, in 1935, renowned mathematicians Pavel Aleksandrov and Andrey Kolmogorov, alongside the entire faculty of the Mathematics Department at Moscow State University, organized the first Moscow Mathematical Olympiad (MMO).

Subsequently, various regions throughout the Soviet Union started hosting their own Mathematical Olympiads, ultimately laying the foundation for the All-Russian Mathematical Olympiad, which was first conducted in 1961. In 1967, the responsibility for organizing All-Russian Mathematical Olympiad was assumed by the Ministry of Education of the Soviet Union, leading to a renaming of the All-Russian Mathematical Olympiad as the All-Soviet-Union Mathematical Olympiad.

In fact, almost all of the best mathematicians born in the Soviet Union after 1930 had participated in Mathematical Olympiads, usually achieving first prizes. This distinguished group includes Fields Medal awardees such as Sergei Novikov, Grigory Margulis, Vladimir Drinfeld, Maxim Kontsevich, Grigori Perelman, and Stanislav Smirnov. Although having claimed that he was never particularly interested in Mathematical Olympiads, Sergei Novikov did secure a second prize in the MMO when he was in eighth grade.

While the United States of America Mathematical Olympiad (USAMO) was first held in 1972, the United States had a longstanding tradition of organizing mathematics competitions prior to that. In 1921, William Lowell Putnam published an article in the *Harvard Graduates' Magazine*, proposing the idea of conducting a university-level mathematics team competition. Following his passing, the Putnam family established the William Lowell Putnam Intercollegiate Memorial Fund to support the organization of the William Lowell Putnam Mathematical Competition (Putnam Competition), administered by the Mathematical Association of America.

With the assistance of George David Birkhoff, the first Putnam Competition took place in 1938, and it has been held annually since then; the top five ranking participants are designated as Putnam Fellows. Due to wartime conditions, the competition was not held from 1943 to 1945. In 1946, George Pólya, Tibor Radó, and Irving Kaplansky (Putnam Fellow in 1938) formed the Putnam Competition Committee, thus reestablishing the competition, but the responsibility for administration was undertaken by

Garrett Birkhoff, the son of George David Birkhoff, and his colleagues in the Harvard University Department of Mathematics.

Many participants in the Putnam Competition have gone on to become prominent mathematicians and scientists. John Milnor, David Mumford, Daniel Quillen, Paul Cohen, John G. Thompson, and Manjul Bhargava have been recipients of Fields Medal. Richard Feynman, Kenneth Geddes Wilson, Steven Weinberg, and Murray Gell-Mann have received Nobel Prize in Physics, while John Nash was awarded the Nobel Prize in Economic Sciences. Additionally, numerous Putnam Fellows have been elected as members of the National Academy of Sciences in the United States.

Building upon the foundation of existing mathematics competitions in many countries, particularly in Eastern European nations, Romania proposed in 1956 the organization of an international mathematics competition involving seven Eastern European countries. This proposal led to the inaugural IMO held in 1959.

The first IMO was held in Braşov, Romania, in 1959. As of 2023, the IMO has been successfully held 64 times, with the exception of the year 1980 when it was not held for specific reasons. Apart from the 61st IMO, which was postponed to September in 2020 due to the impact of the COVID-19 pandemic, the IMO typically takes place in July each year.

The IMO has emerged as the most influential secondary school mathematics competition at present. In recent years, the number of countries and regions participating in this event has exceeded 100.

1 Evolution of the IMO

The first IMO, held in 1959, saw the participation of only 52 contestants from seven countries, including Romania, Hungary, Czechoslovakia, Bulgaria, Poland, the Soviet Union, and the German Democratic Republic. Subsequently, new countries and regions gradually joined this prestigious event. By the 20th IMO, also hosted by Romania in 1978, approximately 20 countries and regions participated (with 21 in the 19th, 17 in the 20th, and 23 in the 21st). The number of participating contestants also reached 132. The historical participation trends are depicted in Figure 1.

As the influence of the IMO continued to expand, the number of participating countries and regions, as well as the number of contestants, grew rapidly. The most significant increase in the number of participating countries and regions occurred in the 34th IMO, which was held in Turkey in 1993. In comparison to the 33rd IMO held in Russia in 1992, the number

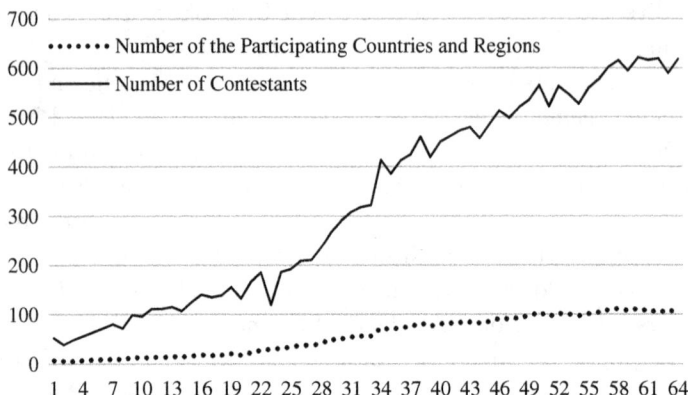

**Figure 1 Numbers of Participating Countries and Regions, as Well as the
Contestants, in the First 64 IMOs**

increased by 17 countries and regions, reaching a total of 73, with 413 contestants. By the 40th IMO, which was still hosted by Romania in 1999, the number of participating countries and regions had reached 81, with 450 contestants.

The first instance of the number of participating countries and regions surpassing one hundred occurred during the 50th IMO, held in Germany in 2009, with a total of 565 contestants. Among the first 64 IMOs, the biggest number of participating countries and regions, as well as the largest number of contestants, was observed in the 60th IMO, hosted by the United Kingdom in 2019, where 621 contestants from 112 countries and regions took part. In the 64th IMO held in Japan in 2023, there were 618 contestants from 112 countries and regions.

As evident from Appendix A, the IMO is primarily hosted by European countries. Moreover, as the number of participating countries and regions in the IMO has increased, it is no longer confined to the seven founding member countries, and many new participating countries and regions have also begun to organize the IMO.

2 Problems in the IMO

The IMO is scheduled to take place annually in July. Each participating country or region officially sends a delegation consisting of six contestants, along with one team leader and one deputy leader. The official competition spans two days, with each day featuring three problems to be solved within a four-and-a-half-hour timeframe. Each problem carries a maximum score of 7 points, resulting in a total maximum score of 42 points, while the total maximum score of the team is 252 points.

In early IMOs, the number of problems and their individual point values varied from one session to another. For instance, the 2nd and 4th IMOs featured seven problems, while all other IMOs had six problems each. Additionally, in the 13th IMO, although the total score remained at 42 points, the six problems were allocated point values of 5, 7, 9, 6, 7, and 8, respectively. It was only from the 22nd IMO, held in the United States in 1981, that the IMO problems have become standardized, with each problem carrying 7 points and a total of six problems.

The number of contestants in each delegation has also become stable at six individuals starting from the 24th IMO held in France in 1983.

2.1 *The number of problems*

The mathematical domains covered by IMO problems encompass four major topics: algebra, combinatorics, geometry, and number theory. These are also the primary focus in various national mathematics competitions.

Across the 1st–64th IMOs, a total of 386 problems have been featured. Among them, geometry problems are the most numerous, with 123 problems, while number theory problems are the least, with 75 problems. Algebra problems account for approximately one-quarter of the total, comprising 101 problems. Furthermore, as indicated in Table 1, the quantity of algebra problems has remained relatively stable over each span of 10 IMOs.

Table 1 Numbers of Problems with Different Topics in the First 64 IMOs

Session	Topic			
	Algebra	Combinatorics	Geometry	Number Theory
1–10	20	6	29	7
11–20	20	12	18	10
21–30	14	16	18	12
31–40	13	16	15	16
41–50	15	11	20	14
51–60	13	19	18	11
61–64	6	7	6	5
Total	101	87	123	75

Remarkably, in the first 64 IMOs, there were five sessions when three algebra problems were presented, specifically in the 3rd, 5th, 7th, 15th, and 18th IMOs. In 29 IMOs, two algebra problems were featured, while in 28 IMOs, only one algebra problem was included. In two IMOs, algebra problem was absent, namely the 30th and 39th IMOs.

2.2 The difficulty level of problems

Typically, the difficulty of a problem is correlated with its problem number in the IMO.

Starting from the 24th IMO, the point value of each problem and the number of contestants per team have become standardized. Therefore, an analysis of the average scores of the 246 problems in the 24th–64th IMOs is presented in Table 2. It can be observed that the first and fourth problems in each IMO are relatively easy, with average scores generally exceeding 3 points. The second and fifth problems are relatively challenging, with average scores mainly ranging from 1 to 4 points. The third and sixth problems are exceptionally difficult, with average scores generally falling below 2 points.

Table 2 Numbers of Problems with Different Average Scores in the 24th–64th IMOs

Problem Number	Problem Mean				
	0–1	1–2	2–3	3–4	4–7
Problem 1	0	0	5	13	23
Problem 2	2	6	18	9	6
Problem 3	22	12	4	3	0
Problem 4	0	0	7	17	17
Problem 5	4	10	14	10	3
Problem 6	24	11	6	0	0
Total	52	39	54	52	49

The 246 problems are categorized into four topics: algebra, combinatorics, geometry, and number theory, as shown in Table 3. Notably, there is a relatively large representation of combinatorics and geometry problems. Combining this information with Table 1, it is evident that in the first 23 IMOs, algebra and geometry problems were predominant.

Furthermore, in early IMOs, geometry problems predominantly appeared in the 1st/4th and 2nd/5th positions. However, starting from

Table 3 Numbers of Problems with Different Topics in the 24th–64th IMOs

	Topic											
	Algebra			Combinatorics			Geometry			Number Theory		
Session	1, 4	2, 5	3, 6	1, 4	2, 5	3, 6	1, 4	2, 5	3, 6	1, 4	2, 5	3, 6
24–30	4	1	5	4	3	5	5	7	0	1	3	4
31–40	4	5	4	5	4	7	6	9	0	5	2	9
41–50	3	8	4	3	2	6	10	6	5	5	4	5
51–60	5	6	2	4	7	8	7	3	6	3	4	4
61–64	1	3	2	2	2	3	3	1	2	2	2	1
Total	17	23	17	18	18	29	31	26	13	16	15	23
		57			65			70			54	

the 41st to 50th IMOs, geometry problems were more commonly found in the 1st/4th and 3rd/6th positions. Similarly, algebra problems were more frequent in the 1st/4th and 2nd/5th positions, combinatorics problems were more prevalent in the 2nd/5th and 3rd/6th positions, while the quantity of number theory problems across different problem numbers does not differ significantly.

From Table 4, it can be observed that among the four topics, the numbers of problems with an average score ranging from 2 to 4 points are quite similar. However, in the combinatorics topic, there is a higher quantity of challenging problems, with 31 problems having an average score between 0 and 2 points. Conversely, the geometry topic has the largest number of relatively easy problems, with 23 problems scoring above 4 points. This discrepancy is largely due to the fact that there are 29 combinatorics problems in the 3rd/6th positions, and 31 geometry problems in the 1st/4th positions.

Table 4 Numbers of Problems with Different Average Scores by Topics in the 24th–64th IMOs

	Problem Mean					
Topic	0–1	1–2	2–3	3–4	4–7	Total
Algebra	7	14	19	6	11	57
Combinatorics	20	11	12	14	8	65
Geometry	13	5	14	15	23	70
Number theory	12	9	9	17	7	54
Total	52	39	54	52	49	246

Furthermore, when considering Table 3 and Table 4, it becomes apparent that among the four topics, the number of problems with an average score ranging from 0 to 2 points closely aligns with the number of problems in the 3rd/6th positions. There are slightly more problems with an average score between 2 to 4 points compared to those in the 2nd/5th positions, and slightly fewer problems with an average score exceeding 4 points compared to those in the 1st/4th positions. This indicates that even the seemingly easier problems in the IMO are not as straightforward as they might appear.

Notably, among these 246 problems, the lowest average score is attributed to IMO 58-3 (Combinatorics, proposed by Austria):

A hunter and an invisible rabbit play a game in a Euclidean plane. The rabbit's starting point, A_0, and the hunter's starting point, B_0, are the same. After $n - 1$ rounds of the game, the rabbit is at point A_{n-1} and the hunter is at point B_{n-1}. In the nth round of the game, three things occur in order.

(i) The rabbit moves invisibly to a point A_n such that the distance between A_{n-1} and A_n is exactly 1.

(ii) A tracking device reports a point P_n to the hunter. The only guarantee provided by the tracking device to the hunter is that the distance between P_n and A_n is at most 1.

(iii) The hunter moves visibly to a point B_n such that the distance between B_{n-1} and B_n is exactly 1.

Is it always possible, no matter how the rabbit moves, and no matter what points are reported by the tracking device, for the hunter to choose her moves so that after 10^9 rounds she can ensure that the distance between her and the rabbit is at most 100?

This unconventional problem received an average score of only 0.042 points. Only two contestants, Mikhail Ivanov from Russia and Linus Cooper from Australia, achieved a perfect score of 7 points. Joe Benton from the United Kingdom scored 5 points, Pavel Hudec from Czech Republic earned 4 points, Hadyn Ka Ming Tang from Australia, Yahor Dubovik from Belarus, and Jeonghyun Ahn from South Korea each scored 1 point.

Furthermore, among the 20 lowest-scoring problems in the 24th–64th IMOs, nearly all of them appeared in the 3rd/6th positions, as indicated in Table 5. There were three algebra problems, eight combinatorics problems, six geometry problems, and three number theory problems among them.

Table 5 The 20 Problems with the Lowest Average Scores in the 24th–64th IMOs

Problem	Mean	Topic	Problem	Mean	Topic
IMO 58-3	0.042	Combinatorics	IMO 54-6	0.296	Combinatorics
IMO 48-6	0.152	Algebra	IMO 48-3	0.304	Combinatorics
IMO 50-6	0.168	Combinatorics	IMO 52-6	0.318	Geometry
IMO 47-6	0.187	Geometry	IMO 53-6	0.336	Number Theory
IMO 57-3	0.251	Number Theory	IMO 55-6	0.339	Combinatorics
IMO 49-6	0.260	Geometry	IMO 56-6	0.355	Combinatorics
IMO 64-6	0.275	Geometry	IMO 51-6	0.368	Algebra
IMO 59-3	0.278	Combinatorics	IMO 62-3	0.372	Geometry
IMO 61-6	0.282	Combinatorics	IMO 62-2	0.375	Algebra
IMO 58-6	0.294	Number Theory	IMO 60-6	0.403	Geometry

These three algebra problems were:

- **(IMO 48-6, proposed by the Netherland).** Let n be a positive integer. Consider

$$S = \{(x,y,z) : x,y,z \in \{0,1,\ldots,n\}, x+y+z > 0\}$$

as a set of $(n+1)^3 - 1$ points in a three-dimensional space. Determine the smallest possible number of planes, the union of which contains S but does not include $(0,0,0)$.

- **(IMO 51-6, proposed by Iran).** Let $a_1, a_2, a_3, \ldots$ be a sequence of positive real numbers. Suppose that for some positive integer s,

$$a_n = \max\{a_k + a_{n-k} | 1 \le k \le n-1\}$$

for all $n > s$. Prove that there exist positive integers l and N with $l \le s$, such that $a_n = a_l + a_{n-l}$ for all $n \ge N$.

- **(IMO 62-2, proposed by Canada).** Show that the inequality

$$\sum_{i=1}^{n}\sum_{j=1}^{n}\sqrt{|x_i - x_j|} \le \sum_{i=1}^{n}\sum_{j=1}^{n}\sqrt{|x_i + x_j|}$$

holds for all real numbers $x_1, x_2, \ldots, x_n$.

2.3 *The classification of problems*

In the 1st–64th IMOs, there were 101 algebra problems, which can be categorized into five specialized subjects: equation problems, function problems, sequence problems, inequality problems, and other algebra problems.

There were 87 combinatorics problems, which can be categorized into six specialized subjects: enumerative combinatorics problems, existence problems, extremal combinatorial problems, operation and logical reasoning problems, combinatorial geometry problems, and graph theory problems.

There were 123 geometry problems, which can be categorized into seven specialized subjects: similarity and congruence problems, circle problems, power-of-point problems, special point and line problems, trigonometry problems, solid geometry problems, and geometric inequality problems.

There were 75 number theory problems, which can be categorized into three specialized subjects: divisibility-of-integer problems, modular arithmetic problems, and indeterminate equation problems.

For the 101 algebra problems categorized and analyzed, as shown in Table 6, it can be observed that the subjects of functions and inequalities had the largest number of problems, with 28 and 33 problems, respectively.

In the first 10 IMOs, algebra problems primarily focused on equations and inequalities. Subsequently, the number of equation problems decreased rapidly, while problems related to functions and sequences gradually increased and maintained a certain frequency.

In the 41st–50th IMOs, algebra problems were primarily centered around inequality problems, whereas in the 51st–60th IMOs, algebra problems primarily focused on function problems.

Table 6   Numbers of Algebra Problems in the First 64 IMOs

Session	Subject				
	Equation	Function	Sequence	Inequality	Others
1–10	11	1	1	5	2
11–20	3	5	3	7	2
21–30	1	6	2	4	1
31–40	0	5	3	4	1
41–50	0	3	1	8	3
51–60	1	7	3	2	0
61–64	0	1	1	3	1
Total	16	28	14	33	10

In the first 64 IMOs, there had been a total of 16 equation problems, accounting for approximately 15.8% of all algebra problems. These problems can be primarily categorized into three types: (1) finding solutions of equations and systems of equations, totaling seven problems; (2) proving relationships satisfied by equations and systems of equations, totaling four problems; (3) investigating conditions under which equations and systems of equations have solutions, totaling five problems.

As shown in Table 7, in the 24th–64th IMOs, there was only one equation problem, which is IMO 57-5 (proposed by Russia). This indicates that equation problems mainly appeared in the first 23 IMOs.

Table 7 Numbers of Equation Problems in the 24th–64th IMOs

Equation Problem	Problem Number			Number of Problems in the First 64 IMOs
	1, 4	2, 5	3, 6	
Finding solutions	0	0	0	7
Proving relationships	0	0	0	4
Investigating conditions	0	1	0	5
Total	0	1	0	16

In the first 64 IMOs, there had been a total of 28 function problems, approximately accounting for 27.7% of all algebra problems. These problems can be primarily categorized into three types: (1) proving properties of functions, totaling nine problems; (2) determining numerical values of function variables or outputs, totaling four problems; (3) deriving expressions for functions that meet specific conditions, totaling 15 problems.

As shown in Table 8, in the 24th–64th IMOs, there were a total of 20 function problems. These problems were predominantly present in the 1st/4th positions as well as in the 2nd/5th positions, with the main focus

Table 8 Numbers of Function Problems in the 24th–64th IMOs

Function Problem	Problem Number			Number of Problems in the First 64 IMOs
	1, 4	2, 5	3, 6	
Proving properties	1	2	1	9
Determining values	0	0	1	4
Deriving expressions	6	8	1	15
Total	7	10	3	28

being on deriving expressions for functions that meet specific conditions. The other two types of function problems appeared less frequently in the last 40 IMOs.

In the first 64 IMOs, there had been a total of 14 sequence problems, approximately accounting for 13.8% of all algebra problems. These problems can be primarily categorized into three types: (1) determining the value of a specific term or the number of terms, totaling four problems; (2) addressing problems related to the existence of sequences, totaling five problems; (3) proving quantitative relationships satisfied by sequences, totaling five problems.

As shown in Table 9, in the 24th–64th IMOs, there were a total of nine sequence problems. These problems were predominantly found in the 3rd/6th positions, with a relatively balanced distribution among different problem types.

Table 9 Numbers of Sequence Problems in the 24th–64th IMOs

Sequence Problem	Problem Number			Number of Problems in the First 64 IMOs
	1, 4	2, 5	3, 6	
Determining values	2	1	0	4
Existence problems	0	0	2	5
Proving quantitative relationships	0	0	4	5
Total	2	1	6	14

In the first 64 IMOs, there had been a total of 33 inequality problems, approximately accounting for 32.7% of all algebra problems. These problems can be primarily categorized into three types: (1) solving inequalities, totaling five problems; (2) proving inequalities, totaling 25 problems; (3) determining value ranges, totaling three problems.

As shown in Table 10, in the 24th–64th IMOs, there were a total of 21 inequality problems. These problems were predominantly found in the 1st/4th positions as well as in the 2nd/5th positions. The primary focus of these problems is to prove inequalities, with the other two types of inequality problems appearing less frequently in the last 40 IMOs.

In the first 64 IMOs, there had been a total of 10 other algebra problems, approximately accounting for 9.9% of all algebra problems. These problems can be primarily categorized into three types: (1) proving trigonometric identities, totaling two problems; (2) finding polynomials, totaling

Table 10 Numbers of Inequality Problems in the 24th–64th IMOs

Inequality Problem	Problem Number			Number of Problems in the First 64 IMOs
	1, 4	2, 5	3, 6	
Solving inequalities	1	0	0	5
Proving inequalities	6	8	4	25
Determining value ranges	0	1	1	3
Total	7	9	5	33

two problems; (3) proving properties of polynomials and sets, totaling six problems.

As shown in Table 11, in the 24th–64th IMOs, there were a total of six other algebra problems, primarily focusing on proving properties of polynomials and sets.

Table 11 Numbers of Other Algebra Problems in the 24th–64th IMOs

Others	Problem Number			Number of Problems in the First 64 IMOs
	1, 4	2, 5	3, 6	
Proving trigonometric identities	0	0	0	2
Finding polynomials	0	1	0	2
Proving properties	1	1	3	6
Total	1	2	3	10

As shown in Table 12, among the algebra problems in the 24th–64th IMOs, the largest proportion of problems is attributed to function and inequality. However, the average scores of function problems are primarily concentrated between 2 and 7 points, while for inequality problems, the average scores are primarily concentrated between 0 and 3 points. This suggests that inequality problems tend to be more challenging, while function problems include a greater number of relatively simple problems.

Overall, the numbers of algebra problems in the 1st/4th, 2nd/5th, and 3rd/6th positions are roughly equal to the numbers of algebra problems with average scores of 4–7, 2–4, and 0–2, respectively. However, upon a further comparison of the distribution of problem numbers among the five subjects, it is evident that inequality problems tend to be more challenging. There are five inequality problems in the 2nd/5th positions, with average

Table 12 Numbers of Algebra Problems with Different Average Scores in the 24th–64th IMOs

Algebra Problem	Problem Mean					Total
	0–1	1–2	2–3	3–4	4–7	
Equation Problems	0	1	0	0	0	1
Function problems	0	4	6	5	5	20
Sequence problems	1	3	3	0	2	9
Inequality problems	3	5	8	1	4	21
Others	3	1	2	0	0	6
Total	7	14	19	6	11	57
		21		25	11	

scores ranging from 0 to 2, and three inequality problems in the 1st/4th positions, with average scores ranging from 2 to 4.

Similarly, there are four function problems in the 1st/4th positions, with average scores ranging from 2 to 4, and one function problem in the 2nd/5th positions, with an average score of 1–2. This indicates that even the easier function problems are not necessarily easy.

2.4 *The proposal for problems*

Problems in the IMO are proposed by the participating countries and regions, except the host. Usually the team leaders are in charge of submitting problems with a limit of six, and these problems are then subjected to the selection by a selection committee composed of experts organized by the host. Approximately 30 problems are chosen as shortlist problems, with around eight problems in each of the four topics: algebra, geometry, combinatorics, and number theory. Subsequently, these problems are submitted to the Jury, which is comprised of team leaders from each participating country or region. The problems are discussed and voted upon to select the official examination problems. Once the problems are finalized, they are translated into five working languages: English, French, German, Russian, and Spanish. Each leader then translates the problems into their respective national languages, and contestants can choose from two languages in which to answer the problems.

Among the first 64 IMOs, algebra problems were contributed by 35 different countries and regions. The United Kingdom had the highest number of problems proposed, with a total of 10. The Netherlands proposed eight problems, followed by Hungary with six problems. Poland, Bulgaria, the Soviet Union, France, Sweden, and South Korea each proposed five problems, while Ireland contributed four problems. These 10 countries collectively provided 58 problems. Remarkably, only Sweden (in the 15th IMO in 1973) and South Korea (in the 46th IMO in 2005) are the two countries that have proposed two algebra problems in the same IMO session.

Furthermore, as indicated in Appendix B, algebra problems in the first 15 IMOs were primarily proposed by the seven founding countries of the IMO. From the 16th–35th IMOs, algebra problems were predominantly contributed by the United Kingdom, with supplementary contributions from other countries and regions. However, in the 36th–64th IMOs, algebra problems demonstrated a more diverse range of proposing countries and regions. This, to some extent, correlates with the expansion of the IMO's influence and the growth in the number of participating countries and regions.

3 Awards in the IMO

In addition to selecting problems, the Jury has several other responsibilities, including: establishing grading criteria, resolving discrepancies in grading between leaders and coordinators, and determining the number of gold, silver, and bronze medals, as well as the score thresholds. In each IMO, approximately 1/12 of the contestants receive a gold medal, 2/12 receive silver, and 3/12 receive bronze.

Apart from the gold, silver, and bronze medals, contestants who do not receive medals but attain a score of 7 on at least one problem in the IMO will receive an Honorable Mention. Contestants who deliver exceptionally elegant solutions to specific problems in the IMO will receive a Special Prize.

As depicted in Figure 2, starting from the 24th IMO, the cutoff scores for gold, silver, and bronze medals have gradually stabilized. The gold medal cutoff is approximately 29 points, the silver medal cutoff is around 22 points, and the bronze medal cutoff is roughly 15 points. Furthermore, the average score of all contestants closely aligns with the bronze medal cutoff. This indicates that the problem difficulty is well-balanced.

Interestingly, in the first 64 IMOs, there were three occasions where the gold medal cutoff was a perfect score, meaning only those who scored full marks could earn a gold medal. These three IMOs were: the 11th IMO (1969, Romania) with a perfect score of 40 points and three gold medalists;

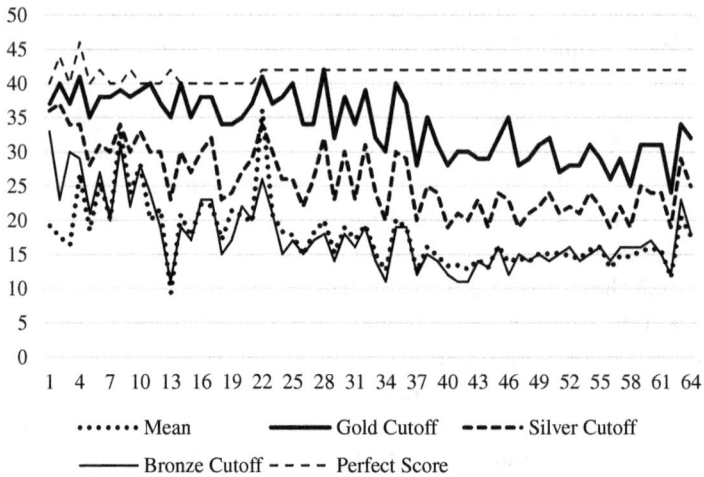

●●●●●● Mean ——— Gold Cutoff — — —· Silver Cutoff

——— Bronze Cutoff — — — — Perfect Score

Figure 2 Medal Cutoff Scores in the First 64 IMOs

the 14th IMO (1972, Poland) with a perfect score of 40 points and eight gold medalists; and the 28th IMO (1988, Cuba) with a perfect score of 42 points and 22 gold medalists.

3.1 *Participation*

In the first 64 IMOs, a total of 269 contestants took part in four or more IMOs. Among them, two contestants attended seven IMOs, four contestants attended six IMOs, 42 contestants attended five IMOs, and 221 contestants attended four IMOs, as shown in Table 13.

Table 13 Contestants with Six or More Participations in the First 64 IMOs

Contestant	Country	Participation Year	Gold	Silver	Bronze	Honorable Mention	Perfect Score
David Kunszenti-Kovács	Norway	1997–2003	1	3	1	1	
Yeoh Zi Song	Malaysia	2014–2020	1	1	4	1	
Zhuo Qun (Alex) Song	Canada	2010–2015	5	0	1	0	1
Teodor von Burg	Serbia	2007–2012	4	1	1	0	
Alexey Entin	Israel	2000–2005	1	3	1	0	
Tan Li Xuan	Malaysia	2016–2021	0	2	2	1	

Coincidentally, in the 43rd IMO held in 2002, the gold medal cutoff was set at 29 points, and David Kunszenti-Kovács achieved a total score of exactly 29 points.

In the 44th IMO held in 2003, the gold medal cutoff was set at 29 points, whereas Alexey Entin attained a total score of exactly 28 points.

In the 51st IMO held in 2010, the bronze medal cutoff was set at 15 points, and Zhuo Qun (Alex) Song achieved a total score of exactly 15 points.

In the 58th IMO held in 2017, the silver medal cutoff was set at 19 points, and Yeoh Zi Song achieved a total score of exactly 19 points. In the 59th IMO held in 2018, the silver medal cutoff was set at 25 points, and Yeoh Zi Song's total score was exactly 24 points. In the 61st IMO held in 2020, the gold medal cutoff was set at 31 points, and Yeoh Zi Song's total score was exactly 31 points.

Additionally, from 2002 to 2005 and in 2007, Sherry Gong participated in the IMO, earning one bronze, two silver, and one gold medal. In the 48th IMO held in 2007, she ranked 7th individually. Notably, from 2002 to 2004, she was a member of the Puerto Rico IMO team, while in 2005 and 2007, she was a member of the United States IMO team.

From 2020 to 2023, Alex Chui participated in the IMO, securing two gold and two silver medals. However, in 2020 and 2021, he was a member of the Chinese Hong Kong IMO team, while in 2022 and 2023, he was a member of the United Kingdom IMO team.

Other than Sherry Gong and Alex Chui, the remaining 267 contestants hailed from 75 different countries and regions. Among them, there were 12 contestants from Cyprus and Moldova each, 11 from Malaysia, eight from Trinidad and Tobago, seven from each of Estonia, Germany, Sri Lanka, and North Macedonia, and six from Japan and Philippines each.

3.2 *Gold medals*

In the first 64 IMOs, a total of 49 contestants achieved three or more gold medals. Among them, one contestant earned five gold medals, six contestants earned four gold medals, and 42 contestants earned three gold medals.

As shown in Table 14, Zhuo Qun (Alex) Song, Reid Barton, and Lisa Sauermann have all achieved perfect scores.

Coincidentally, in the 54th IMO held in 2013, the gold medal cutoff was 31 points, and Nipun Pitimanaaree achieved a total score of exactly 31 points.

Table 14 Contestants with Four or More Gold Medals in the First 64 IMOs

Contestant	Country	Participation Year	Gold Year	Perfect Score Year
Zhuo Qun (Alex) Song	Canada	2010–2015	2011–2015	2015
Reid Barton	The United States of America	1998–2001	1998–2001	2001
Christian Reiher	Germany	1999–2003	2000–2003	
Lisa Sauermann	Germany	2007–2011	2008–2011	2011
Teodor von Burg	Serbia	2007–2012	2009–2012	
Nipun Pitimanaaree	Thailand	2009–2013	2010–2013	
Luke Robitaille	The United States of America	2019–2022	2019–2022	

In the 41st IMO held in 2000, the top four contestants all achieved perfect scores, and Reid Barton ranked fifth with a total score of 39. In the 43rd IMO held in 2002, the top three contestants all achieved perfect scores, and Christian Reiher ranked fourth with a total score of 36. Furthermore, in the 50th IMO held in 2009, Lisa Sauermann achieved a total score of 41, securing the third position.

Moreover, Oleg Golberg participated in the IMO from 2002 to 2004, achieving three gold medals. He consistently ranked within the top 10 in terms of total scores. Notably, in 2002 and 2003, he was a member of the Russia IMO team, and in 2004, he was a member of the United States IMO team.

Apart from Oleg Golberg, the remaining 48 contestants hailed from 22 different countries and regions. Among them, there were four contestants from each of Russia, Bulgaria, Germany, Hungary, Romania, and the United States. Both South Korea and the United Kingdom had three contestants, while Canada, Japan, Singapore, and the Soviet Union were represented by two contestants each.

3.3 *Special prizes*

In the first 64 IMOs, only 44 contestants have received special prizes. Among them, one contestant has received the special prize three times, seven contestants have earned twice, and 36 contestants have achieved once. It indicates that achieving a special prize is even more challenging than securing a gold medal.

As shown in Table 15, John Rickard, Imre Ruzsa, and Marc van Leeuwen all achieved two special prizes in one IMO for their elegant solutions. Furthermore, John Rickard, Imre Ruzsa, and László Lovász have all earned a perfect score twice.

Table 15 Contestants with Multiple Special Prizes in the First 64 IMOs

Contestant	Country	Participation Year	Special Prize Year	Gold Year	Perfect Score Year
John Rickard	The United Kingdom	1975–1977	1976, 1977 (2)	1975–1977	1975, 1977
József Pelikán	Hungary	1963–1966	1965, 1966	1964–1966	1966
László Lovász	Hungary	1963–1966	1965, 1966	1964–1966	1965, 1966
László Babai	Hungary	1966–1968	1966, 1968	1968	1968
Simon Phillips Norton	The United Kingdom	1967–1969	1967, 1969	1967–1969	1969
Wolfgang Burmeister	The German Democratic Republic	1967–1971	1970, 1971	1968, 1970, 1971	1970
Imre Ruzsa	Hungary	1969–1971	1971 (2)	1970, 1971	1970, 1971
Marc van Leeuwen	The Netherlands	1977–1978	1978 (2)		

Coincidentally, in the 11th IMO held in 1969, only three gold medals were awarded, with Imre Ruzsa ranking 4th and receiving a silver medal. Similarly, in the 19th IMO held in 1977, which also resulted in only 13 gold medals, Marc van Leeuwen ranked 14th and earned a silver medal.

Additionally, these 44 contestants hailed from 16 different countries and regions. Among them, there were seven contestants from each of Hungary and the United Kingdom, five from the German Democratic Republic, four from Bulgaria and Poland each, three from each of Czechoslovakia and the Soviet Union, and two from Finland and the United States each.

Special prizes were more frequently granted in the first 20 IMOs, with a total of 27 special prizes earned by 22 contestants from the 11th–20th IMOs. Subsequently, the frequency of special prize presentations declined. Since Moldovan contestant Iurie Boreico received a special prize for his brilliant solution to IMO 46-3 (Algebra, proposed by South Korea) in 2005, no contestant has achieved this accolade to date.

From 2003 to 2007, Iurie Boreico consistently participated in the IMO, earning three gold and two silver medals. He achieved a perfect score in 2005 and 2006. It's noteworthy that the 44th IMO held in 2003 only yielded

37 gold medals, with Iurie Boreico placing 38th individually and receiving a silver medal.

• **(IMO 46-3, proposed by South Korea).** Let x, y, z be three positive reals such that $xyz \geq 1$. Prove that

$$\frac{x^5 - x^2}{x^5 + y^2 + z^2} + \frac{y^5 - y^2}{y^5 + z^2 + x^2} + \frac{z^5 - z^2}{z^5 + x^2 + y^2} \geq 0.$$

4   Summary

The IMO stands as a distinguished intellectual competition for young minds. According to a study by Agarwal R. and Gaule P., statistical analysis reveals that among contestants in the IMO (including those who did not secure medals), 22% choose to pursue further studies in mathematics, ultimately obtaining doctoral degrees in the field. Additionally, 1% of these contestants become presenters at the International Congress of Mathematicians, and 0.2% attain the Fields Medal. These statistics underscore the vital role of the IMO in identifying and nurturing mathematical talent.

It's essential not to perceive the IMO as a mere selection exam. Rather than focusing solely on the brief two-day competition, the crucial aspect lies in the learning and preparation undertaken before participating. As the mathematician Paul Halmos aptly put it, what mathematics really consists of is problems and solutions. Contestants, through their exploration of Olympiad problems, not only enhance their mathematical abilities but also experience the joy and satisfaction of problem-solving. This experience plants the seeds of a future career in mathematics.

However, it's important to acknowledge that Olympiad problems and research problems in mathematics differ. Research problems often lack readily available answers and may require the investment of countless days and nights. Hence, the IMO is just one pathway in the growth of mathematical talents, and success in the IMO is not the sole qualification for becoming an outstanding mathematician.

Although every contestant aims for a gold medal, their aspirations go far beyond accolades. On this stage, they have the opportunity to showcase their intellectual capabilities, revel in the mathematical exploration, and relish competing with talented young minds from around the world, all without the narrow goal of proving their superiority over others. While the competition results may vary, each contestant stands as a victor in their own right and becomes a companion and witness to one another's life journeys.

In contrast to the Olympics, where athletes' careers are closely intertwined with the Games, the IMO is merely a chapter in the growth of these gifted young individuals. Following the IMO, the door to a new mathematical world has already swung wide open for them.

Chapter 1

Equation Problems

Equations, in layman's terms, are mathematical statements containing unknowns that demonstrate the equality relationship between two mathematical expressions.

Ancient scholars had early investigations into equations, including mathematicians from Babylon, Egypt, Greece, and China. By the time the Indian mathematician Brahmagupta found a general solution, most ancient civilizations already knew and understood solutions to quadratic equations. Moreover, the 5th-century Chinese mathematician Chongzhi Zu, 7th-century Chinese mathematician Xiaotong Wang, and 13th-century Italian mathematician Fibonacci all approximated the positive roots of cubic equations in the form $x^3 + px^2 + qx = r$, where $p, q, r > 0$.

However, before the 16th century, mathematicians had not discovered a general formula for the roots of cubic equations. In fact, the history of cubic equation roots is closely linked to the mathematics competition among Italian mathematicians in the 16th century.

During that time, Italian mathematicians often openly challenged each other to gain honors and benefits. The discoverer of the cubic equation root formula, Nicolo Tartaglia, triumphed over his opponents with a general method for solving the three equations:

$$x^3 + px^2 = q, \quad x^3 = px + q, \quad x^3 + q = px^2.$$

However, he kept his discovery undisclosed for over two decades, leading to unfortunate consequences. Shortly afterward, Ludovico Ferrari found a solution to quartic equations.

Equations of the fifth degree and higher do not have radical solutions, a fact proven by the 19th-century French mathematician Evariste Galois, who introduced the concept of groups and thereby initiated a new branch of mathematics called "Abstract Algebra." This not only elevated the study of algebra to new heights but also had a significant impact on modern mathematics, physics, chemistry, and other fields.

Today, the knowledge and applications of equations are integral to all stages of mathematical education, ranging from elementary puzzles like the classic chicken-rabbit cage problem to complex university-level differential equations. Equations have evolved to become essential tools in mathematics, enabling the exploration and understanding of relationships between various mathematical elements and uncovering the fundamental principles that govern these relationships.

In the first 64 IMOs, there had been a total of 16 equation problems, accounting for approximately 15.8% of all algebra problems. These problems can be primarily categorized into three types: (1) finding solutions of equations and systems of equations, totaling seven problems; (2) proving relationships satisfied by equations and systems of equations, totaling four problems; (3) investigating conditions under which equations and systems of equations have solutions, totaling five problems. The statistical distribution of these three types of problems in the previous IMOs is presented in Table 1.1.

Table 1.1 Numbers of Equation Problems in the First 64 IMOs

Content	Session							Total
	1–10	11–20	21–30	31–40	41–50	51–60	61–64	
Finding solutions	7	0	0	0	0	0	0	7
Proving relationships	2	2	0	0	0	0	0	4
Investigating conditions	2	1	1	0	0	1	0	5
Algebra problems	20	20	14	13	15	13	6	101
The percentage of equation problems among the algebra problems	55.0%	15.0%	7.1%	0.0%	0.0%	7.7%	0.0%	15.8%

It is evident that equation problems predominantly appeared in early IMOs, with a majority focusing on finding solutions of equations and systems of equations. Particularly, in the 1st–10th IMOs, the proportion of equation problems among all algebra problems was very high, exceeding 50%, but rapidly decreased thereafter.

This can partly be attributed to the fact that the study of equations spans various stages of primary and secondary school mathematics education. The ability to formulate and solve appropriate equations has become an essential skill and method in dealing with mathematical problems. Therefore, in recent years, algebra problems rarely specifically address equation problems, with the focus shifting towards more flexible topics such as solving functional equations and proving inequalities.

Problems involving finding solutions are generally not very difficult, and commonly used methods include factoring, the method of undetermined coefficients, trial and error, etc. It is important to note the verification of extraneous roots during the solving process. As for problems about proving relationships, the relationship between the unknowns and coefficients of the equations or systems is somehow clear, and the direction from conditions to conclusions is often more apparent. In contrast, problems investigating conditions tend to be more open-ended.

This chapter will be divided into three parts. The first part introduces common equations and methods, followed by presentations of important theorems related to equations, mainly focusing on roots and coefficients. These theorems are highly applicable when solving equations or proving relationships satisfied by equations. Polynomials and equations are closely related, with additional contents related to polynomials planned for the fifth chapter, "Other Algebra Problems."

The second part revolves around three types of problems: "finding solutions of equations and systems of equations," "proving relationships satisfied by equations and systems of equations," and "investigating conditions under which equations and systems of equations have solutions." These problems are presented in chronological order, and some problems include various solutions, generalizations, and similar problems.

It is important to note that for each problem, the solutions are followed by information on the scores, including the number of contestants in each score range, the average score, and the scores of the top five teams. However, early IMOs often lacked information on contestant scores, so the number of contestants in each score range only represents the counted number of contestants, and some problems lack scores of the top five teams.

The third part provides a brief summary of this chapter.

Some of the equation problems are quite innovative, and readers are encouraged to contemplate the solutions before referring to the answers.

1.1   Common Theorems, Formulas, and Methods

1.1.1   *Common equations*

(1) *Linear equations of one variable*

The standard form of a linear equation of one variable is $ax = b$, where $a \neq 0$, and the equation has a unique solution $x = \frac{b}{a}$.

(2) *Quadratic equations of one variable*

The standard form of a quadratic equation of one variable is $ax^2 + bx + c = 0$, where $a \neq 0$, and a, b, and c are real numbers, with the discriminant $\Delta = b^2 - 4ac$.

  (i) When $\Delta > 0$, the equation has two distinct real roots.
 (ii) When $\Delta = 0$, the equation has two identical real roots.
(iii) When $\Delta < 0$, the equation has no real roots but two complex conjugate
      roots.

The quadratic formula is given by

$$x = \frac{-b \pm \sqrt{b^2 - 4ac}}{2a}.$$

(3) *Higher-degree equations*

In general, polynomial equations with the highest power of the variable greater than 2 are referred to as higher-degree equations. Their general form is

$$a_n x^n + \cdots + a_i x^i + \cdots + a_1 x + a_0 = 0,$$

where $a_n \neq 0$, and $a_i (i = 0, 1, 2, \ldots, n)$ are real numbers.

In particular, for a cubic equation of one variable $x^3 + px + q = 0$, the discriminant is

$$\Delta = \left(\frac{q}{2}\right)^2 + \left(\frac{p}{3}\right)^3.$$

 (i) When $\Delta > 0$, there is one real root and two complex roots.
(ii) When $\Delta = 0$, there are three real roots. If $p = q = 0$, then there is a
     triple root; if $pq \neq 0$, then two of the three real roots are equal.

(iii) When $\Delta < 0$, there are three distinct real roots given by

$$\begin{cases} x_1 = 2\sqrt[3]{r}\cos\theta, \\ x_2 = 2\sqrt[3]{r}\cos(\theta + 120°), \\ x_3 = 2\sqrt[3]{r}\cos(\theta + 240°), \end{cases}$$

where $r = \sqrt{-\left(\frac{p}{3}\right)^3}$ and $\theta = \frac{1}{3}\arccos\left(-\frac{q}{2r}\right)$.

(4) *Fractional equations*

Equations with unknowns in the numerators and denominators are termed fractional (or rational) equations. Typically, methods such as clearing denominators, substitution, and basic identity transformations are employed to transform fractional equations into polynomial equations for solving. It is important to note that this process may introduce extraneous solutions.

(5) *Irrational equations*

An irrational equation, also known as a radical equation, is an equation containing an unknown under a radical root symbol. The fundamental approach to solving irrational equations involves transforming the "irrational" into "rational." Typically, techniques such as factoring, substitution, and completing the square are employed to convert the equation. When dealing with irrational equations, caution must be exercised regarding the possibility of extraneous solutions during the process.

(6) *Systems of equations*

A system of equations refers to a combination of two or more equations with multiple unknowns. In secondary education, the most commonly encountered type is the n-variable linear system:

$$\begin{cases} a_{11}x_1 + a_{12}x_2 + \cdots + a_{1n}x_n = b_1, \\ a_{21}x_1 + a_{22}x_2 + \cdots + a_{2n}x_n = b_2, \\ \cdots\cdots\cdots\cdots\cdots\cdots\cdots\cdots\cdots\cdots \\ a_{s1}x_1 + a_{s2}x_2 + \cdots + a_{sn}x_n = b_s, \end{cases}$$

where $a_{11}, a_{12}, \ldots, a_{sn}$ are coefficients, and $b_1, b_2, \ldots, b_s$ are constants. The number of equations may be equal to, less than, or greater than the number of unknowns. There are three possible scenarios for the solutions of an n-variable linear system: (i) no solution; (ii) a unique solution; (iii) an infinite number of solutions.

1.1.2 *Common methods for solving equations*

(1) *Completing the square*

This method involves transforming an expression or part of it through identity transformations into a perfect square or the sum of several perfect squares. It is frequently used in manipulating identities to uncover implicit conditions in the problem. In particular, if there are nested square roots, then one might consider completing the square to eliminate one of the square roots.

For example, for solving the equation

$$\sqrt{x} + \sqrt{y-1} + \sqrt{z-2} = \frac{1}{2}(x+y+z)$$

within the real number domain, completing the square can be employed to transform the equation into

$$\left(\sqrt{x}-1\right)^2 + \left(\sqrt{y-1}-1\right)^2 + \left(\sqrt{z-2}-1\right)^2 = 0.$$

(2) *Elimination method*

The elimination method involves eliminating certain elements from multiple relations through finite transformations to simplify the problem.

For example, for solving the system of linear equations in two variables:

$$\begin{cases} 3x + 7y = 2, \\ 2x + y = 7, \end{cases}$$

using the elimination method, the system can be transformed into $3x + 7(7-2x) = 2$, significantly simplifying the problem.

(3) *Method of undetermined coefficients*

This method involves expressing a polynomial in another form containing undetermined coefficients, resulting in an identity. Subsequently, based on the properties of the identity, equations or systems of equations can be derived to determine the values of the coefficients. This method is commonly applied in factorization.

For example, given the polynomial $2x^4 - 3x^3 + ax^2 + 7x + b$ divisible by $x^2 + x - 2$, the method of undetermined coefficients allows us to assume

$$2x^4 - 3x^3 + ax^2 + 7x + b = (x^2 + x - 2)(mx^2 + nx + k),$$

where solving for m, n, k allows us to determine the values of a, b.

(4) Formula method

For specific types of equations, applying corresponding formulas or conclusions directly facilitates the solution of the equation.

For example, for solving the equation $2x^2 + 7x + 5 = 0$, using the quadratic formula reveals the real roots as

$$x_1 = \frac{-7 + \sqrt{7^2 - 4 \times 2 \times 5}}{2 \times 2} \quad \text{and} \quad x_2 = \frac{-7 - \sqrt{7^2 - 4 \times 2 \times 5}}{2 \times 2}.$$

(5) Trial-and-error method

Also known as the trial root method, this involves making speculative guesses for roots. For instance, in solving a polynomial equation with integer coefficients

$$a_n x^n + \cdots + a_1 x + a_0 = 0 (a_n \neq 0),$$

by listing the divisors of the leading coefficient a_n and the constant term a_0, denoted as q and p, respectively, and dividing them separately, potential rational roots $\frac{p}{q}$ for the equation can be obtained. Subsequently, these values are substituted back into the equation for verification.

For example, for solving the equation $2x^3 - 6x^2 + 3x + 2 = 0$, using the trial-and-error method yields potential rational roots as $\pm 1, \pm 2, \pm \frac{1}{2}$. Upon testing, $x = 2$ is found to be a rational root, leading to the factorization of $2x^3 - 6x^2 + 3x + 2 = (x - 2)(2x^2 - 2x - 1)$.

(6) Substitution method

When faced with a complex equation structure, considering certain parts as a whole and then replacing them with a new variable is a common approach. This method is often employed to reduce the number of unknowns or decrease the degree of the equation, simplifying complex problems.

For example, for solving the equation $(x^2 - 2x)^2 - 3(x^2 - 2x) - 4 = 0$, using substitution, let $y = x^2 - 2x$. The original equation is then simplified to $y^2 - 3y - 4 = 0$, and solving this quadratic equation yields $y_1 = 4$ and $y_2 = -1$. Substituting the value of y into $y = x^2 - 2x$ provides the corresponding values for x.

1.1.3   *Other important theorems*

Theorem 1.1 (Vieta's Formulas). For a real-coefficient quadratic equation $ax^2 + bx + c = 0$ with the roots x_1 and x_2, where $a \neq 0$, the following relations hold:

$$x_1 + x_2 = -\frac{b}{a}, \quad x_1 x_2 = \frac{c}{a}.$$

For a real-coefficient polynomial equation of degree n given by $a_n x^n + \cdots + a_1 x + a_0 = 0$ with the roots $x_1, x_2, \ldots, x_n$, where $a_n \neq 0$, the following relations hold:

$$x_1 + x_2 + \cdots + x_n = -\frac{a_{n-1}}{a_n},$$

$$\sum_{1 \leq i < j \leq n} x_i x_j = \frac{a_{n-2}}{a_n},$$

$$\sum_{1 \leq i < j < k \leq n} x_i x_j x_k = -\frac{a_{n-3}}{a_n},$$

$$\ldots\ldots\ldots\ldots\ldots\ldots\ldots\ldots\ldots\ldots$$

$$x_1 x_2 \cdots x_n = \frac{(-1)^n a_0}{a_n}.$$

Theorem 1.2 (Inverse Vieta's Formulas). If real numbers α and β satisfy the following relations:

$$\alpha + \beta = -\frac{b}{a}, \quad \alpha\beta = \frac{c}{a},$$

then α and β are the roots of the quadratic equation $ax^2 + bx + c = 0$, where a, b, c are real numbers and $a \neq 0$.

Theorem 1.3 (Rational Root Theorem). For a polynomial equation with integer coefficients

$$a_n x^n + \cdots + a_1 x + a_0 = 0 (a_n \neq 0),$$

let $\frac{p}{q}$ be a rational root, where p and q are relatively prime. Then p is an integer factor of the constant term a_0 and q is an integer factor of the leading coefficient a_n.

In particular, for a quadratic equation $ax^2 + bx + c = 0$, where a, b, c are integers,

(i) when $a = 1$, the rational roots of the equation must be integers;
(ii) integer roots must be divisors of c;
(iii) the discriminant Δ of the equation is a perfect square.

1.2 Problems and Solutions

1.2.1 *Finding solutions (without parameters)*

Problem 1.1 (IMO 1-2, proposed by Romania). For what real values of x is

$$\sqrt{x + \sqrt{2x-1}} + \sqrt{x - \sqrt{2x-1}} = A,$$

given (a) $A = \sqrt{2}$, (b) $A = 1$, (c) $A = 2$, where only non-negative real numbers are admitted for square roots?

Solution 1. It is easy to see that $x \geq \frac{1}{2}$, and we also have

$$\sqrt{x + \sqrt{2x-1}} = \frac{1}{\sqrt{2}}|\sqrt{2x-1}+1|$$

$$= \frac{1}{\sqrt{2}}(\sqrt{2x-1}+1),$$

$$\sqrt{x - \sqrt{2x-1}} = \frac{1}{\sqrt{2}}|\sqrt{2x-1}-1|.$$

It is evident that

$$|\sqrt{2x-1}-1| = 1 - \sqrt{2x-1}, \quad \frac{1}{2} \leq x \leq 1,$$

$$|\sqrt{2x-1}-1| = \sqrt{2x-1}-1, \quad x \geq 1.$$

Let

$$y = \sqrt{x + \sqrt{2x-1}} + \sqrt{x - \sqrt{2x-1}}$$

$$= \frac{1}{\sqrt{2}}(\sqrt{2x-1}+1+|\sqrt{2x-1}-1|).$$

For $\frac{1}{2} \leq x \leq 1$, we have $y = \sqrt{2}$; for $x \geq 1$, we have $y = \sqrt{2}\sqrt{2x-1} \geq \sqrt{2}$.

Therefore, the result for this problem is (a) $\frac{1}{2} \leq x \leq 1$; (b) no solution; (c) $x = \frac{3}{2}$.

Solution 2. Let $y = \sqrt{x + \sqrt{2x-1}} + \sqrt{x - \sqrt{2x-1}}$. Squaring both sides and simplifying, we get

$$\frac{y^2}{2} = x + |x - 1|, \quad x \geq \frac{1}{2}.$$

So, when $\frac{1}{2} \leq x \leq 1$, we have $1 = \frac{y^2}{2}$; when $1 \leq x$, we have $2x - 1 = \frac{y^2}{2}$. By substituting $y = \sqrt{2}, 1, 2$, we can obtain the same result as Solution 1.

Note. $\sqrt{a + \sqrt{b}} = \sqrt{\frac{a+c}{2}} + \sqrt{\frac{a-c}{2}}$, where $c^2 = a^2 - b$.

【Score Situation】 This particular problem saw the following distribution of scores among contestants: 1 contestant scored 8 points, 2 contestants scored 7 points, 4 contestants scored 6 points, 1 contestant scored 5 points, 1 contestant scored 4 points, no contestant scored 3 points, no contestant scored 2 points, no contestant scored 1 point, and no contestant scored 0 point. The average score of this problem is 6.111, indicating that it was simple.

Among the top five teams in the team scores, the Romania team achieved a total score of 249 points, the Hungary team achieved a total score of 233 points, the Czechoslovakia team achieved a total score of 192 points, the Bulgaria team achieved a total score of 131 points, and the Poland team achieved a total score of 122 points.

The gold medal cutoff for this IMO was set at 37 points (with 3 contestants earning gold medals), the silver medal cutoff was 36 points (with 3 contestants earning silver medals), and the bronze medal cutoff was 33 points (with 5 contestants earning bronze medals).

In this IMO, only one contestant achieved a perfect score of 40 points, namely Bohuslav Diviš from Czechoslovakia.

Problem 1.2 (IMO 4-4, proposed by Romania). Solve the equation $\cos^2 x + \cos^2 2x + \cos^2 3x = 1$.

Solution 1. Substituting $\cos^2 x = \frac{1}{2}(1 + \cos 2x)$ and $\cos^2 3x = \frac{1}{2}(1 + \cos 6x)$ into the given equation, we get

$$0 = \cos 2x + 2\cos^2 2x + \cos 6x = 2\cos 2x(\cos 4x + \cos 2x)$$

$$= 4\cos x \cos 2x \cos 3x.$$

For this, we obtain

$$x_1 = k\pi + \frac{\pi}{2}, \quad x_2 = \frac{k\pi}{2} + \frac{\pi}{4}, \quad x_3 = \frac{k\pi}{3} + \frac{\pi}{6}.$$

Here, k takes all integers, and it is easy to see that the first solution x_1 is included in the third solution x_3, so it can be omitted.

Solution 2. Let the trigonometric form of the complex number $a + bi$ be $z = r(\cos\theta + i\sin\theta)$, where $r = \sqrt{a^2 + b^2}$ and $\tan\theta = \frac{b}{a}$.

From De Moivre's formula, if $r = 1$, then $z^n = \cos n\theta + i\sin n\theta$ and $z^{-n} = \cos n\theta - i\sin n\theta$. So $\cos n\theta = \frac{1}{2}(z^n + z^{-n})$ and $\sin n\theta = \frac{1}{2i}(z^n - z^{-n})$. Replacing the unknown angle x with θ in the original equation, we have

$$(z + z^{-1})^2 + (z^2 + z^{-2})^2 + (z^3 + z^{-3})^2 = 4.$$

Simplifying the above gives $z^{-6} + z^{-4} + z^{-2} + 1 + z^2 + z^4 + z^6 = -1$. It's worth noting that $z^2 \neq 1$, otherwise, θ would be a multiple of π, and substituting it into the original equation wouldn't make it valid. So, $\frac{z^{-6}-z^8}{1-z^2} = -1$, and this leads to $z^7 - z^{-7} = -(z - z^{-1})$, or

$$\sin 7\theta = -\sin\theta = \sin(-\theta),$$

which implies $7\theta = n\pi + (-1)^n(-\theta)$.

If $n = 2k$, then $8\theta = 2k\pi$, so $\theta = \frac{k\pi}{4}$, where k is an integer that cannot be divisible by 4.

If $n = 2k+1$, then $6\theta = (2k+1)\pi$, so $\theta = \frac{(2k+1)\pi}{6}$, where k is an integer.

Note. The solutions $\theta = \frac{k\pi}{4}$ in Solution 2, where k cannot be divisible by 4, are equivalent to the solutions in Solution 1, $x = \frac{k\pi}{2} + \frac{\pi}{4}$ and $x = k\pi + \frac{\pi}{2}$, where k is an integer.

【Score Situation】This particular problem saw the following distribution of scores among contestants: 8 contestants scored 5 points, 5 contestants scored 4 points, no contestant scored 3 points, 3 contestants scored 2 points, 1 contestant scored 1 point, and no contestant scored 0 point. The average score of this problem is 3.941, indicating that it was relatively straightforward.

Among the top five teams in the team scores, the Hungary team achieved a total score of 289 points, the Soviet Union team achieved a total score of 263 points, the Romania team achieved a total score of 257 points, the Czechoslovakia team achieved a total score of 212 points, and the Poland team achieved a total score of 212 points.

The gold medal cutoff for this IMO was set at 41 points (with 4 contestants earning gold medals), the silver medal cutoff was 34 points (with 12 contestants earning silver medals), and the bronze medal cutoff was 29 points (with 15 contestants earning bronze medals).

In this IMO, only one contestant achieved a perfect score of 46 points, namely Iosif Bernstein from the Soviet Union.

Problem 1.3 (IMO 7-4, proposed by the Soviet Union). Find all sets of four real numbers x_1, x_2, x_3, x_4 such that the sum of any one and the product of the other three is equal to 2.

Solution 1. For the system of equations

$$\begin{cases} x_1 + x_2 x_3 x_4 = 2, & (1) \\ x_2 + x_3 x_4 x_1 = 2, & (2) \\ x_3 + x_4 x_1 x_2 = 2, & (3) \\ x_4 + x_1 x_2 x_3 = 2, & (4) \end{cases}$$

it is evident that $x_i \neq 0 (i = 1, 2, 3, 4)$. Otherwise, for example, if $x_1 = 0$, then from (2), (3), and (4), we have $x_2 = x_3 = x_4 = 2$. However, $x_2 x_3 x_4 = 2$ from (1), which leads to a contradiction.

Now, from (1) and (2),

$$x_3 x_4 = \frac{2 - x_1}{x_2} = \frac{2 - x_2}{x_1},$$

so $2x_1 - x_1^2 = 2x_2 - x_2^2$, which implies $(x_1 - 1)^2 = (x_2 - 1)^2$.

Therefore,

$$|x_1 - 1| = |x_2 - 1| = |x_3 - 1| = |x_4 - 1|. \tag{5}$$

Next, we'll discuss five cases:

Case 1: All $x_i \geq 1$. Then from (5),

$$x_1 = x_2 = x_3 = x_4.$$

From (1), we have $x_1 + x_1^3 = 2$, so $(x_1 - 1)(x_1^2 + x_1 + 2) = 0$. Since $x_1^2 + x_1 + 2$ is always greater than 0, we see that $x_1 = x_2 = x_3 = x_4 = 1$.

Case 2: Exactly three $x_i \geq 1$. Without loss of generality, assume $x_1 < 1$ and $x_2 = x_3 = x_4 \geq 1$. Then

$$1 - x_1 = x_2 - 1 = x_3 - 1 = x_4 - 1.$$

Therefore, $x_2 = x_3 = x_4 = 2 - x_1$. From (1),

$$2 - x_2 + x_2^3 = 2,$$

which simplifies to $x_2(x_2^2 - 1) = 0$. Solving this gives $x_2 = 1 = x_1$, leading to a contradiction.

Case 3: Exactly two $x_i \geq 1$. Without loss of generality, assume $x_1, x_2 < 1 \leq x_3, x_4$. From (5), we can deduce $x_1 = x_2$ and $x_3 = x_4 = 2 - x_1$. From (3),

$$2 - x_1 + x_1^2(2 - x_1) = 2,$$

which can be further simplified to $x_1(x_1 - 1)^2 = 0$. Solving this equation yields $x_1 = 1$, leading to a contradiction.

Case 4: Exactly one $x_i \geq 1$. Without loss of generality, assume $x_1, x_2, x_3 < 1 \leq x_4$. Then from (5),

$$x_1 = x_2 = x_3 = 2 - x_4.$$

Now from (4),

$$2 - x_1 + x_1^3 = 2,$$

which simplifies to $x_1(x_1^2 - 1) = 0$. Solving this equation gives $x_1 = x_2 = x_3 = -1$ and $x_4 = 3$. Similarly, there are four sets of solutions: $(-1, -1, -1, 3)$, $(-1, -1, 3, -1)$, $(-1, 3, -1, -1)$, and $(3, -1, -1, -1)$.

Case 5: All $x_i < 1$. Then from (5),

$$x_1 = x_2 = x_3 = x_4.$$

It is easy to see that $x_1 = 1$, which contradicts the condition that $x_1 < 1$.

In summary, there are totally five solutions: $(1, 1, 1, 1)$, $(-1, -1, -1, 3)$, $(-1, -1, 3, -1)$, $(-1, 3, -1, -1)$, and $(3, -1, -1, -1)$.

Solution 2. Using P to represent $x_1 x_2 x_3 x_4$, we see that each equation has the same form (as shown in Solution 1, $P \neq 0$):

$$x_i + \frac{P}{x_i} = 2 \ (i = 1, 2, 3, 4),$$

which implies $x_i = 1 \pm \sqrt{1 - P}(i = 1, 2, 3, 4)$. It is easy to see that $P \leq 1$.

Case 1: If $P = 1$, then $x_1 = x_2 = x_3 = x_4 = 1$.

Case 2: If $P < 1$ and x_1, x_2, x_3, x_4 are not all equal, then we discuss the following three cases.

(i) $x_i (i = 1, 2, 3, 4)$ include two $1 + \sqrt{1 - P}$ and two $1 - \sqrt{1 - P}$. Then

$$P = x_1 x_2 x_3 x_4$$
$$= (1 + \sqrt{1 - P})^2 (1 - \sqrt{1 - P})^2$$
$$= P^2.$$

This leads to $P = 1$, which is a contradiction.

(ii) $x_i (i = 1, 2, 3, 4)$ include three $1 + \sqrt{1 - P}$ and one $1 - \sqrt{1 - P}$. Then

$$P = x_1 x_2 x_3 x_4$$
$$= (1 + \sqrt{1 - P})^3 (1 - \sqrt{1 - P})$$
$$= P(1 + \sqrt{1 - P})^2.$$

We still have $P = 1$, which is a contradiction.

(iii) $x_i(i = 1, 2, 3, 4)$ include one $1 + \sqrt{1 - P}$ and three $1 - \sqrt{1 - P}$. Then

$$P = x_1 x_2 x_3 x_4$$
$$= (1 + \sqrt{1 - P})(1 - \sqrt{1 - P})^3$$
$$= P(1 - \sqrt{1 - P})^2.$$

We can solve out $P = -3$, and consequently, one of the $x_i(i = 1, 2, 3, 4)$ is 3, while the other three are -1.

Note. This problem is the same as the seventh problem in 1967 British Mathematical Olympiad Round 1. Furthermore, there are several similar problems:

- **(German Mathematical Olympiad 2023, Final Round, Problem 4).** Determine all triples (a, b, c) of real numbers with

$$a + \frac{4}{b} = b + \frac{4}{c} = c + \frac{4}{a}.$$

- **(Junior Balkan Mathematical Olympiad 2018, Problem 3).** Let $k > 1$ be a positive integer and $n > 2018$ be an odd positive integer. Suppose that nonzero rational numbers $x_1, x_2, \ldots, x_n$ are not all equal and satisfy

$$x_1 + \frac{k}{x_2} = x_2 + \frac{k}{x_3} = x_3 + \frac{k}{x_4} = \cdots = x_{n-1} + \frac{k}{x_n} = x_n + \frac{k}{x_1}.$$

  Find:
  (a) the product $x_1 x_2 \cdots x_n$ as a function of k and n;
  (b) the least value of k, such that there exist $n, x_1, x_2, \ldots, x_n$ satisfying the given conditions.

- **(Vojtěch Jarník International Mathematical Competition 2017, Category II, Problem 3).** Let $n \geq 2$ be an integer. Consider the system of equations

$$x_1 + \frac{2}{x_2} = x_2 + \frac{2}{x_3} = \cdots = x_{n-1} + \frac{2}{x_n} = x_n + \frac{2}{x_1}. \qquad (1)$$

  (a) Prove that (1) has infinitely many real solutions $(x_1, x_2, \ldots, x_n)$ such that the numbers $x_1, x_2, \ldots, x_n$ are distinct.
  (b) Prove that every solution $(x_1, x_2, \ldots, x_n)$ of (1), such that the numbers $x_1, x_2, \ldots, x_n$ are not all equal, satisfies $|x_1 x_2 \cdots x_n| = 2^{\frac{n}{2}}$.

- **(Irish Mathematical Olympiad 2014, Problem 4).** Three different nonzero real numbers a, b, c satisfy the equations

$$a + \frac{2}{b} = b + \frac{2}{c} = c + \frac{2}{a} = p,$$

where p is a real number. Prove that $abc + 2p = 0$.

- **(British Mathematical Olympiad 2008, 1st Round, Problem 2).** Find all real values of x, y, and z such that

$$(x+1)yz = 12, \quad (y+1)zx = 4, \quad \text{and} \quad (z+1)xy = 4.$$

- **(All-Russian Mathematical Olympiad 1993, 4th Round, Grade 10, Problem 3).** Solve for positive numbers the system

$$x_1 + \frac{1}{x_2} = 4, \quad x_2 + \frac{1}{x_3} = 1, \quad x_3 + \frac{1}{x_4} = 4,$$

$$x_4 + \frac{1}{x_5} = 1, \ldots, \quad x_{99} + \frac{1}{x_{100}} = 4, \quad x_{100} + \frac{1}{x_1} = 1.$$

- **(Canadian Mathematical Olympiad 1970, Problem 1).** Find all number triples (x, y, z) such that when any one of these numbers is added to the product of the other two, the result is 2.

【Score Situation】 This particular problem saw the following distribution of scores among contestants: 37 contestants scored 6 points, 6 contestants scored 5 points, 7 contestants scored 4 points, 4 contestants scored 3 points, 19 contestants scored 2 points, 5 contestants scored 1 point, and 2 contestants scored 0 point. The average score for this problem is 4.188, indicating that it was simple.

Among the top five teams in the team scores, the scores of this problem are as follows: the Soviet Union team scored 43 points (with a total team score of 281 points), the Hungary team scored 38 points (with a total team score of 244 points), the Romania team scored 39 points (with a total team score of 222 points), the Poland team scored 44 points (with a total team score of 178 points), and the German Democratic Republic team scored 34 points (with a total team score of 175 points).

The gold medal cutoff for this IMO was set at 38 points (with 8 contestants earning gold medals), the silver medal cutoff was 30 points (with 12 contestants earning silver medals), and the bronze medal cutoff was 20 points (with 17 contestants earning bronze medals).

In this IMO, only two contestants achieved a perfect score of 40 points, namely László Lovász from Hungary and Pavel Bleher from the Soviet Union.

1.2.2 *Finding solutions (with parameters)*

Problem 1.4 (IMO 3-3, proposed by Bulgaria). Solve the equation $\cos^n x - \sin^n x = 1$, where n is a positive integer.

Solution. When n is even, denote $n = 2m$. We have

$$\cos^{2m} x = 1 + \sin^{2m} x.$$

Since $\cos^{2m} x \leq 1 \leq 1 + \sin^{2m} x$, it can only be the case that

$$\sin x = 0, \quad \cos x = \pm 1.$$

In this situation, x is of the form $k\pi$, where k is an integer.

When n is odd and $n \geq 3$, we can observe from

$$1 = \cos^n x - \sin^n x \leq \cos^2 x + \sin^2 x = 1$$

that the equality holds if and only if $\cos x = 1$ or $\sin x = -1$. Consequently, $x = 2k\pi$ or $x = 2k\pi - \frac{\pi}{2}$, where k is an integer.

When $n = 1$, the original equation becomes $\cos x - \sin x = 1$, which can be rewritten as $\cos(x + \frac{\pi}{4}) = \frac{\sqrt{2}}{2}$. This also leads to $x = 2k\pi$ or $x = 2k\pi - \frac{\pi}{2}$, where k is an integer.

Note. There is a similar problem:

- **(All-Russian Mathematical Olympiad 2002, Final Round, Grade 11, Problem 3).** Prove that if $0 < x < \frac{\pi}{2}$ and $n > m$, where n, m are positive integers, then

$$2|\sin^n x - \cos^n x| \leq 3|\sin^m x - \cos^m x|.$$

【Score Situation】 This particular problem saw the following distribution of scores among contestants: 1 contestant scored 7 points, 1 contestant scored 6 points, 1 contestant scored 5 points, 1 contestant scored 4 points, no contestant scored 3 points, no contestant scored 2 points, 1 contestant scored 1 point, and 4 contestants scored 0 point. The average score of this problem is 2.556, indicating that it had a certain level of difficulty.

Among the top five teams in the team scores, the Hungary team achieved a total score of 270 points, the Poland team achieved a total score of 203 points, the Romania team achieved a total score of 197 points, the Czechoslovakia team achieved a total score of 159 points, and the German Democratic Republic team achieved a total score of 146 points.

The gold medal cutoff for this IMO was set at 37 points (with 3 contestants earning gold medals), the silver medal cutoff was 34 points (with 4 contestants earning silver medals), and the bronze medal cutoff was 30 points (with 4 contestants earning bronze medals).

In this IMO, only one contestant achieved a perfect score of 40 points, namely Béla Bollobás from Hungary.

Problem 1.5 (IMO 5-1, proposed by Czechoslovakia). Find all real solutions of the equation

$$\sqrt{x^2 - p} + 2\sqrt{x^2 - 1} = x,$$

where p is a real parameter.

Solution. If $p < 0$, then

$$\sqrt{x^2 - p} + 2\sqrt{x^2 - 1} \geq \sqrt{x^2 - p} > x.$$

In this case, the equation has no solution. Therefore, we can assume $p \geq 0$, $x \geq 1$, and $x^2 \geq p$.

The original equation can be rewritten as

$$2\sqrt{x^2 - 1} = x - \sqrt{x^2 - p}.$$

Squaring and rearranging, we get

$$2x^2 + p - 4 = -2x\sqrt{x^2 - p},$$

and squaring both sides again, we obtain

$$8(2 - p)x^2 = (p - 4)^2.$$

It's easy to see that p must satisfy $0 \leq p < 2$, and in this case, there is a solution

$$x = \frac{4 - p}{\sqrt{8(2 - p)}}.$$

Substituting this into the original equation and simplifying, we get $|3p - 4| = 4 - 3p$, which implies $p \leq \frac{4}{3}$.

In conclusion, the equation has a unique solution if and only if $0 \leq p \leq \frac{4}{3}$, and the solution is $x = \frac{(4-p)\sqrt{4-2p}}{8-4p}$.

Note. There is a similar problem:

- **(Canadian Mathematical Olympiad 1998, Problem 2).** Find all real numbers x such that

$$x = \sqrt{x - \frac{1}{x}} + \sqrt{1 - \frac{1}{x}}.$$

【Score Situation】 This particular problem saw the following distribution of scores among contestants: 2 contestants scored 6 points, 3 contestants scored 5 points, 4 contestants scored 4 points, 1 contestant scored 3 points, 5 contestants scored 2 points, no contestant scored 1 point, and 1 contestant scored 0 point. The average score of this problem is 3.500, indicating that it was relatively straightforward.

Among the top five teams in the team scores, the Soviet Union team achieved a total score of 271 points, the Hungary team achieved a total score of 234 points, the Romania team achieved a total score of 191 points, the Yugoslavia team achieved a total score of 162 points, and the Czechoslovakia team achieved a total score of 151 points.

The gold medal cutoff for this IMO was set at 35 points (with 7 contestants earning gold medals), the silver medal cutoff was 28 points (with 11 contestants earning silver medals), and the bronze medal cutoff was 21 points (with 17 contestants earning bronze medals).

In this IMO, no contestant achieved a perfect score of 40 points.

Problem 1.6 (IMO 5-4, proposed by the Soviet Union). Find all solutions x_1, x_2, x_3, x_4, x_5 of the system

$$\begin{cases} x_5 + x_2 = yx_1, \\ x_1 + x_3 = yx_2, \\ x_2 + x_4 = yx_3, \\ x_3 + x_5 = yx_4, \\ x_4 + x_1 = yx_5, \end{cases}$$

where y is a parameter.

Solution. First, we sum up all five equations to obtain

$$(x_1 + x_2 + x_3 + x_4 + x_5)(y - 2) = 0,$$

which implies that either $x_1 + x_2 + x_3 + x_4 + x_5 = 0$ or $y = 2$.

If $y = 2$, then we can deduce from the original equation:

$$(x_1 + x_3) + (x_2 + x_4) = 2x_2 + 2x_3 \Rightarrow x_1 + x_4 = x_2 + x_3,$$

and $x_1 + x_4 = 2x_5$. Consequently,

$$x_1 + x_2 + x_3 + x_4 + x_5 = 5x_5.$$

Furthermore, $x_1 + x_2 + x_3 + x_4 + x_5 = 5x_i (i = 1, 2, 3, 4, 5)$, which implies that

$$x_1 = x_2 = x_3 = x_4 = x_5 = k,$$

where k can be any real number.

If $y \neq 2$, then

$$(x_5 + x_2) - (x_3 + x_5) = yx_1 - yx_4 \Rightarrow x_2 - x_3 = y(x_1 - x_4)$$

and $x_1 + x_4 = yx_5 = y(yx_1 - x_2)$. These two equations, along with $x_1 + x_3 = yx_2$ and $x_2 + x_4 = yx_3$, can be combined to eliminate x_4, resulting in the following system:

$$\begin{cases} yx_1 - x_2 + x_3 - y^2 x_3 + yx_2 = 0, \\ (1 - y^2)x_1 + yx_2 + yx_3 - x_2 = 0, \end{cases}$$

and substituting $x_3 = yx_2 - x_1$ and simplifying, we have

$$\begin{cases} (y^2 + y - 1)x_1 + (-y^3 + 2y - 1)x_2 = 0, \\ (1 - y - y^2)x_1 + (y^2 + y - 1)x_2 = 0. \end{cases}$$

Since $-y^3 + 2y - 1 = (y^2 + y - 1)(1 - y)$, we can rewrite the above equations as

$$\begin{cases} (y^2 + y - 1)(x_1 + (1 - y)x_2) = 0, \\ (y^2 + y - 1)(x_2 - x_1) = 0. \end{cases}$$

If $y^2 + y - 1 \neq 0$, then $x_1 = x_2$ and $x_1 + (1 - y)x_2 = (2 - y)x_1 = 0$.

Assuming $y \neq 2$, we find $x_1 = x_2 = 0$, and it can be further shown that $x_3 = x_4 = x_5 = 0$, which is the trivial solution.

If $y^2 + y - 1 = 0$, then x_1 and x_2 can take any values. And x_3, x_4, x_5 can be expressed in terms of x_1, x_2, y as

$$x_3 = yx_2 - x_1, \quad x_4 = yx_3 - x_2 = y(yx_2 - x_1) - x_2 = -yx_2 - yx_1,$$

$$x_5 = yx_1 - x_2,$$

where $y = \frac{1}{2}(-1 \pm \sqrt{5})$.

【Score Situation】 This particular problem saw the following distribution of scores among contestants: no contestant scored 6 points, 1 contestant scored 5 points, no contestant scored 4 points, 1 contestant scored 3 points, 4 contestants scored 2 points, 6 contestants scored 1 point, and 4 contestants scored 0 point. The average score of this problem is 1.375, indicating that it was relatively challenging.

Among the top five teams in the team scores, the Soviet Union team achieved a total score of 271 points, the Hungary team achieved a total score of 234 points, the Romania team achieved a total score of 191 points, the Yugoslavia team achieved a total score of 162 points, and the Czechoslovakia team achieved a total score of 151 points.

The gold medal cutoff for this IMO was set at 35 points (with 7 contestants earning gold medals), the silver medal cutoff was 28 points (with 11 contestants earning silver medals), and the bronze medal cutoff was 21 points (with 17 contestants earning bronze medals).

In this IMO, no contestant achieved a perfect score of 40 points.

Problem 1.7 (IMO 8-5, proposed by Czechoslovakia). Solve the system of equations

$$\begin{cases} |a_1 - a_2|x_2 + |a_1 - a_3|x_3 + |a_1 - a_4|x_4 = 1, \\ |a_2 - a_1|x_1 + |a_2 - a_3|x_3 + |a_2 - a_4|x_4 = 1, \\ |a_3 - a_1|x_1 + |a_3 - a_2|x_2 + |a_3 - a_4|x_4 = 1, \\ |a_4 - a_1|x_1 + |a_4 - a_2|x_2 + |a_4 - a_3|x_3 = 1, \end{cases}$$

where a_1, a_2, a_3, a_4 are four different real numbers.

Solution. If we swap the subscripts of a_i and a_j, then the original system remains unchanged. Therefore, we can assume $a_1 > a_2 > a_3 > a_4$, and the original system becomes

$$\begin{cases} (a_1 - a_2)x_2 + (a_1 - a_3)x_3 + (a_1 - a_4)x_4 = 1, & (1) \\ (a_1 - a_2)x_1 + (a_2 - a_3)x_3 + (a_2 - a_4)x_4 = 1, & (2) \\ (a_1 - a_3)x_1 + (a_2 - a_3)x_2 + (a_3 - a_4)x_4 = 1, & (3) \\ (a_1 - a_4)x_1 + (a_2 - a_4)x_2 + (a_3 - a_4)x_3 = 1. & (4) \end{cases}$$

(1)–(2) yields $(a_1 - a_2)(x_3 + x_2 + x_4 - x_1) = 0$.

(2)–(3) yields $(a_2 - a_3)(x_3 + x_4 - x_1 - x_2) = 0$.

(3)–(4) yields $(a_3 - a_4)(x_4 - x_1 - x_2 - x_3) = 0$.

Since a_i are all distinct, we have

$$\begin{cases} x_2 + x_3 + x_4 - x_1 = 0, \\ x_3 + x_4 - x_1 - x_2 = 0, \\ x_4 - x_1 - x_2 - x_3 = 0. \end{cases}$$

Solving this system results in $x_1 = x_4 = \frac{1}{a_1 - a_4}$ and $x_2 = x_3 = 0$.

Note. Using the same method, we can solve a system of n equations:

$$\begin{cases} |a_1 - a_2|x_2 + |a_1 - a_3|x_3 + \cdots + |a_1 - a_n|x_n = 1, \\ |a_2 - a_1|x_1 + |a_2 - a_3|x_3 + \cdots + |a_2 - a_n|x_n = 1, \\ \cdots\cdots\cdots\cdots\cdots\cdots\cdots\cdots\cdots\cdots\cdots\cdots\cdots\cdots\cdots \\ |a_n - a_1|x_1 + |a_n - a_2|x_2 + \cdots + |a_n - a_{n-1}|x_{n-1} = 1. \end{cases}$$

If $a_1 > a_2 > \cdots > a_n$, then the solution is $x_1 = x_n = \frac{1}{a_1 - a_n}$ and $x_2 = x_3 = \cdots = x_{n-1} = 0$.

【Score Situation】 This particular problem saw the following distribution of scores among contestants: 15 contestants scored 7 points, no contestant scored 6 points, 1 contestant scored 5 points, 3 contestants scored 4 points, 1 contestant scored 3 points, 4 contestants scored 2 points, 2 contestants scored 1 point, and 1 contestant scored 0 point. The average score of this problem is 5.000, indicating that it was simple.

Among the top five teams in the team scores, the Soviet Union team achieved a total score of 293 points, the Hungary team achieved a total score of 281 points, the German Democratic Republic team achieved a total score of 280 points, the Poland team achieved a total score of 269 points, and the Romania team achieved a total score of 257 points.

The gold medal cutoff for this IMO was set at 39 points (with 13 contestants earning gold medals), the silver medal cutoff was 34 points (with 15 contestants earning silver medals), and the bronze medal cutoff was 31 points (with 11 contestants earning bronze medals).

In this IMO, a total of 11 contestants achieved a perfect score of 40 points.

1.2.3 *Proving relationships*

Problem 1.8 (IMO 7-2, proposed by Poland). Consider the system of equations

$$\begin{cases} a_{11}x_1 + a_{12}x_2 + a_{13}x_3 = 0, \\ a_{21}x_1 + a_{22}x_2 + a_{23}x_3 = 0, \\ a_{31}x_1 + a_{32}x_2 + a_{33}x_3 = 0, \end{cases}$$

with unknowns x_1, x_2, x_3. The coefficients satisfy the conditions:

(a) a_{11}, a_{22}, a_{33} are positive numbers;
(b) the remaining coefficients are negative numbers;
(c) in each equation, the sum of the coefficients is positive.

Prove that the given system has only the solution $x_1 = x_2 = x_3 = 0$.

Proof 1. Assume that the given system of equations has a non-trivial solution, with some $x_i (i = 1, 2, 3)$ having the largest absolute value, say $i = 2$, so that $x_2 \neq 0$, $|x_2| \geq |x_1|$, and $|x_2| \geq |x_3|$.

According to the given conditions,

$$a_{22}|x_2| = |-a_{21}x_1 - a_{23}x_3|$$

$$\leq -a_{21}|x_1| - a_{23}|x_3|$$

$$\leq -a_{21}|x_2| - a_{23}|x_2|.$$

Since $x_2 \neq 0$, we have $a_{22} \leq -a_{21} - a_{23}$, which contradicts the condition $a_{22} + a_{21} + a_{23} > 0$. Thus, the original system of equations has the unique solution $x_1 = x_2 = x_3 = 0$.

Proof 2. A linear homogeneous system of equations has the unique zero solution only when the determinant of the coefficient matrix is non-zero. Let's prove that

$$D = \begin{vmatrix} a_{11} & a_{12} & a_{13} \\ a_{21} & a_{22} & a_{23} \\ a_{31} & a_{32} & a_{33} \end{vmatrix} \neq 0.$$

The last column can be replaced with the sum of all three columns, resulting in

$$D = \begin{vmatrix} a_{11} & a_{12} & S_1 \\ a_{21} & a_{22} & S_2 \\ a_{31} & a_{32} & S_3 \end{vmatrix},$$

where $S_i = a_{i1} + a_{i2} + a_{i3} > 0 \ (i = 1, 2, 3)$.

Using a cofactor expansion, we have

$$D = S_1 \begin{vmatrix} a_{21} & a_{22} \\ a_{31} & a_{32} \end{vmatrix} - S_2 \begin{vmatrix} a_{11} & a_{12} \\ a_{31} & a_{32} \end{vmatrix} + S_3 \begin{vmatrix} a_{11} & a_{12} \\ a_{21} & a_{22} \end{vmatrix}$$

$$= S_1(a_{21}a_{32} - a_{22}a_{31}) + S_2(a_{12}a_{31} - a_{11}a_{32}) + S_3(a_{11}a_{22} - a_{12}a_{21}).$$

Note that the first two terms in this expression are positive, from conditions (a) and (b). It is also apparent that from conditions (b) and (c),

$$a_{11} + a_{12} > a_{11} + a_{12} + a_{13} > 0 \Rightarrow a_{11} > -a_{12} = |a_{12}|,$$

$$a_{22} + a_{21} > a_{22} + a_{21} + a_{23} > 0 \Rightarrow a_{22} > -a_{21} = |a_{21}|.$$

Therefore $a_{11}a_{22} > a_{12}a_{21}$, making the third term in the above expression positive. Hence, $D > 0$, and as a result, the original system of equations has only the trivial solution $x_1 = x_2 = x_3 = 0$.

Note. This problem is a special case of Hadamard's theorem.

◎ **Hadamard's Theorem.** For an $n \times n$ matrix $\boldsymbol{A} = (a_{ij})_{n \times n}$, if the following conditions are satisfied:

(i) a_{ij} are complex numbers, where $i, j \in \{1, 2, \ldots, n\}$;

(ii) $|a_{ii}| > \sum_{\substack{j=1 \\ j \neq i}}^{n} |a_{ij}|$ for $i = 1, 2, \ldots, n$,

then the determinant of $\boldsymbol{A}$ satisfies $|\boldsymbol{A}| \neq 0$.

Furthermore, there is a similar problem:

- **(Germany Team Selection Test 2004, Problem 7).** Let a_{ij}, where $i, j \in \{1, 2, 3\}$, be real numbers such that a_{ij} is positive for $i = j$ and negative for $i \neq j$. Prove the existence of positive real numbers c_1, c_2, c_3 such that the numbers

$$a_{11}c_1 + a_{12}c_2 + a_{13}c_3, \quad a_{21}c_1 + a_{22}c_2 + a_{23}c_3, \quad a_{31}c_1 + a_{32}c_2 + a_{33}c_3$$

are either all negative, all positive, or all zero.

【Score Situation】 This particular problem saw the following distribution of scores among contestants: 19 contestants scored 6 points, 6 contestants scored 5 points, 4 contestants scored 4 points, 6 contestants scored 3 points, 10 contestants scored 2 points, 18 contestants scored 1 point, and 17 contestants scored 0 point. The average score for this problem is 2.700, indicating that it had a certain level of difficulty.

Among the top five teams in the team scores, the scores of this problem are as follows: the Soviet Union team scored 46 points (with a total team score of 281 points), the Hungary team scored 36 points (with a total team score of 244 points), the Romania team scored 35 points (with a total team score of 222 points), the Poland team scored 25 points (with a total team score of 178 points), and the German Democratic Republic team scored 17 points (with a total team score of 175 points).

The gold medal cutoff for this IMO was set at 38 points (with 8 contestants earning gold medals), the silver medal cutoff was 30 points (with 12 contestants earning silver medals), and the bronze medal cutoff was 20 points (with 17 contestants earning bronze medals).

In this IMO, only two contestants achieved a perfect score of 40 points, namely László Lovász from Hungary and Pavel Bleher from the Soviet Union.

Problem 1.9 (IMO 10-3, proposed by Bulgaria). Consider the system of equations

$$
\begin{cases}
ax_1^2 + bx_1 + c = x_2, \\
ax_2^2 + bx_2 + c = x_3, \\
\dots\dots\dots\dots\dots\dots \\
ax_{n-1}^2 + bx_{n-1} + c = x_n, \\
ax_n^2 + bx_n + c = x_1,
\end{cases}
$$

with unknowns $x_1, x_2, \ldots, x_n$, where a, b, c are real and $a \neq 0$. Let $\Delta = (b-1)^2 - 4ac$.

Prove that for this system,

(a) if $\Delta < 0$, then there is no solution;
(b) if $\Delta = 0$, then there is exactly one solution;
(c) if $\Delta > 0$, then there is more than one solution.

Proof. Consider the quadratic function $Q(x) = ax^2 + (b-1)x + c$. It is clear that the discriminant of $Q(x)$ is $\Delta = (b-1)^2 - 4ac$, and its graph is a parabola.

Summing up the equations of the original system, we get

$$\sum_{i=1}^{n} Q(x_i) = 0.$$

 (i) If $\Delta < 0$, then $Q(x) > 0(a > 0)$ or $Q(x) < 0(a < 0)$, implying $Q(x_i) > 0(a > 0)$ or $Q(x_i) < 0(a < 0)$. It is impossible to have $\sum_{i=1}^{n} Q(x_i) = 0$. Therefore, there is a contradiction, and the original system has no solution.

 (ii) If $\Delta = 0$, then $Q(x) \geq 0(a > 0)$ or $Q(x) \leq 0(a < 0)$, implying $Q(x_i) \geq 0(a > 0)$ or $Q(x_i) \leq 0(a < 0)$. To have $\sum_{i=1}^{n} Q(x_i) = 0$, we must have $Q(x_i) = 0$, meaning $x_1 = x_2 = \cdots = x_n = r$. Here, r is the unique root of $Q(x) = 0$.

(iii) If $\Delta > 0$, then $Q(x) = 0$ has two roots r_1 and r_2. It is evident that $x_1 = x_2 = \cdots = x_n = r_1$ and $x_1 = x_2 = \cdots = x_n = r_2$ are both solutions of the original system.

Note. There are several similar problems:

- **(All-Russian Mathematical Olympiad 2011, Grade 10, Problem 2).** Nine quadratics are written on the board:
$$x^2 + a_1 x + b_1, x^2 + a_2 x + b_2, \ldots, x^2 + a_9 x + b_9.$$
It is known that $a_1, a_2, \ldots, a_9$ and $b_1, b_2, \ldots, b_9$ are both arithmetic sequences, and the sum of these nine quadratics has at least one real root. What is the maximum number of these original nine quadratics that can have no real roots?

- **(Mexican Mathematical Olympiad 2011, Problem 3).** Let n be a positive integer. Find all real solutions $(a_1, a_2, \ldots, a_n)$ to the system:
$$\begin{cases} a_1^2 + a_1 - 1 = a_2, \\ a_2^2 + a_2 - 1 = a_3, \\ \cdots\cdots\cdots\cdots\cdots \\ a_n^2 + a_n - 1 = a_1. \end{cases}$$

- (**All-Russian Mathematical Olympiad 2009, Regional Round, Grade 10, Problem 7**). Let $x_1, x_2, \ldots, x_{2009}$ be positive real numbers satisfying

$$x_1^2 - x_1 x_2 + x_2^2 = x_2^2 - x_2 x_3 + x_3^2 = \cdots = x_{2008}^2 - x_{2008} x_{2009} + x_{2009}^2$$
$$= x_{2009}^2 - x_{2009} x_1 + x_1^2.$$

Prove that $x_1 = x_2 = \cdots = x_{2009}$.

- (**Austrian–Polish Mathematical Competition 2001, Problem 2**). Given an integer $n > 2$, solve in non-negative real numbers the system

$$x_k + x_{k+1} = x_{k+2}^2, \quad k = 1, 2, \ldots, n,$$

where $x_{n+i} = x_i$.

- (**Polish Mathematical Olympiad 2000, Problem 1**). For a given integer $n \geq 2$, find the number of non-negative real solutions of the system of equations:

$$\begin{cases} x_1 + x_n^2 = 4x_n, \\ x_2 + x_1^2 = 4x_1, \\ x_3 + x_2^2 = 4x_2, \\ \cdots\cdots\cdots\cdots \\ x_n + x_{n-1}^2 = 4x_{n-1}. \end{cases}$$

- (**Canadian Mathematical Olympiad 1996, Problem 2**). Find all real solutions to the following system of equations:

$$\begin{cases} \frac{4x^2}{1+4x^2} = y, \\ \frac{4y^2}{1+4y^2} = z, \\ \frac{4z^2}{1+4z^2} = x. \end{cases}$$

- (**Turkey Team Selection Test 1995, Problem 1**). Given real numbers $b \geq a > 0$, find all solutions of the system

$$\begin{cases} x_1^2 + 2ax_1 + b^2 = x_2, \\ x_2^2 + 2ax_2 + b^2 = x_3, \\ \cdots\cdots\cdots\cdots\cdots\cdots \\ x_{n-1}^2 + 2ax_{n-1} + b^2 = x_n, \\ x_n^2 + 2ax_n + b^2 = x_1. \end{cases}$$

【Score Situation】 This particular problem saw the following distribution of scores among contestants: 38 contestants scored 7 points, 9 contestants scored 6 points, 1 contestant scored 5 points, 2 contestants scored 4 points, no contestant scored 3 points, 7 contestants scored 2 points, 21 contestants scored 1 point, and 13 contestants scored 0 point. The average score of this problem is 4.044, indicating that it was simple.

Among the top five teams in the team scores, the German Democratic Republic team achieved a total score of 304 points, the Soviet Union team achieved a total score of 298 points, the Hungary team achieved a total score of 291 points, the United Kingdom team achieved a total score of 263 points, and the Poland team achieved a total score of 262 points.

The gold medal cutoff for this IMO was set at 39 points (with 22 contestants earning gold medals), the silver medal cutoff was 33 points (with 22 contestants earning silver medals), and the bronze medal cutoff was 26 points (with 20 contestants earning bronze medals).

In this IMO, a total of 16 contestants achieved a perfect score of 40 points.

Problem 1.10 (IMO 18-2, proposed by Finland). Let $p_1(x) = x^2 - 2$ and $P_j(x) = P_1(P_{j-1}(x))$ for $j = 2, 3, \ldots$. Show that, for any positive integer n, the roots of the equation $P_n(x) = x$ are real and distinct.

Proof 1. It is evident that $p_n(x) = x$ is a polynomial of degree 2^n, implying it has at most 2^n distinct roots. Moreover, $p_1(x) > x$ when $x > 2$, and

$$p_2(x) = p_1(p_1(x)) > p_1(x) > x.$$

Similarly, $p_n(x) > x > 2$, and $p_n(x) > 2 > x$ when $x < -2$. Thus the real roots of $p_n(x) = x$ lie within the closed interval $[-2, 2]$. Therefore, we can assume that the roots of the equation $p_n(x) = x$ take the form $x = 2\cos t$. Consequently,

$$p_1(2\cos t) = 4\cos^2 t - 2 = 2\cos 2t,$$

$$p_2(2\cos t) = p_1(2\cos 2t) = 2\cos 4t.$$

In general, $p_n(2\cos t) = 2\cos 2^n t$. For the equation $2\cos 2^n t = 2\cos t$, the following solutions exist:

$$t_k = \frac{2k\pi}{2^n - 1}, \quad k = 0, 1, \ldots, 2^{n-1} - 1,$$

$$s_l = \frac{2l\pi}{2^n + 1}, \quad l = 1, 2, \ldots, 2^{n-1},$$

leading to $0 = t_0 < t_1 < \cdots < t_{2^{n-1}-1} < \pi$, and $0 < s_1 < s_2 < \cdots < s_{2^{n-1}} < \pi$.

Since the cosine function is strictly monotonic on $[0, \pi]$, the values

$$2 \cos t_k (k = 0, 1, \ldots, 2^{n-1} - 1)$$

are 2^{n-1} distinct real numbers. Similarly, $2 \cos s_l (l = 1, 2, \ldots, 2^{n-1})$ also constitute 2^{n-1} distinct real numbers.

Moreover, for any k and l within the aforementioned range, it is impossible to have $t_k = s_l$. Otherwise, there exists the equation

$$\frac{2^n - 1}{2^n + 1} = \frac{k}{l}.$$

Since $k < 2^n - 1$ and $l < 2^n + 1$, it indicates that the fraction on the left side of the equation has not been reduced to its simplest form. However, $2^n - 1$ and $2^n + 1$ are consecutive odd numbers and coprime, meaning they cannot have a common factor greater than 1. This implies that $2 \cos t_k$ and $2 \cos s_l$ are all 2^n distinct real roots of the equation $p_n(x) = x$.

Proof 2. It is easy to observe that the polynomial

$$p_n(x) = (\cdots ((x^2 - 2)^2 - 2)^2 - \cdots)^2 - 2$$

is of degree 2^n and an even function, i.e., $p_n(-x) = p_n(x)$.

As shown in Figure 1.1, p_1 maps the interval $[-2, 2]$ onto the interval $[-2, 2]$. When x increases from -2 to 0, we see that $p_1(x)$ decreases from 2 to -2. Furthermore, since $p_1(x)$ is an even function, $p_1(x)$ increases from -2 to 2 when x increases from 0 to 2.

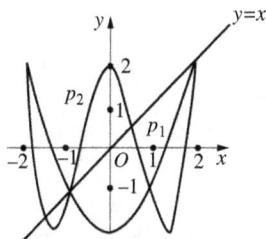

Figure 1.1 Graph of the Function $P_n(x)$

Next, consider $p_2(x) = p_1(p_1(x))$. When x increases from -2 to 0, clearly $p_1(x)$ decreases from 2 to -2. Subsequently, $p_1(p_1(x))$ maps these values onto the interval $[-2, 2]$, and the graph for the interval $-2 \leq x \leq 0$ is symmetric about the y-axis with the graph for the interval $0 \leq x \leq 2$.

Similarly, in the graph of $p_3(x)$, the number of times its ordinate decreases from 2 to -2 or increases from -2 to 2 is twice that of $p_2(x)$, and so forth. Therefore, in the graph of $p_n(x)$, the curve of $p_n(x)$ repeats between -2 and 2 2^n times.

Furthermore, the solutions of $p_n(x) = x$ correspond to the abscissas of the intersection points of $p_n(x)$ and the line $y = x$. When $-2 < x < 2$, this line precisely corresponds to the diagonal of the 4×4 square. As $p_n(x)$ oscillates 2^n times, there are 2^n distinct intersection points, i.e., 2^n distinct real roots.

Moreover, since the degree of $p_n(x)$ is 2^n, the number of roots of $p_n(x) = x$ is no more than 2^n. Therefore, all its roots are distinct real numbers.

Note. The nth Chebyshev polynomial $T_n(x)$, to be presented in Chapter 5, is defined by $T_n(\cos \theta) = \cos n\theta$. It is easy to verify that $p_n(x) = 2T_{2^n}\left(\frac{x}{2}\right)$.

【Score Situation】 This particular problem saw the following distribution of scores among contestants: 25 contestants scored 7 points, 5 contestants scored 6 points, 4 contestants scored 5 points, no contestant scored 4 points, 4 contestants scored 3 points, 5 contestants scored 2 points, 22 contestants scored 1 point, and 74 contestants scored 0 point. The average score for this problem is 1.935, indicating that it was relatively challenging.

Among the top five teams in the team scores, the scores of this problem are as follows: the Soviet Union team scored 46 points (with a total team score of 250 points), the United Kingdom team scored 35 points (with a total team score of 214 points), the United States team scored 18 points (with a total team score of 188 points), the Bulgaria team scored 18 points (with a total team score of 174 points), and the Austria team scored 16 points (with a total team score of 167 points).

The gold medal cutoff for this IMO was set at 34 points (with 9 contestants earning gold medals), the silver medal cutoff was 23 points (with 28 contestants earning silver medals), and the bronze medal cutoff was 15 points (with 45 contestants earning bronze medals).

In this IMO, only one contestant achieved a perfect score of 40 points, namely Laurent Pierre from France.

Problem 1.11 (IMO 18-5, proposed by the Netherlands). Consider the system of p equations in $q = 2p$ unknowns $x_1, x_2, \ldots, x_q$:

$$
\begin{cases}
a_{11}x_1 + a_{12}x_2 + \cdots + a_{1q}x_q = 0, \\
a_{21}x_1 + a_{22}x_2 + \cdots + a_{2q}x_q = 0, \\
\cdots\cdots\cdots\cdots\cdots\cdots\cdots\cdots\cdots\cdots \\
a_{p1}x_1 + a_{p2}x_2 + \cdots + a_{pq}x_q = 0,
\end{cases}
$$

with every coefficient a_{ij} a member of the set $\{-1, 0, 1\}$. Prove that the system has a solution $(x_1, x_2, \ldots, x_q)$ satisfying:

(a) all $x_j (j = 1, 2, \ldots, q)$ are integers;
(b) there is at least one value of j for which $x_j \neq 0$;
(c) $|x_j| \leq q (j = 1, 2, \ldots, q)$.

Proof. Consider the arrays $(x_1, x_2, \ldots, x_q)$ substituted into the system, satisfying $|x_j| \leq p$. According to the given conditions,

$$
|b_i| = |a_{i1}x_1 + a_{i2}x_2 + \cdots + a_{iq}x_q| \leq pq.
$$

Thus, $a_{i1}x_1 + a_{i2}x_2 + \cdots + a_{iq}x_q$ can take at most $2pq + 1$ possible values. Consequently, the values corresponding to the left side of the system can take at most $(2pq + 1)^p$ possible sets of integers $(b_1, b_2, \ldots, b_p)$.

For each x_j, let it take $2p + 1$ values ranging from $-p$ to p, resulting in $(2p + 1)^q$ possible sets of integers $(b_1, b_2, \ldots, b_p)$. Since

$$
\begin{aligned}
(2p + 1)^q &= (2p + 1)^{2p} \\
&= (4p^2 + 4p + 1)^p \\
&> (4p^2 + 1)^p \\
&= (2pq + 1)^p,
\end{aligned}
$$

by the Pigeonhole Principle, there must exist two distinct arrays $(x_1', x_2', \ldots, x_q')$ and $(x_1'', x_2'', \ldots, x_q'')$ such that they correspond to the same set of integers $(b_1', b_2', \ldots, b_p')$ on the left side of the system.

Consider the array $(x_1' - x_1'', x_2' - x_2'', \ldots, x_q' - x_q'')$, which evidently satisfies the conditions (a) and (b). Moreover, since $|x_j'| \leq p$ and $|x_j''| \leq p$, it follows that $|x_j' - x_j''| \leq 2p = q$, and thus condition (c) is also satisfied. Therefore, the proposition holds.

Note. This problem can also be solved from a matrix perspective. Let A represent the coefficient matrix of the system of equations, $\overline{0}$ represent the constant term vector, and X the unknown vector, that is, let

$$A = \begin{pmatrix} a_{11} & a_{12} & \cdots & a_{1q} \\ a_{21} & a_{22} & \cdots & a_{2q} \\ \vdots & \vdots & & \vdots \\ a_{p1} & a_{p2} & \cdots & a_{pq} \end{pmatrix}, \quad \overline{0} = \begin{pmatrix} 0 \\ 0 \\ \vdots \\ 0 \end{pmatrix}, \quad X = \begin{pmatrix} x_1 \\ x_2 \\ \vdots \\ x_q \end{pmatrix}.$$

Then the original system of equations can be succinctly written as $AX = \overline{0}$.

It is easily known that there are $(2p+1)^q$ different integer vectors X that satisfy $|x_j| \le p (j = 1, 2, \ldots, q)$. For such X, it is evident that

$$|a_{i1}x_1 + a_{i2}x_2 + \cdots + a_{iq}x_q| \le |a_{i1}x_1| + |a_{i2}x_2| + \cdots + |a_{iq}x_q|$$

$$\le |x_1| + |x_2| + \cdots + |x_q|$$

$$\le pq.$$

Thus, the vector AX has at most $(2pq+1)^p$ different values.

Since $(2p+1)^q > (2pq+1)^p$, there must exist two distinct vectors X_1 and X_2 such that $AX_1 = AX_2$. Hence $A(X_1 - X_2) = \overline{0}$. Consequently, the components of $X_1 - X_2$ are solutions.

【Score Situation】 This particular problem saw the following distribution of scores among contestants: 15 contestants scored 7 points, no contestant scored 6 points, no contestant scored 5 points, 3 contestants scored 4 points, 2 contestants scored 3 points, 3 contestants scored 2 points, 23 contestants scored 1 point, and 93 contestants scored 0 point. The average score for this problem is 1.094, indicating that it was relatively challenging.

Among the top five teams in the team scores, the scores of this problem are as follows: the Soviet Union team scored 31 points (with a total team score of 250 points), the United Kingdom team scored 11 points (with a total team score of 214 points), the United States team scored 26 points (with a total team score of 188 points), the Bulgaria team scored 6 points (with a total team score of 174 points), and the Austria team scored 8 points (with a total team score of 167 points).

The gold medal cutoff for this IMO was set at 34 points (with 9 contestants earning gold medals), the silver medal cutoff was 23 points (with 28 contestants earning silver medals), and the bronze medal cutoff was 15 points (with 45 contestants earning bronze medals).

In this IMO, only one contestant achieved a perfect score of 40 points, namely Laurent Pierre from France.

1.2.4 *Investigating conditions*

Problem 1.12 (IMO 1-3, proposed by Hungary). Let a, b, c be real numbers. Consider the quadratic equation in $\cos x$:

$$a \cos^2 x + b \cos x + c = 0.$$

Using the numbers a, b, c, form a quadratic equation in $\cos 2x$, whose roots are the same as those of the original equation. Compare the equations in $\cos x$ and $\cos 2x$ for $a = 4$, $b = 2$, and $c = -1$.

Solution 1. Expressing the original equation as

$$a \cos^2 x + c = -b \cos x,$$

squaring both sides, and multiplying by 4, we obtain

$$(a(1 + \cos 2x) + 2c)^2 = 2b^2(1 + \cos 2x),$$

simplifying to

$$a^2 \cos^2 2x + (2a^2 + 4ac - 2b^2) \cos 2x + (a^2 + 4ac - 2b^2 + 4c^2) = 0.$$

For the specific values $a = 4$, $b = 2$, and $c = -1$, the equation concerning $\cos 2x$ becomes

$$4 \cos^2 2x + 2 \cos 2x - 1 = 0.$$

This quadratic equation shares the same coefficients as the given equation involving $\cos x$.

Solution 2. According to Vieta's formulas, for the roots (which may be identical) of the given equation, $\cos x_1 + \cos x_2 = -\frac{b}{a}$ and $\cos x_1 \cdot \cos x_2 = \frac{c}{a}$. Thus,

$$\cos 2x_1 + \cos 2x_2 = 2(\cos^2 x_1 + \cos^2 x_2 - 1)$$
$$= 2((\cos x_1 + \cos x_2)^2 - 2 \cos x_1 \cdot \cos x_2 - 1)$$
$$= 2\left(\left(-\frac{b}{a}\right)^2 - 2 \cdot \frac{c}{a} - 1\right)$$
$$= -\frac{2a^2 + 4ac - 2b^2}{a^2},$$

$$\cos 2x_1 \cdot \cos 2x_2 = (2\cos^2 x_1 - 1)(2\cos^2 x_2 - 1)$$
$$= 4(\cos x_1 \cdot \cos x_2)^2 - 2(\cos^2 x_1 + \cos^2 x_2) + 1$$
$$= 4\left(\frac{c}{a}\right)^2 - 2\left(\left(-\frac{b}{a}\right)^2 - 2\cdot\frac{c}{a}\right) + 1$$
$$= \frac{a^2 + 4ac - 2b^2 + 4c^2}{a^2}.$$

Therefore, the required quadratic equation satisfied by $\cos 2x$ is

$$a^2 \cos^2 2x + (2a^2 + 4ac - 2b^2)\cos 2x + (a^2 + 4ac - 2b^2 + 4c^2) = 0.$$

When $a = 4$, $b = 2$, and $c = -1$, this quadratic equation shares the same coefficients as the given equation $a\cos^2 x + b\cos x + c = 0$.

【Score Situation】This particular problem saw the following distribution of scores among contestants: 4 contestants scored 7 points, no contestant scored 6 points, 2 contestants scored 5 points, no contestant scored 4 points, no contestant scored 3 points, no contestant scored 2 points, no contestant scored 1 point, and 3 contestants scored 0 point. The average score of this problem is 4.222, indicating that it was simple.

Among the top five teams in the team scores, the Romania team achieved a total score of 249 points, the Hungary team achieved a total score of 233 points, the Czechoslovakia team achieved a total score of 192 points, the Bulgaria team achieved a total score of 131 points, and the Poland team achieved a total score of 122 points.

The gold medal cutoff for this IMO was set at 37 points (with 3 contestants earning gold medals), the silver medal cutoff was 36 points (with 3 contestants earning silver medals), and the bronze medal cutoff was 33 points (with 5 contestants earning bronze medals).

In this IMO, only one contestant achieved a perfect score of 40 points, namely Bohuslav Diviš from Czechoslovakia.

Problem 1.13 (IMO 3-1, proposed by Hungary). Solve the system of equations:

$$\begin{cases} x + y + z = a, \\ x^2 + y^2 + z^2 = b^2, \\ xy = z^2, \end{cases}$$

where a and b are constants. Give conditions that a and b must satisfy, so that x, y, z (the solutions of the system) are distinct positive numbers.

Solution. Since the solutions of the system of equations are distinct positive numbers, it is evident that $a > 0$. Additionally, it is easy to observe that

$$b^2 = x^2 + y^2 + z^2 = x^2 + y^2 + 2xy - z^2 = (x + y + z)(x + y - z),$$

which leads to

$$x + y - z = \frac{b^2}{a},$$

$$z = \frac{(x + y + z) - (x + y - z)}{2} = \frac{a^2 - b^2}{2a},$$

$$x + y = \frac{a^2 + b^2}{2a},$$

$$xy = \frac{(a^2 - b^2)^2}{4a^2}.$$

Consider the quadratic equation $f(x) = x^2 - \frac{a^2 + b^2}{2a}x + \frac{(a^2 - b^2)^2}{4a^2}$. Then,

$$\Delta = \frac{1}{4a^2}((a^2 + b^2)^2 - 4(a^2 - b^2)^2)$$

$$= \frac{1}{4a^2}(3a^2 - b^2)(3b^2 - a^2) \geq 0,$$

and in this case, the system of equations has solutions

$$\begin{cases} x = \frac{1}{4a}(a^2 + b^2 \pm \sqrt{(3a^2 - b^2)(3b^2 - a^2)}), \\ y = \frac{1}{4a}(a^2 + b^2 \mp \sqrt{(3a^2 - b^2)(3b^2 - a^2)}), \\ z = \frac{1}{2a}(a^2 - b^2). \end{cases}$$

To ensure that the solutions are positive, $z > 0$, implying $a^2 > b^2$.

Since $x \neq y$, we have $\Delta > 0$. Additionally, as $3a^2 - b^2 > a^2 - b^2 > 0$, there holds $3b^2 > a^2$.

Therefore,

$$0 < |b| < a < \sqrt{3}|b|.$$

It is evident that $\frac{a^2 + b^2}{4a} > \frac{\sqrt{\Delta}}{2}$. Thus, $x > 0$ and $y > 0$. Furthermore, since x, y are distinct positive numbers, and $z = \sqrt{xy}$, it is clear that x, y, z are distinct positive numbers.

Note. From the given conditions, it is known that the distance from the plane $x + y + z = a > 0$ to the origin is $\frac{a}{\sqrt{3}} > 0$. The radius of the sphere centered at the origin with the equation $x^2 + y^2 + z^2 = b^2$ is $|b|$. The plane and the sphere have a common point only when $|b| \geq \frac{a}{\sqrt{3}}$. However, when $\sqrt{3}|b| = a$, the plane and the sphere touch at a single point, and the coordinates of this point are equal. Therefore, $0 < a < \sqrt{3}|b|$ is a necessary condition for x, y, z to be distinct positive numbers.

Furthermore, by using the AM-GM inequality

$$\frac{x+y}{2} > \sqrt{xy}(x \neq y),$$

it is evident that

$$\frac{x+y}{2} = \frac{1}{2}(a - z) = \frac{1}{2}\left(a - \frac{a^2 - b^2}{2a}\right) > \frac{a^2 - b^2}{2a} = z = \sqrt{xy}.$$

This implies $a^2 < 3b^2$. Given that $a > 0$, the condition $0 < a < \sqrt{3}|b|$ still holds.

There are several similar problems:

- **(British Mathematical Olympiad 2007, 1st Round, Problem 2).**
  Find all solutions in positive integers x, y, z to the simultaneous equations

  $$x + y - z = 12 \quad \text{and} \quad x^2 + y^2 - z^2 = 12.$$

- **(British Mathematical Olympiad 1998, 2nd Round, Problem 4).**
  Find a solution of the simultaneous equations

  $$xy + yz + zx = 12 \quad \text{and} \quad xyz = 2 + x + y + z,$$

  in which all of x, y, z are positive, and prove that it is the only such solution.

  Show that a solution exists in which x, y, z are real and distinct.

- **(British Mathematical Olympiad 1996, 2nd Round, Problem 4).**
  Let a, b, c, and d be positive real numbers such that

  $$a + b + c + d = 12 \quad \text{and} \quad abcd = 27 + ab + ac + ad + bc + bd + cd.$$

  Find all possible values of a, b, c, d satisfying these equations.

【Score Situation】 This particular problem saw the following distribution of scores among contestants: 1 contestant scored 6 points, 2 contestants scored 5 points, 2 contestants scored 4 points, 1 contestant scored 3 points, 1 contestant scored 2 points, no contestant scored 1 point, and 2 contestants scored 0 point. The average score of this problem is 3.222, indicating that it was relatively straightforward.

Among the top five teams in the team scores, the Hungary team achieved a total score of 270 points, the Poland team achieved a total score of 203 points, the Romania team achieved a total score of 197 points, the Czechoslovakia team achieved a total score of 159 points, and the German Democratic Republic team achieved a total score of 146 points.

The gold medal cutoff for this IMO was set at 37 points (with 3 contestants earning gold medals), the silver medal cutoff was 34 points (with 4 contestants earning silver medals), and the bronze medal cutoff was 30 points (with 4 contestants earning bronze medals).

In this IMO, only one contestant achieved a perfect score of 40 points, namely Béla Bollobás from Hungary.

Problem 1.14 (IMO 15-3, proposed by Sweden). Let a and b be real numbers for which the equation

$$x^4 + ax^3 + bx^2 + ax + 1 = 0$$

has at least one real solution. For all such pairs (a, b), find the minimum value of $a^2 + b^2$.

Solution 1. It is evident that $x = 0$ is not a root of the original equation. Therefore, the original equation can be transformed into

$$\left(x + \frac{1}{x}\right)^2 + a\left(x + \frac{1}{x}\right) + (b - 2) = 0.$$

Let $y = x + \frac{1}{x}$. Then it is easy to observe that $|y| \geq 2$. Solving the equation $y^2 + ay + (b - 2) = 0$ yields

$$y = \frac{-a \pm \sqrt{a^2 - 4(b - 2)}}{2},$$

where at least one of the roots has an absolute value not less than 2. This is equivalent to

$$|a| + \sqrt{a^2 - 4(b - 2)} \geq 4,$$

or

$$\sqrt{a^2 - 4(b - 2)} \geq 4 - |a|.$$

(i) When $|a| \geq 4$, it follows that $a^2 + b^2 \geq 16$.

(ii) When $|a| < 4$, squaring both sides and simplifying, we obtain

$$2|a| \geq 2 + b.$$

If $b \leq -2$, then $a^2 + b^2 \geq 4$. Considering $b > -2$ and squaring both sides again, we get

$$4a^2 \geq 4 + 4b + b^2,$$

$$4a^2 + 4b^2 \geq 5b^2 + 4b + 4$$

$$= 5\left(b + \frac{2}{5}\right)^2 + \frac{16}{5}$$

$$\geq \frac{16}{5}.$$

Therefore, when $b = -\frac{2}{5}$ and $a = \pm\frac{4}{5}$, the expression $a^2 + b^2$ attains the minimum value of $\frac{4}{5}$. Upon verification, the original equation has real roots $x = \mp 1$.

Solution 2. It is evident that $x = 0$ is not a root of the original equation. By the Cauchy–Schwarz inequality, we have

$$\left(\frac{1}{2}a^2 + b^2 + \frac{1}{2}a^2\right)(2x^6 + x^4 + 2x^2) \geq (ax^3 + bx^2 + ax)^2 = (x^4 + 1)^2.$$

Therefore,

$$a^2 + b^2 \geq \frac{(x^4 + 1)^2}{2x^6 + x^4 + 2x^2} = \frac{\left(x^2 + \frac{1}{x^2}\right)^2}{2\left(x^2 + \frac{1}{x^2}\right) + 1}$$

$$= \frac{1}{2}\left(x^2 + \frac{1}{x^2} + \frac{1}{2} + \frac{\frac{1}{4}}{x^2 + \frac{1}{x^2} + \frac{1}{2}} - 1\right).$$

Since $x^2 + \frac{1}{x^2} + \frac{1}{2} \geq 2 + \frac{1}{2} = \frac{5}{2}$ and $a^2 + b^2 \geq \frac{1}{2}(\frac{5}{2} + \frac{1}{10} - 1) = \frac{4}{5}$, the equality holds when $x = \mp 1$, $b = -\frac{2}{5}$, and $a = \pm\frac{4}{5}$.

Note. In Solution 1, from $|y| \geq 2$, it follows that

$$\sqrt{a^2 - 4(b - 2)} \geq 4 - |a|.$$

Squaring both sides and rearranging yield $b \leq -2a - 2$ or $b \leq 2a - 2$. Consequently, the pairs of numbers (a, b) that lead to solutions of the original equation lie below the lines $y = -2x - 2$ or $y = 2x - 2$. The distance from the origin to the lines $y = -2x - 2$ or $y = 2x - 2$ is exactly $\frac{2}{\sqrt{5}}$. Thus, $\sqrt{a^2 + b^2} \geq \frac{2}{\sqrt{5}}$.

【Score Situation】This particular problem saw the following distribution of scores among contestants: 42 contestants scored 8 points, 4 contestants scored 7 points, 8 contestants scored 6 points, 4 contestants scored 5 points, 4 contestants scored 4 points, 6 contestants scored

3 points, 9 contestants scored 2 points, 15 contestants scored 1 point, and 33 contestants scored 0 point. The average score of this problem is 3.992, indicating that it was relatively straightforward.

Among the top five teams in the team scores, the scores of this problem are as follows: the Soviet Union team scored 53 points (with a total team score of 254 points), the Hungary team scored 50 points (with a total team score of 215 points), the German Democratic Republic team scored 31 points (with a total team score of 188 points), the Poland team scored 16 points (with a total team score of 174 points), and the United Kingdom team scored 38 points (with a total team score of 164 points).

The gold medal cutoff for this IMO was set at 35 points (with 5 contestants earning gold medals), the silver medal cutoff was 27 points (with 15 contestants earning silver medals), and the bronze medal cutoff was 17 points (with 48 contestants earning bronze medals).

In this IMO, only one contestant achieved a perfect score of 40 points, namely Sergei Konyagin from the Soviet Union.

Problem 1.15 (IMO 21-5, proposed by Israel). Find all real numbers a for which there exist non-negative real numbers x_1, x_2, x_3, x_4, x_5 satisfying the relations

$$\sum_{k=1}^{5} k x_k = a, \quad \sum_{k=1}^{5} k^3 x_k = a^2, \quad \sum_{k=1}^{5} k^5 x_k = a^3.$$

Solution 1. The relations can be rewritten as

$$a^2 \sum_{k=1}^{5} k x_k + \sum_{k=1}^{5} k^5 x_k = 2a \sum_{k=1}^{5} k^3 x_k,$$

and hence

$$\sum_{k=1}^{5} k(a - k^2)^2 x_k = 0.$$

This implies that each of the five non-negative terms is zero, i.e.,

$$k(a - k^2)^2 x_k = 0, \quad k = 1, 2, 3, 4, 5.$$

If $x_1 = x_2 = x_3 = x_4 = x_5 = 0$, then $a = 0$; otherwise, $x_k \neq 0$ and $a = k^2$ for some k.

Thus, the possible values for a are 0, 1, 4, 9, 16, 25.

Solution 2. Using the Cauchy–Schwarz inequality, we have

$$(a^2)^2 = \left(\sum_{k=1}^{5} k^3 x_k \right)^2$$

$$\leq \left(\sum_{k=1}^{5} k x_k \right) \left(\sum_{k=1}^{5} k^5 x_k \right)$$

$$= a^4.$$

The equality implies the existence of a real number λ such that $\sqrt{k^5 x_k} = \lambda \sqrt{k x_k}$, or $k^4 x_k = \lambda^2 x_k$. In other words,

$$(k^4 - \lambda^2) x_k = 0, \quad k = 1, 2, 3, 4, 5.$$

If $x_k \neq 0$ for some k, then $\lambda = k^2$. The remaining x_i are all zero. Replacing them into the original expression yields $k x_k = a$ and $k^3 x_k = a^2$, implying $a = k^2 = \lambda$, while also satisfying $k^5 x_k = a^3$.

The other case is $x_k = 0$ for $k = 1, 2, 3, 4, 5$, leading to $a = 0$.

Thus, the possible values for a are 0, 1, 4, 9, 16, 25.

【Score Situation】 This particular problem saw the following distribution of scores among contestants: 46 contestants scored 7 points, 2 contestants scored 6 points, 7 contestants scored 5 points, 4 contestants scored 4 points, 10 contestants scored 3 points, 31 contestants scored 2 points, 42 contestants scored 1 point, and 24 contestants scored 0 point. The average score for this problem is 3.127, indicating that it was relatively straightforward.

Among the top five teams in the team scores, the scores of this problem are as follows: the Soviet Union team scored 41 points (with a total team score of 267 points), the Romania team scored 40 points (with a total team score of 240 points), the Germany team scored 17 points (with a total team score of 235 points), the United Kingdom team scored 35 points (with a total team score of 218 points), and the United States team scored 38 points (with a total team score of 199 points).

The gold medal cutoff for this IMO was set at 37 points (with 8 contestants earning gold medals), the silver medal cutoff was 29 points (with 32 contestants earning silver medals), and the bronze medal cutoff was 20 points (with 42 contestants earning bronze medals).

In this IMO, a total of four contestants achieved a perfect score of 40 points.

Problem 1.16 (IMO 57-5, proposed by Russia). The equation

$$(x-1)(x-2)\cdots(x-2016) = (x-1)(x-2)\cdots(x-2016)$$

is written on the board, with 2016 linear factors on each side. What is the least possible value of k for which it is possible to erase exactly k of these 4032 linear factors so that at least one factor remains on each side and the resulting equation has no real solutions?

Solution. In order to ensure that the resulting equation has no real roots, the same linear factor cannot appear on both sides of the equation. At least one of them needs to be removed, requiring the removal of at least 2016 linear factors.

Next, it is explained that if all linear factors $(x-k)$, where $k \equiv 2,3 \pmod 4$, are removed from the left side, and all linear factors $(x-m)$, where $m \equiv 0,1 \pmod 4$, are removed from the right side, then the resulting equation

$$\prod_{j=0}^{503}(x-4j-1)(x-4j-4) = \prod_{j=0}^{503}(x-4j-2)(x-4j-3) \qquad (1)$$

will have no real roots.

Case 1: $x \in \{1,2,\ldots,2016\}$.

In this case, one side of (1) is zero, and the other side is non-zero, making (1) false.

Case 2: $x \in (4k+1, 4k+2) \cup (4k+3, 4k+4)$, where $k \in \{0,1,2,\ldots,503\}$.

For $j \in \{0,1,2,\ldots,503\}$ and $j \neq k$, the inequalities

$$(x-4j-1)(x-4j-4) > 0 \quad \text{and} \quad (x-4j-2)(x-4j-3) > 0$$

hold.

If $j = k$, then

$$(x-4j-1)(x-4j-4) < 0 \quad \text{and} \quad (x-4j-2)(x-4j-3) > 0.$$

Multiplying these inequalities results in the left side of (1) being negative and the right side being positive, making (1) false.

Case 3: $x < 1$ or $x > 2016$ or $x \in (4k, 4k+1)$, where $k \in \{1,2,\ldots,503\}$.

For $j \in \{0,1,2,\ldots,503\}$, the inequalities

$$0 < (x-4j-1)(x-4j-4) < (x-4j-2)(x-4j-3)$$

are valid.

Multiplying these inequalities results in the left side of (1) being less than the right side, making (1) false.

Case 4: $x \in (4k+2, 4k+3)$, where $k \in \{0, 1, 2, \ldots, 503\}$.

For $j \in \{1, 2, \ldots, 503\}$, the inequalities

$$0 < (x - 4j + 1)(x - 4j - 2) < (x - 4j)(x - 4j - 1)$$

are true.

Additionally, $x - 1 > x - 2 > 0 > x - 2015 > x - 2016$. Multiplying these inequalities results in

$$\prod_{j=0}^{503} (x - 4j - 1)(x - 4j - 4) < \prod_{j=0}^{503} (x - 4j - 2)(x - 4j - 3) < 0.$$

Thus, (1) is false.

In conclusion, the minimum number of linear factors that need to be removed is 2016.

Note. There are several similar problems:

- **(Turkey Tean Selection Test 2023, Problem 8).** Initially, the equation

$$\bigstar \frac{1}{x-1} \bigstar \frac{1}{x-2} \bigstar \frac{1}{x-4} \bigstar \cdots \bigstar \frac{1}{x - 2^{2023}} = 0$$

  is written on the board. In each turn Asli and Zehra deletes one of the stars in the equation and writes $+$ or $-$ instead. The first move is performed by Asli and continues in order. What is the maximum number of real solutions Asli can guarantee after all the stars have been replaced by signs?

- **(Chinese Team Selection Test 2014, Problem 15).** Show that there are no 2-tuples (x, y) of positive integers satisfying the equation

$$(x+1)(x+2) \cdots (x+2014) = (y+1)(y+2) \cdots (y+4028).$$

- **(All-Russian Mathematical Olympiad 2010, Grade 11, Problem 1).** Do there exist non-zero real numbers $a_1, a_2, \ldots, a_{10}$ for which

$$\left(a_1 + \frac{1}{a_1}\right)\left(a_2 + \frac{1}{a_2}\right) \cdots \left(a_{10} + \frac{1}{a_{10}}\right) = \left(a_1 - \frac{1}{a_1}\right)\left(a_2 - \frac{1}{a_2}\right) \cdots$$

$$\left(a_{10} - \frac{1}{a_{10}}\right)?$$

- **(All-Russian Mathematical Olympiad 2005, Final Round, Grade 11, Problem 1).** Assume that the number of solutions N of the equation

$$|x - a_1| + |x - a_2| + \cdots + |x - a_{50}| = |x - b_1| + |x - b_2| + \cdots + |x - b_{50}|$$

is finite, where a_i, b_j are distinct numbers. What is the greatest possible value of N?

- **(All-Russian Mathematical Olympiad 2000, 4th Round, Grade 11, Problem 1).** Prove that there exist distinct real numbers $a_1, a_2, \ldots, a_{10}$ such that the equation

$$(x - a_1)(x - a_2) \cdots (x - a_{10}) = (x + a_1)(x + a_2) \cdots (x + a_{10})$$

has exactly five distinct real roots.

【Score Situation】 This particular problem saw the following distribution of scores among contestants: 81 contestants scored 7 points, 4 contestants scored 6 points, 2 contestants scored 5 points, 50 contestants scored 4 points, 21 contestants scored 3 points, 55 contestants scored 2 points, 36 contestants scored 1 point, and 353 contestants scored 0 point. The average score for this problem is 1.678, indicating that it was relatively challenging.

Among the top five teams in the team scores, the scores of this problem are as follows: the United States team scored 41 points (with a total team score of 214 points), the South Korea team scored 33 points (with a total team score of 207 points), the China team scored 42 points (with a total team score of 204 points), the Singapore team scored 42 points (with a total team score of 196 points), and the Chinese Taiwan team scored 30 points (with a total team score of 175 points).

The gold medal cutoff for this IMO was set at 29 points (with 44 contestants earning gold medals), the silver medal cutoff was 22 points (with 101 contestants earning silver medals), and the bronze medal cutoff was 16 points (with 135 contestants earning bronze medals).

In this IMO, a total of six contestants achieved a perfect score of 42 points.

1.3 Summary

In the first 64 IMOs, there were a total of 16 equation problems. These problems can be broadly categorized into three types, as depicted in Figure 1.2. The score details for these problems are presented in Table 1.2. Due to the smaller number of participating teams and missing contestant score information in early IMOs, there are several blanks in Table 1.2.

Problems 1.1–1.7 focus on "finding solutions of equations and systems of equations;" among these seven problems, the one with the lowest average

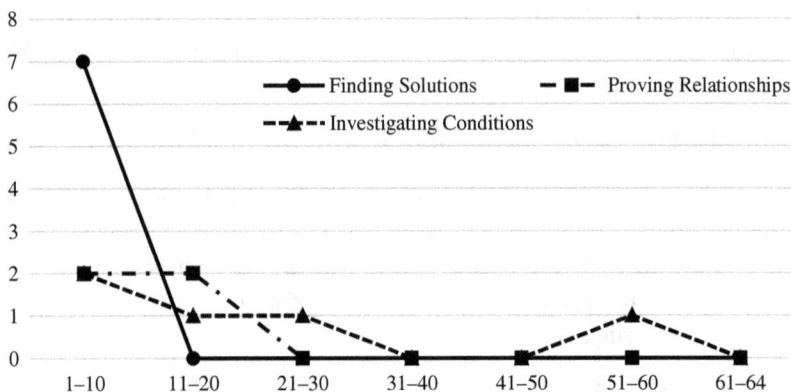

Figure 1.2 Numbers of Equation Problems in the First 64 IMOs

score is Problem 1.6 (IMO 5-4), proposed by the Soviet Union. Problems 1.8–1.11 deal with "proving relationships satisfied by equations and systems of equations;" among these four problems, the one with the lowest average score is Problem 1.11 (IMO 18-5), proposed by the Netherlands. Problems 1.12–1.16 are about "investigating conditions under which equations and systems of equations have solutions;" among these five problems, the one with the lowest average score is Problem 1.16 (IMO 57-5), proposed by Russia.

These 16 problems were proposed by 11 countries. The Soviet Union, Romania, Bulgaria, Czechoslovakia, and Hungary each contributed two problems.

From Table 1.2, it can be observed that in the first 64 IMOs, there were four equation problems with an average score of 1–2 points; two problems with an average score of 2–3 points; five problems with an average score of 3–4 points; five problems with an average score above 4 points. Overall, the equation problems were relatively simple, and the contestants scored relatively high.

In the 24th–64th IMOs, there was only one equation problem, specifically Problem 1.16 (IMO 57-5), which had an average score of 1.678 points. This indicates that equation problems were primarily featured in the first 23 IMOs, as shown in Table 1.3.

From Table 1.2, it can be observed that, excluding Problem 1.16 (IMO 57-5), the average score of the top five teams is typically about 1 point higher than the average score of the problem. However, the average score

Table 1.2 Score Details of Equation Problems in the First 64 IMOs

Problem	1.1	1.2	1.3	1.4	1.5	1.6	1.7	1.8
Full points	8.000	5.000	6.000	7.000	6.000	6.000	7.000	6.000
Average score	6.111	3.941	4.188	2.556	3.500	1.375	5.000	2.700
Top five mean			4.950					3.975
6th–15th mean								
16th–25th mean								
Problem number in IMO	1-2	4-4	7-4	3-3	5-1	5-4	8-5	7-2
Proposing country	Romania	Romania	The Soviet Union	Bulgaria	Czecho-slovakia	The Soviet Union	Czecho-slovakia	Poland

Problem	1.9	1.10	1.11	1.12	1.13	1.14	1.15	1.16
Full points	7.000	7.000	7.000	7.000	6.000	8.000	7.000	7.000
Average score	4.044	1.935	1.094	4.222	3.222	3.992	3.127	1.678
Top five mean		3.325	2.050			4.700	4.275	6.267
6th–15th mean		1.563	0.863			3.850	3.434	4.485
16th–25th mean								2.636
Problem number in IMO	10-3	18-2	18-5	1-3	3-1	15-3	21-5	57-5
Proposing country	Bulgaria	Finland	The Netherlands	Hungary	Hungary	Sweden	Israel	Russia

Note. Top five mean = Total score of the top five teams ÷ Total number of contestants from the top five teams,
6th–15th mean = Total score of the 6th–15th teams ÷ Total number of contestants from the 6th–15th teams,
16th–25th mean = Total score of the 16th–25th teams ÷ Total number of contestants from the 16th–25th teams.

Table 1.3 Numbers of Equation Problems in the 24th–64th IMOs

Equation Problem	Problem Number			Number of Problems in the First 64 IMOs
	1, 4	2, 5	3, 6	
Finding solutions	0	0	0	7
Proving relationships	0	0	0	4
Investigating conditions	0	1	0	5
Total	0	1	0	16

of the 6th–15th teams tends to be close to or below the average score of the problem, as seen in Problem 1.10 (IMO 18-2), Problem 1.11 (IMO 18-5), Problem 1.14 (IMO 15-3), and Problem 1.15 (IMO 21-5).

This phenomenon is due to the smaller number of participating teams in early IMOs. It was not until the 22nd IMO in 1981 that the number of participating teams exceeded 25. Furthermore, the relatively low level of difficulty associated with equation problems during this period suggests a minimal variation in the average scores between the overall, top five, and 6th–15th teams.

Chapter 2

Function Problems

The concept of functions represents a type of correspondence between sets. Since this concept was introduced, it has permeated various levels of mathematics, and the theory surrounding it has become quite comprehensive. However, there are still many issues that remain worthy of exploration.

In mathematics competitions, problems often arise where the specific form of a function is not given. Instead, certain properties, relations, or functional equations are provided, and the task is to determine the function's expression, calculate its values, or prove properties it possesses.

In the first 64 IMOs, there had been a total of 28 function problems, accounting for approximately 27.7% of all algebra problems. These problems can be primarily categorized into three types: (1) proving properties of functions, totaling nine problems; (2) determining numerical values of function variables or outputs, totaling four problems; (3) deriving expressions for functions that meet specific conditions, totaling 15 problems. The statistical distribution of these three types of problems in the previous IMOs is presented in Table 2.1.

From 21st–40th and 51st–60th IMOs, function problems accounted for a high proportion, close to or exceeding 40%. Initially, most problems involved proving properties of functions, requiring some mathematical analysis methods in advanced mathematics. Subsequently, problems on determining function values and solving functional equations appeared, with the former being relatively straightforward when the properties of the function were clear. Consequently, such problems gradually faded, while functional equations became a focal point.

Table 2.1　Numbers of Function Problems in the First 64 IMOs

Content	Session							Total
	1–10	11–20	21–30	31–40	41–50	51–60	61–64	
Proving properties	1	4	1	1	0	2	0	9
Determining values	0	1	3	0	0	0	0	4
Deriving expressions	0	0	2	4	3	5	1	15
Algebra problems	20	20	14	13	15	13	6	101
The percentage of function problems among the algebra problems	5.0%	25.0%	42.9%	38.5%	20.0%	53.8%	16.7%	27.7%

The study of functional equations dates back to the 14th-century mathematician Nicola Oresme. In 1352, in his paper "Tractatus de configurationibus qualitatum et motuum," Oresme employed the functional equation

$$\frac{y - x}{z - y} = \frac{f(y) - f(x)}{f(z) - f(y)}$$

to indirectly provide a definition of linear functions and use graphs to represent such functional relationships, predating René Descartes.

Although Oresme's work on linear functions can be considered an early example of functional equations, the field's theoretical foundation is more aptly marked by Cauchy's work. In 1676, Newton extended the binomial theorem to

$$(1 + x)^z = 1 + C_z^1 x + C_z^2 x^2 + C_z^3 x^3 + \cdots,$$

with z as any real number, but his derivation was not rigorous. Cauchy further refined Newton's work. On this basis, Cauchy defined the function

$$f(z) = 1 + C_z^1 x + C_z^2 x^2 + C_z^3 x^3 + \cdots$$

and proved that $f(z+w) = f(z)f(w)$ for any real numbers z and w, known as Cauchy's exponential functional equation.

Later, many renowned mathematicians, such as Jean Le Rond d'Alembert, Léonard Euler, Carl Friedrich Gauss, Adrien-Marie Legendre, Niels Henrik Abel, and David Hilbert, contributed to the study of functional equations, leading to equations named after them. Today, functional equations remain an actively developing branch of mathematics, with widespread applications in various scientific and technical fields.

Furthermore, functional equations frequently appeared in publications like the Mathematical Association of America's *Mathematics Magazine* and the Canadian Mathematical Society's *Crux Mathematicorum*, often solvable with elementary mathematics.

Functional equations also appeared in various secondary and tertiary mathematics competitions. For instance, the Putnam Mathematical Competition has featured numerous such problems, the earliest being from the 4th Putnam Mathematical Competition in 1941:

- **(B14).** Show that any solution $f(t)$ of the functional equation

$$f(x+y)f(x-y) = (f(x))^2 + (f(y))^2 - 1, \quad (x, y \text{ are real}),$$

is such that

$$f''(t) = \pm m^2 f(t), \quad (m \text{ is a constant and } m \geq 0),$$

assuming the existence and continuity of the second derivative. Deduce that $f(t)$ is one of the functions $\pm \cos mt$ and $\pm \cosh mt$.

These problems typically involve advanced mathematics and are thus more common in university-level competitions.

The development of functional equations has also influenced Mathematical Olympiad problems, such as the 5th problem of the 14th IMO in 1972:

- **(IMO 14-5).** Let f and g be real-valued functions defined for all real values of x and y, and satisfying the equation

$$f(x+y) + f(x-y) = 2f(x)g(y)$$

for all $x, y \in \mathbf{R}$. Prove that if $f(x)$ is not identically zero and $|f(x)| \leq 1$ for all $x \in \mathbf{R}$, then $|g(y)| \leq 1$ for all $y \in \mathbf{R}$.

This problem represents a special case of functional equations of interest to applied mathematicians in the 18th century.

Over recent decades, problems involving deriving expressions have appeared frequently in the IMO, almost every two to three years. This is due to the fact that many significant problems in natural sciences can ultimately be reduced to solving specific or classes of functional equations. However, a complete and systematic theory of functional equations has not yet been established, and general methods for solving them are scarce. Thus, solving functional equations itself demands a good mathematical foundation.

Function iterations, a special case of function compositions, often appear in functional equations. Specifically, in cases where the recurrence relation

of a sequence $\{a_n\}$ is known, a_n can be similarly regarded as the nth iteration of the initial value a_0 under the recurrence relation. Moreover, function
iterations can sometimes serve as a problem-solving approach, such as using
fixed-point properties to solve functional equations.

This chapter will be divided into three parts. The first part introduces
basic properties of functions and common functional equations, followed
by presentations of common methods for solving function iterations and
functional equations. These methods include the Cauchy method, method
of undetermined coefficients, fixed point method, and substitution method,
all of which find applications in the IMO.

The second part revolves around three types of problems: "proving
properties of functions," "determining numerical values of function variables or outputs," and "deriving expressions for functions that meet specific conditions." These problems are presented in chronological order, and
some problems include various solutions, generalizations, and similar problems. Problems combining functional equations with number theory are
presented in *IMO Problems, Theorems, and Methods: Number Theory*.

It is important to note that for each problem, the solutions are followed by information on the scores, including the number of contestants in
each score range, the average score, and the scores of the top five teams.
However, early IMOs often lacked information on contestant scores, so the
number of contestants in each score range only represents the counted number of contestants, and some problems lack scores of the top five teams.

The third part provides a brief summary of this chapter.

It should be noted that even functional equations with similar structures
may require entirely different approaches. In such cases, more intuitive
ideas can prove to be more effective. These include considering whether the
function is injective (one-to-one) or surjective (onto), determining whether
the function is odd or even, examining the set of points where the function equals the target function, and proving that it constitutes the entire
set. Although these ideas are straightforward and can be easily overlooked,
continuous exploration may outline the entire problem.

2.1 Common Theorems, Formulas, and Methods

2.1.1 *Common elementary functions*

Let A and B be non-empty sets. If for every element x in set A, there
exists a unique corresponding element y in set B based on a certain defined
correspondence f, then $f : A \to B$ is called a function from set A to set B,

denoted as

$$y = f(x), \quad x \in A.$$

Here, x is referred to as the independent variable, and its range A is called the domain of the function. The corresponding value y of x is known as the function value, and the set of function values $\{f(x)|x \in A\}$ is termed the range of the function. If for every y in the range, there exists a unique x in the domain such that $f(x) = y$, then the function mapping y back to x is called the inverse function of f, denoted as:

$$f^{-1} : \{f(x)|x \in A\} \to A,$$

or

$$x = f^{-1}(y), \quad y \in \{f(x)|x \in A\}.$$

It is important to note that the domain of function f is precisely the range of its inverse function f^{-1}, and vice versa. Moreover, the inverse function notation "f^{-1}" should not be confused with the reciprocal $\frac{1}{f}$.

Common elementary functions include:

(1) *Power functions*

The function $y = x^a (a \in \mathbb{R})$ is called a power function, where the base x is the independent variable and the exponent a is a constant, such as $y = x$ and $y = \frac{1}{x}(x \neq 0)$.

(2) *Exponential functions*

The function $y = a^x (a > 0, a \neq 1)$ is called an exponential function, where the exponent x is the independent variable and the base a is a constant. For $a > 1$, the function $y = a^x$ is monotonically increasing; for $0 < a < 1$, the function $y = a^x$ is monotonically decreasing.

(3) *Logarithmic functions*

The function $y = \log_a x (a > 0, a \neq 1)$ is called a logarithmic function, where the positive real number x is the independent variable and the base a is a constant.

A logarithmic function is the inverse function of an exponential function, and their graphs in the Cartesian coordinate system are symmetric about the line $y = x$. For $a > 1$, the function $y = \log_a x$ is monotonically increasing; for $0 < a < 1$, the function $y = \log_a x$ is monotonically decreasing.

(4) *Trigonometric functions*

These include the sine function $\sin x$, cosine function $\cos x$, tangent function $\tan x$, cotangent function $\cot x$, secant function $\sec x$, and cosecant function $\csc x$.

In secondary mathematics, the sine, cosine, and tangent functions are frequently encountered, and like other elementary functions, they exhibit basic properties such as monotonicity, periodicity, and odd-even symmetry.

(5) *Inverse trigonometric functions*

Inverse trigonometric functions include the arcsine function $\arcsin x$, arccosine function $\arccos x$, arctangent function $\arctan x$, arccotangent function $\text{arccot}\, x$, arcsecant function $\text{arcsec}\, x$, and arccosecant function $\text{arccsc}\, x$.

There is a difference between these functions and the inverse functions of other common elementary functions, i.e., they are multivalued.

(6) *Constant functions*

Constant functions are those whose function values remain unchanged, i.e., the function values are a constant.

Consider two functions

$$z = f(y), \quad \text{where } y \in E,$$

$$y = g(x), \quad \text{where } x \in D.$$

If the set $D^* = \{x \in D | g(x) \in E\}$ is non-empty, then a new function can be defined with elements from D^* as the independent variable, passing through the successive actions of g and f, and elements from the range of f as the function value, denoted as

$$z = f(g(x)), \quad x \in D^*.$$

This function is usually represented as $f \circ g$ and called the function composition of f and g.

Function compositions follow the associative law, i.e., $(f \circ g) \circ h = f \circ (g \circ h)$, but not the commutative law, as $f \circ g$ and $g \circ f$ are generally not the same. There is a related problem:

- **(Romanian Master of Mathematics 2013, Problem 2).** Does there exist a pair (g, h) of functions $g, h : \mathbf{R} \to \mathbf{R}$ such that the only function $f : \mathbf{R} \to \mathbf{R}$ satisfying $f(g(x)) = g(f(x))$ and $f(h(x)) = h(f(x))$ for all $x \in \mathbf{R}$ is the identity function $f(x) \equiv x$?

2.1.2 *Basic properties of functions*

In mathematics, an injective function, also known as injection or one-to-one function, is a function $f : A \to B$ that maps distinct elements of A to distinct elements, i.e., if $x_1 \neq x_2$, then $f(x_1) \neq f(x_2)$.

A surjective function, also known as surjection or onto function, is a function $f : A \to B$ such that, for every element $y \in B$, there exists at least one element $x \in A$ satisfying $y = f(x)$.

A bijective function, also known as bijection, is a function that is both injective and surjective.

Actually, every bijective function $f : \mathbf{Z} \to \mathbf{Z}$ can be written in the way $f = u + v$, where $u, v : \mathbf{Z} \to \mathbf{Z}$ are bijective functions (Romanian Master of Mathematics 2008, Problem 2).

(1) *Odd–even symmetry*

Suppose the domain D of a function $f(x)$ is symmetric about the origin. If $f(-x) = -f(x)$ for every $x \in D$, then $f(x)$ is called an odd function. If $f(-x) = f(x)$ for every $x \in D$, then $f(x)$ is called an even function.

Actually, every function f can be written in the way $f = g + h$, where g is odd and h is even. Furthermore, the following propositions are relatively straightforward to prove by using definitions.

Proposition 2.1. *Only the zero function can be both odd and even.*

Proposition 2.2. *The sum of an odd function and an even function is neither odd nor even, unless one of them is identically zero.*

Proposition 2.3. *The sum and difference of two odd (even) functions are odd (even) functions, and an odd (even) function multiplied by a non-zero constant remains odd (even).*

Proposition 2.4. *The product and quotient of two odd (even) functions are even functions, while the product and quotient of an odd function and an even function are odd functions.*

Proposition 2.5. *The composition of two odd functions is an odd function, the composition of an even function and an odd function (the even function acts second) is an even function, and the composition of any function and an even function (the even function acts first) is an even function.*

More generally, for a function $f(x)$ with domain $\mathbf{R}$, the following holds:

Proposition 2.6. *If $f(a+x) = f(b-x)$ is always true, then the graph of $f(x)$ is symmetric about the line $x = \frac{a+b}{2}$.*

Proposition 2.7. *If $f(a+x) = c - f(b-x)$ is always true, then the graph of $f(x)$ is symmetric about the point $\left(\frac{a+b}{2}, \frac{c}{2}\right)$.*

Proposition 2.8. *The function $g(x) = f(a+x) + f(b-x)$ is symmetric about the line $x = \frac{b-a}{2}$.*

Proposition 2.9. *The function $g(x) = f(a+x) - f(b-x)$ is symmetric about the point $\left(\frac{b-a}{2}, 0\right)$.*

(2) *Monotonicity*

Let the domain of a function $f(x)$ be D. If for any $x_1, x_2 \in D_1 \subset D$, it holds that $f(x_1) < f(x_2)$ whenever $x_1 < x_2$, then $f(x)$ is said to be (strictly) increasing on D_1; if $f(x_1) > f(x_2)$ whenever $x_1 < x_2$, then $f(x)$ is said to be (strictly) decreasing on D_1.

If the function $f(x)$ is increasing or decreasing on a subset of its domain (usually an interval), then $f(x)$ is said to be (strictly) monotonic on this subset. This subset is referred to as the monotonic interval of $f(x)$.

It should be noted that the monotonicity of a function pertains to a subset of its domain and is a "local" property. Even if $f(x)$ is increasing (decreasing) on subintervals D_1 and D_2 of its domain, it cannot be directly inferred that $f(x)$ is increasing (decreasing) on $D_1 \cup D_2$.

Furthermore, in advanced mathematics, the definition of monotonicity is slightly weaker, indicating that $f(x_1) \leq f(x_2)$ or $f(x_1) \geq f(x_2)$ for $x_1 < x_2$.

The following propositions are relatively straightforward to prove by using definitions.

Proposition 2.10. *If functions $f(x)$ and $g(x)$ are both increasing (decreasing) on a common interval I, then the function $f(x) + g(x)$ is also increasing (decreasing) on I.*

Proposition 2.11. *If two positive-valued functions $f(x)$ and $g(x)$ are both increasing (decreasing) on a common interval I, then the function $f(x)g(x)$ is also increasing (decreasing) on I.*

Proposition 2.12. *If function $f(x)$ is increasing (decreasing) on interval I with the range $f(I)$, then the inverse function $f^{-1}(x)$ certainly exists and is increasing (decreasing) on $f(I)$.*

Proposition 2.13. *If functions $f(x)$ and $g(x)$ are both monotonic on their respective intervals, then the function $f(g(x))$ is also monotonic on the monotonic interval of $g(x)$. If $f(x)$ and $g(x)$ have the same type of monotonicity, then $f(g(x))$ is increasing. If $f(x)$ and $g(x)$ have opposite types of monotonicity, then $f(g(x))$ is decreasing.*

Moreover, there is a related problem:

- **(Romanian Master of Mathematics 2011, Problem 1).** Prove that there exist two functions $f, g : \mathbf{R} \to \mathbf{R}$ such that $f \circ g$ is strictly decreasing and $g \circ f$ is strictly increasing.

(3) *Periodicity*

Let the domain of a function $f(x)$ be D. If there exists a non-zero constant T such that $x + T \in D$ and $f(x + T) = f(x)$ for every $x \in D$, then $f(x)$ is termed a periodic function, and T is known as a period of the function $f(x)$.

If among all positive periods of the function $f(x)$, there exists the smallest value T_0, then T_0 is called the fundamental period of the periodic function $f(x)$.

However, not every periodic function has a fundamental period. For instance, a constant function can have any real number as a period, and the function

$$d(x) = \begin{cases} 1, & x \in \mathbf{Q}, \\ 0, & x \notin \mathbf{Q} \end{cases}$$

can have any rational number as a period.

The following propositions are relatively straightforward to prove by using definitions.

Proposition 2.14. *If a function $f(x)$ with domain $\mathbf{R}$ is symmetric about the lines $x = a$ and $x = b$, where $a \neq b$, then $f(x)$ is periodic with a period $T = 2|a - b|$.*

In particular, $f(x)$ being an even function is equivalent to $f(x)$ being symmetric about the line $x = 0$.

Proposition 2.15. *If a function $f(x)$ with domain $\mathbf{R}$ is symmetric about the points $(a, 0)$ and $(b, 0)$, where $a \neq b$, then $f(x)$ is periodic with a period $T = 2|a - b|$.*

In particular, $f(x)$ being an odd function is equivalent to $f(x)$ being symmetric about the origin.

Proposition 2.16. *If a function $f(x)$ with domain $\mathbf{R}$ is symmetric about the line $x = a$ and the point $(b, 0)$, where $a \neq b$, then $f(x)$ is periodic with a period $T = 4|a - b|$.*

(4) *Convexity and concavity*

Let $f(x)$ be a function defined on an interval I. If for any $x, y \in I$ and any $t \in [0, 1]$,

$$f(tx + (1 - t)y) \leq tf(x) + (1 - t)f(y),$$

then $f(x)$ is called a convex function on I. Conversely, if for any $x, y \in I$ and any $t \in [0, 1]$,

$$f(tx + (1 - t)y) \geq tf(x) + (1 - t)f(y),$$

then $f(x)$ is called a concave function on I.

It is important to note that the terminology for convex and concave functions has varied considerably in different periods of the literature, what is defined as a convex function in one text might be defined as a concave function in another, necessitating a careful distinction.

The following propositions are relatively straightforward to prove by using definitions.

Proposition 2.17. *If $f(x)$ is a convex function defined on I and c is a positive real number, then $cf(x)$ is also a convex function.*

Proposition 2.18. *If $f(x)$ and $g(x)$ are both convex functions defined on I, then $h_1(x) = f(x) + g(x)$ and $h_2(x) = \max\{f(x), g(x)\}$ are both convex functions.*

Proposition 2.19. *If $f(x)$ and $g(x)$ are both convex functions and $g(x)$ is increasing, then $g(f(x))$ is a convex function.*

Proposition 2.20. *If $f(x)$ is a convex function defined on I, then $q(x) = f(ax + b)$ is also a convex function, where a, b are real numbers, with $a \neq 0$.*

Convex and concave functions are closely related to Jensen's inequality, which will be introduced in Chapter 4 on inequalities.

(5) *Continuity*

A continuous function is one in which a sufficiently small change in the independent variable results in a correspondingly small change in the function value. However, this is not the rigorous definition of a continuous function, which relies on the concept of limits in advanced mathematics.

Let x_0 be a point in the domain D of the function $f(x)$. If for any sequence $\{x_n\} \subset D$ converging to x_0, the sequence $\{f(x_n)\}$ tends to $f(x_0)$, then $f(x)$ is said to be continuous at x_0.

Suppose the domain of the function $f(x)$ is D, and the interval I is a subset of D. If $f(x)$ is continuous at every point x in I, then $f(x)$ is said to be continuous on the interval I.

Intuitively, functions whose graphs are unbroken curves are continuous. For example, polynomial function, power function, exponential function, logarithmic function, trigonometric function, and the absolute value function are all continuous on subintervals of their domains.

The following propositions are relatively straightforward to prove by using definitions.

Proposition 2.21. *If f and g are both continuous functions, then $f \pm g$, fg, and $\frac{f}{g}$ are all continuous functions; note that the domain of $\frac{f}{g}$ excludes points x where $g(x) = 0$.*

Proposition 2.22. *The composition $f \circ g$ of two continuous functions f and g is a continuous function.*

It is not an exaggeration to say that the concept of continuous functions has widespread applications in almost all disciplines related to mathematics.

2.1.3 *Common function iterations*

Let $f : D \to D$ be a function. For any $x \in D$, denote

$$f^{(0)}(x) = x, \quad f^{(n+1)}(x) = f(f^{(n)}(x)) \quad \text{for } n \in \mathbf{N}_+.$$

Then $f^{(n)}(x)$ is called the nth iteration of the function $f(x)$ over the domain D, and n is called the iteration index.

It's important to note that $f^{(n)}(x)$ is also commonly used to represent the nth derivative of $f(x)$. Therefore, it's crucial to discern the context of a problem for a proper understanding.

In the above defined function iteration, the iteration index is always a non-negative integer. If $f(x)$ is a bijection from D to D, negative iteration indices can also be defined as follows.

Let $f(x)$ be a bijection from D to D. Define $f^{(-1)}(x) = f^{-1}(x)$, i.e., the inverse function of $f(x)$. Then, for any positive integer n, define $f^{(-n)}(x)$ as the nth iteration of $f^{(-1)}(x)$.

Common forms of function iterations include:

(i) The nth iteration of the function $f(x) = x + c$ is $f^{(n)}(x) = x + nc$.

(ii) The nth iteration of the function $f(x) = x^m$ is $f^{(n)}(x) = x^{m^n}$.

(iii) The nth iteration of the function $f(x) = ax + b$ is

$$f^{(n)}(x) = a^n \left(x - \frac{b}{1-a} \right) + \frac{b}{1-a}.$$

(iv) The nth iteration of the function $f(x) = \frac{x}{a+bx}$ is

$$f^{(n)}(x) = \frac{x}{a^n + bx \cdot \frac{1-a^n}{1-a}}.$$

(v) The nth iteration of the function $f(x) = ax^2 + bx + c$ (where $a \neq 0$ and $4ac = b^2 - 2b$) is

$$f^{(n)}(x) = a^{2^n - 1} \left(x + \frac{b}{2a} \right)^{2n} - \frac{b}{2a}.$$

Note that some functions' iterations may exhibit a periodic behavior. For example, for the function $f(x) = \frac{x-1}{x}$, it is easy to verify that $f^{(4)}(x) = f^{(1)}(x) = f(x)$.

2.1.4 *Common methods for determining function iterations*

The primary issue in function iterations is to determine the expression for the nth iteration of a given function.

(1) *Mathematical induction*

This approach involves guessing and then proving. Initially, iterate the function $f(x)$ a few times, observe the pattern, conjecture the expression for $f^{(n)}(x)$, and finally prove it using mathematical induction.

Example 2.1. Let $f(x) = ax + b$ and find $f^{(n)}(x)$.

Solution. It is straightforward that

$$f(x) = ax + b,$$

$$f^{(2)}(x) = f(f(x)) = a(ax + b) + b = a^2x + ab + b,$$
$$f^{(3)}(x) = f(f^{(2)}(x)) = a(a^2x + ab + b) + b = a^3x + a^2b + ab + b,$$

leading to the conjecture $f^{(n)}(x) = a^n x + a^{n-1}b + a^{n-2}b + \cdots + ab + b$.

Next, prove it by mathematical induction. For $n = 1$, the proposition holds.

Assume $f^{(k)}(x) = a^k x + a^{k-1}b + a^{k-2}b + \cdots + ab + b$ holds. Then

$$f^{(k+1)}(x) = f(f^{(k)}(x)) = a(a^k x + a^{k-1}b + \cdots + ab + b) + b$$
$$= a^k x + a^{k-1}b + \cdots + ab + b,$$

implying that the proposition also holds for $n = k + 1$, and thus for any positive integer n.

(2) *Recursive method*

This involves solving by constructing a sequence. Let $a_0 = x$ and $a_n = f^{(n)}(x)$ for all positive integers n. From $a_n = f^{(n)}(x) = f(a_{n-1})$, find $a_n = g(a_0)$. Then $g(x) = f^{(n)}(x)$.

Example 2.2. Let $f(x) = 3\sqrt[3]{x}(\sqrt[3]{x} + 1) + x + 1$ and find $f^{(n)}(x)$.

Solution. It is evident that $f(x) = (\sqrt[3]{x} + 1)^3$. Set $a_0 = x$ and $a_n = f^{(n)}(x)$. Then

$$a_n = f(a_{n-1}) = (\sqrt[3]{a_{n-1}} + 1)^3,$$

which implies $\sqrt[3]{a_n} - \sqrt[3]{a_{n-1}} = 1$. Hence, $\sqrt[3]{a_n} = \sqrt[3]{a_0} + n = \sqrt[3]{x} + n$, leading to

$$a_n = (\sqrt[3]{x} + n)^3,$$

i.e., $f^{(n)}(x) = (\sqrt[3]{x} + n)^3$.

(3) *Similarity method*

If there exist a function $\varphi(x)$ and its inverse $\varphi^{-1}(x)$ such that $f(x) = \varphi^{-1}(g(\varphi(x)))$, then the functions $f(x)$ and $g(x)$ are said to be conjugate or similar. We call $\varphi(x)$ the bridge function between $f(x)$ and $g(x)$, and it is denoted as $f \sim g$ or $f \sim_\varphi g$.

The significance of introducing a bridge function lies in its ability to establish an equivalence relation, maintaining the similarity during the process of function iterations. It is easy to observe the following properties:

(i) Reflexive Property: $f \sim f$.

(ii) Symmetric Property: If $f \sim g$, then $g \sim f$.

(iii) Transitive Property: If $f \sim g$ and $g \sim h$, then $f \sim h$.

(iv) If $f \sim g$, then $f^{(n)} \sim g^{(n)}$, i.e., $f^{(n)}(x) = \varphi^{-1}(g^{(n)}(\varphi(x)))$.

This implies that if functions $f(x)$ and $g(x)$ are similar, then finding the nth iteration of $f(x)$ is essentially equivalent to finding the nth iteration of $g(x)$. If the nth iteration of $g(x)$ is easier to determine, then the problem becomes more straightforward.

Example 2.3. Let $f(x) = \frac{x}{1+ax}$ and find $f^{(n)}(x)$.

Solution. Let $g(x) = x + a$ and $\varphi(x) = \frac{1}{x}$. Then $\varphi^{-1}(x) = \frac{1}{x}$. It is easy to verify that $f(x) = \varphi^{-1}(g(\varphi(x)))$, i.e., $f \sim g$. Therefore,

$$f^{(n)}(x) = \varphi^{-1}(g^{(n)}(\varphi(x))) = \varphi^{-1}(\varphi(x) + na)$$

$$= \frac{1}{\frac{1}{x} + na} = \frac{x}{1 + nax}.$$

(4) *Fixed point method*

In the process of deriving function iterations using the similarity method, the key step is finding a bridge function $\varphi(x)$. However, this is not easy in most cases. Here, we introduce a method for identifying a class of bridge functions, i.e., the fixed point method.

A fixed point of a function $f(x)$ is a point that satisfies $f(x) = x$. It is evident that fixed points have the following properties:

(i) If x_0 is a fixed point of $f(x)$, then x_0 is also a fixed point of $f^{(n)}(x)$.

(ii) If $f \sim_\varphi g$ and x_0 is a fixed point of $f(x)$, then $\varphi(x_0)$ is a fixed point of $g(x)$.

Note that when using the similarity method, $g(x)$ is often chosen as $x+1$, ax, ax^2, ax^3, etc., where the fixed points of $g(x)$ are 0 or $+\infty$. Therefore, based on the properties of fixed points, the chosen bridge function $\varphi(x)$ should map the fixed points of $f(x)$ to 0 or $+\infty$.

If $f(x)$ has only one fixed point x_0, then consider $\varphi(x) = x - x_0$ or $\varphi(x) = \frac{1}{x-x_0}$. If $f(x)$ has two distinct fixed points α and β, then consider $\varphi(x) = \frac{x-\alpha}{x-\beta}$.

Example 2.4. Let $f(x) = \frac{x+6}{x+2}$ and find $f^{(n)}(x)$.

Solution. Setting $f(x) = x$, we find that the fixed points of $f(x)$ are 2 and -3. Let $\varphi(x) = \frac{x-2}{x+3}$ and $g(x) = -\frac{1}{4}x$, and it is easy to verify $\varphi(f(x)) = g(\varphi(x))$.

Hence, $f \sim g$ and $f^{(n)} \sim g^{(n)}$, and since $g^{(n)}(x) = \left(-\frac{1}{4}\right)^n x$, we have

$$f^{(n)}(x) = \varphi^{-1}(g^{(n)}(\varphi(x))) = \frac{(2 \cdot (-4)^n + 3)x + 6 \cdot ((-4)^n - 1)}{((-4)^n - 1)x + (3 \cdot (-4)^n + 2)}.$$

In this way, we can deduce a bridge function's expression based on the function's fixed points and then apply the similarity method to find the nth iteration of the function. Similarly, we have:

(i) For functions like $f(x) = a(x-h)^k + h(a \neq 0, k \notin \{0,1\})$, a bridge function can be $\varphi(x) = x - h$.

(ii) For functions like

$$f_1(x) = \sqrt[k]{ax^k + b}, \quad f_2(x) = \frac{x}{\sqrt[k]{ax^k + b}}, \quad f_3(x) = \frac{\sqrt[k]{ax^k + b}}{x},$$

where $ab \neq 0$, a bridge function can be $\varphi(x) = x^k$.

(iii) For functions like

$$f_1(x) = (a\sqrt[k]{x} + b)^k, \quad f_2(x) = \frac{x}{(a\sqrt[k]{x} + b)^k}, \quad f_3(x) = \frac{(a\sqrt[k]{x} + b)^k}{x},$$

where $ab \neq 0$, a bridge function can be $\varphi(x) = \sqrt[k]{x}$.

(iv) For functions like $f(x) = \frac{cx^k}{x^k - c(x-c)^k}$, where $c \neq 0$, a bridge function can be $\varphi(x) = \frac{x-c}{cx}$.

Note that the expression of the nth iteration of most functions $f(x)$ is not easy to determine and can be quite complex. In such cases, it is more common to study the properties of $f^{(n)}(x)$, such as given x_0, examining the sequence

$$x_0, f(x_0), f^{(2)}(x_0), \ldots, f^{(n)}(x_0), \ldots$$

and the quantitative relationships it satisfies. This sequence is called the orbit of x_0, and the function sequence

$$x, f(x), f^{(2)}(x), \ldots, f^{(n)}(x), \ldots$$

is called the Picard sequence of the function $f(x)$. There is a related problem:

- **(Romanian Master in Mathematics 2010, Problem 6).** Given a polynomial $f(x)$ with rational coefficients, of degree $d \geq 2$, we define the sequence of sets $f^{(0)}(\mathbf{Q}), f^{(1)}(\mathbf{Q}), \ldots$ as

$$f^{(0)}(\mathbf{Q}) = \mathbf{Q}, \quad f^{(n+1)}(\mathbf{Q}) = f(f^{(n)}(\mathbf{Q})) \quad \text{for } n \geq 0.$$

  (Given a set S, we write $f(S)$ for the set $\{f(x)|x \in S\}$.)

  Let $f^{(\omega)}(\mathbf{Q}) = \bigcap\limits_{n=0}^{\infty} f^{(n)}(\mathbf{Q})$ be the set of numbers that are in all of the sets $f^{(n)}(\mathbf{Q})$ with $n \geq 0$. Prove that $f^{(\omega)}(\mathbf{Q})$ is a finite set.

2.1.5 *Common functional equations*

The renowned Hungarian mathematician János Aczél defined functional equations as follows: "Functional equations are equations in which both sides contain a finite number of functions, some known and some unknown."

In layman's terms, functional equations are equations where the unknowns are functions, but the number of solutions to a functional equation is not necessarily related to the number of unknown functions.

Furthermore, functional equations can be classified in various ways, such as by the domain of the unknown functions, by the number of variables in the functions, by the number of unknown functions, or by the number of iterations the unknown functions undergo, among others.

Common functional equations include (some of which are named after famous mathematicians):

(1) *Cauchy's functional equation*

Let $f : \mathbf{R} \to \mathbf{R}$ be a continuous function. If for all real numbers x and y, the Cauchy's functional equation

$$f(x + y) = f(x) + f(y)$$

is satisfied, then for all $x \in \mathbf{R}$, one has $f(x) = f(1) \cdot x$.

Similarly, we can deduce the following functional equations:

(i) If $f(x + y) = f(x) + f(y) + a$, then $f(x) = (f(1) + a)x - a$.
(ii) If $f(x + y) = f(x)f(y)$, then $f(x) = (f(1))^x$.
(iii) If $f(xy) = f(x) + f(y)(x, y > 0)$, then $f(x) = f(a) \cdot \log_a x(a > 0, a \neq 1)$.
(iv) If $f(xy) = f(x)f(y)(x, y > 0)$, then $f(x) = (f(a))^{\log_a x}(a > 0, a \neq 1)$.

(v) If $f(x+y) = \frac{f(x)f(y)}{f(x)+f(y)}$ ($f(x) \neq 0$), then $f(x) = \frac{f(1)}{x}$ ($x \neq 0$).

(vi) If $f(x+y) = f(x)+f(y)+2\sqrt{f(x)f(y)}$ ($x, y > 0$), then $f(x) = f(1) \cdot x^2$.

(vii) If $f(x+y) = f(x) + f(y) + kxy$, then $f(x) = \frac{k}{2}x^2 + (f(1) - \frac{k}{2})x$.

(viii) If $(f(x+y))^2 = (f(x))^2 + (f(y))^2$ ($x, y > 0$), then $f(x) = f(1) \cdot \sqrt{x}$.

(ix) If $(f(x+y))^2 + (f(x-y))^2 = 2(f(x))^2 + 2(f(y))^2$, then $f(x) = f(1) \cdot |x|$.

(2) *Jensen's functional equation*

Jensen's functional equation has the form

$$f\left(\frac{x+y}{2}\right) = \frac{f(x) + f(y)}{2},$$

which can be seen as a mean-form of Cauchy's functional equation. The general solution is $f(x) = ax + b$.

(3) *Linear functional equations*

A linear functional equation has the form

$$f(ax + by + c) = pf(x) + qf(x) + r,$$

where a, b, c, p, q, r are constants.

(4) *Cauchy's exponential functional equation*

Cauchy's exponential functional equation has the form

$$f(x + y) = f(x) \cdot f(y),$$

where $f : \mathbf{R} \to \mathbf{R}$ is a continuous function that is not identically zero.

(5) *Pexider's functional equation*

Pexider's functional equation has the form

$$f(x + y) = g(x) + h(y),$$

where $f, g, h : \mathbf{R} \to \mathbf{R}$ are continuous functions. Pexider's functional equation is a natural generalization of Cauchy's functional equation, and its general solution is

$$\begin{cases} f(z) = cz + a + b, \\ g(x) = cx + a, \\ h(y) = cy + b, \end{cases}$$

where a, b, c are arbitrary constants.

(6) *Aczel's functional equation*

Aczel's functional equation has the form

$$f(x + y) = g(x) \cdot k(y) + h(y),$$

which is a generalization of the Pexider's functional equation.

(7) *D'Alembert's functional equation*

D'Alembert's functional equation has the form

$$g(x + y) + g(x - y) = 2g(x) \cdot g(y).$$

Additionally, there are many other famous functional equations.

   (i) Abel's functional equation: $f(g(x)) = f(x) + a$.
  (ii) Euler's functional equation: $f(tx, ty) = t^k f(x, y)$.
 (iii) Wilson's functional equation: $g(x + y) + g(x - y) = 2g(x)f(y)$.
 (iv) Gauss's functional equation: $f(\sqrt{x^2 + y^2}) = f(x)f(y)$.
  (v) Davidson's functional equation: $f(xy) + f(x + y) = f(xy + x) + f(y)$.

2.1.6   *Common methods for solving functional equations*

Solving functional equations involves finding solutions under given conditions or determining that no solutions exist. Typically, functional equations have more than one solution, and sometimes extraneous solutions may arise (akin to extraneous roots). Therefore, it is necessary to substitute the obtained solutions back into the original equations for verification.

    Functional equations are closely related to function iterations, and some methods for solving functional equations are similar to those for finding expressions of the nth iteration of a function.

(1) *Cauchy's method*

First, determine the solution $f(q)$ when the independent variable is a rational number q. Then, under certain assumptions or conditions about the function $f(x)$ (such as monotonicity, continuity, boundedness, etc.), find the solution $f(x)$ for any real number x.

Example 2.5. Suppose a nonzero continuous function $f(x)$ satisfies the functional equation $f(\sqrt{x^2 + y^2}) = f(x)f(y)$, where x, y are real numbers. Prove that $f(x) = (f(1))^{x^2}$.

Proof. It is evident that $f(x)$ is an even function, so it suffices to prove the case for $x \geq 0$.

Setting $x = y = 0$ yields $f(0) = (f(0))^2$. Since $f(x) \neq 0$, it follows that $f(0) = 1$, and thus $f(0) = (f(1))^{0^2}$.

Suppose n is a positive integer, first prove the following proposition:

$$f(\sqrt{n}y) = (f(y))^n \quad \text{for } n \in \mathbf{N}_+, \quad y \in \mathbf{R}_+.$$

For $n = 1$, the proposition is trivially true. Assume it holds for $n = k$. For $n = k + 1$,

$$f(\sqrt{k+1}y) = f(\sqrt{ky^2 + y^2}) = f(\sqrt{k}y)f(y) = (f(y))^k f(y) = (f(y))^{k+1}.$$

Thus, the proposition holds.

Setting $y = 1$ and $y = \sqrt{n}$ in $f(\sqrt{n}y) = (f(y))^n$ gives $f(\sqrt{n}) = (f(1))^n$ and $f(n) = (f(\sqrt{n}))^n$, so $f(n) = (f(1))^{n^2}$.

For $x = \frac{p}{q}$, where p and q are positive integers, the proposition yields

$$f(p) = f\left(\sqrt{q^2} \cdot \frac{p}{q}\right) = \left(f\left(\frac{p}{q}\right)\right)^{q^2} \quad \text{and} \quad f(p) = (f(1))^{p^2}.$$

Hence, $\left(f\left(\frac{p}{q}\right)\right)^{q^2} = (f(1))^{p^2}$, and therefore $f\left(\frac{p}{q}\right) = (f(1))^{\frac{p^2}{q^2}}$.

From the continuity of $f(x)$, it follows that $f(x) = (f(1))^{x^2}$ also holds for irrational numbers x.

(2) *Method of undetermined coefficients*

This method involves assuming a form of the solution based on the structure of the functional equation, and then substituting this form into the functional equation to determine the coefficients it contains.

Example 2.6. Suppose $f(x)$ is a quadratic function satisfying

$$f(x+1) - f(x) = 8x + 3 \quad \text{and} \quad f(0) = 5.$$

Find $f(x)$.

Solution. Assume $f(x) = ax^2 + bx + c$. From $f(0) = 5$, we have $c = 5$. Also,

$$a(x+1)^2 + b(x+1) + c - ax^2 - bx - c = 8x + 3.$$

Comparing coefficients on both sides of the above equation yields $a = 4$ and $b = -1$.

Thus, $f(x) = 4x^2 - x + 5$.

(3) *Recursive method*

The recursive method is a technique that involves studying functional equations through sequences. On the one hand, the function value at a point and its iterated values can be treated as a sequence. By determining its general term and combining the value range, the form of the function can be deduced.

Example 2.7. Find all functions $f : \mathbf{R}_+ \cup \{0\} \to \mathbf{R}_+ \cup \{0\}$ such that for any $x \geq 0$,

$$f(f(x)) = 10x - 3f(x).$$

Solution. For any fixed $x_0 \geq 0$, let $a_n = f^{(n)}(x_0)$, thus obtaining a non-negative real number sequence $\{a_n\}$. This sequence satisfies the recurrence relation $a_{n+2} = -3a_{n+1} + 10a_n$, with its characteristic equation being $x^2 + 3x - 10 = 0$, and roots 2 and -5. Therefore, $a_n = 2^n A + (-5)^n B$, where A, B are constants.

Since $a_n = 2^n \left(A + B \left(-\frac{5}{2} \right)^n \right)$, if $B \neq 0$, then a_n can be negative when n is a sufficiently large odd integer. Hence, B must be zero, and $a_n = 2^n A$.

Given $a_0 = x_0 = A$, we find $f^{(n)}(x_0) = a_n = 2^n x_0$. In particular, $f(x_0) = 2x_0$, which holds true for any $x_0 \geq 0$. Upon verification, the only function satisfying the conditions is $f(x) = 2x$.

On the other hand, the recursive relation can be used to determine a series of function values. For instance, if the values of $f(1), f(2), \ldots, f(n)$ are known and the value of $f(n+1)$ is uniquely determined by the recurrence relation, then one can also employ the characteristic equation.

(4) *Fixed point method*

This method is commonly used to solve functional equations involving function composition. If the function in the functional equation has a fixed point, then the solutions can be derived by using this fixed point and the properties of the function itself.

Example 2.8 (Putnam Mathematical Competition 1971, A2).

Find all polynomial functions $P(x)$ satisfying

$$P(x^2 + 1) = (P(x))^2 + 1 \quad \text{and} \quad P(0) = 0$$

Solution. From the conditions, it is known that $x_0 = 0$ is a fixed point of $P(x)$. If x_k is a fixed point of $P(x)$, then $P(x_k^2 + 1) = (P(x_k))^2 + 1 = x_k^2 + 1$, making $x_{k+1} = x_k^2 + 1$ also a fixed point of $P(x)$. Thus, there exists an infinite sequence

$$0, 1, 2, 5, 26, \ldots$$

such that each term is a fixed point of $P(x)$, i.e., roots of $P(x) - x = 0$.

However, since $P(x)$ is a polynomial, $P(x) - x = 0$ can only have a finite number of roots, implying $P(x) - x$ is identically zero, i.e., $P(x) = x$.

Moreover, one may initially determine a fixed point x_0 of $f(x)$, express it in terms of $f(x)$ and x, and then find the specific value of x_0 to derive the expression of $f(x)$. Examples include IMO 24-1 (Problem 2.14) and IMO 56-5 (Problem 2.25).

If there is more than one fixed point, then the problem becomes much more complex, and the sought function could be a piecewise function.

(5) *Method of equivalent transformations*

Two functional equations (I) and (II) are equivalent if, by making appropriate transformations of the independent variable, (I) can be converted into (II) and vice versa. If one of them is already solved, then the solution for the other can also be determined.

Example 2.9. Solve the functional equation $2f\left(\frac{x+y}{2}\right) = f(x) + f(y)$.

Solution. Let $f(0) = a$. Then

$$f\left(\frac{x}{2}\right) = f\left(\frac{x+0}{2}\right) = \frac{1}{2}(f(x) + f(0)) = \frac{1}{2}(f(x) + a).$$

Also, since $\frac{1}{2}(f(x) + f(y)) = f\left(\frac{x+y}{2}\right) = f\left(\frac{x+y}{2} + 0\right) = \frac{1}{2}(f(x+y) + a)$, it follows that $f(x+y) = f(x) + f(y) - a$.

Let $g(x) = f(x) - a$. Then $g(x+y) = g(x) + g(y)$, which is Cauchy's functional equation with the solution $g(x) = x \cdot g(1)$. Hence,

$$f(x) = g(x) + a = (f(1) - f(0))x + f(0).$$

Similarly, it can be shown that Lobachevsky's functional equation

$$f\left(\frac{x+y}{2}\right) = \sqrt{f(x)f(y)}$$

is equivalent to the functional equation $f(x + y) = f(x)f(y)$, where $f(x) > 0$.

(6) *Estimation method*

Sometimes, deeper relationships of magnitude are implicit in functional equations or inequalities. These relationships are closely related to the solution to the problem. In particular, if it is known that $f(x) \geq g(x)$ and $f(x) \leq g(x)$, then $f(x) = g(x)$.

Example 2.10. Find all functions $f : \mathbf{R}_+ \to \mathbf{R}_+$ such that for any positive real numbers x and y,

$$\frac{x + f(y)}{xf(y)} = f\left(\frac{1}{y} + f\left(\frac{1}{x}\right)\right).$$

Solution. It is easy to see that $f(x) = x$ is a solution of the functional equation.

Suppose there exists $x_0 > 0$ such that $\frac{1}{x_0} > f\left(\frac{1}{x_0}\right)$. Let $x = x_0$ and $y = \frac{1}{\frac{1}{x_0} - f\left(\frac{1}{x_0}\right)}$. Then

$$\frac{x + f(y)}{xf(y)} = \frac{1}{x} + \frac{1}{f(y)} = \frac{1}{x_0} + \frac{1}{f(y)},$$

$$f\left(\frac{1}{y} + f\left(\frac{1}{x}\right)\right) = f\left(\frac{1}{x_0} - f\left(\frac{1}{x_0}\right) + f\left(\frac{1}{x_0}\right)\right) = f\left(\frac{1}{x_0}\right).$$

Thus, $\frac{1}{x_0} + \frac{1}{f(y)} = f\left(\frac{1}{x_0}\right)$, i.e., $\frac{1}{x_0} < f\left(\frac{1}{x_0}\right)$, a contradiction.

Therefore, for any positive real number x_0, there is $\frac{1}{x_0} \leq f\left(\frac{1}{x_0}\right)$, i.e., $x \leq f(x)$.

Also, $\frac{1}{x} + \frac{1}{f(y)} = f\left(\frac{1}{y} + f\left(\frac{1}{x}\right)\right) \geq \frac{1}{y} + f\left(\frac{1}{x}\right) \geq \frac{1}{y} + \frac{1}{x}$, i.e., $y \geq f(y)$.

Consequently, the solution to the functional equation is $f(x) = x$.

(7) *Calculus method*

For some types of functional equations, it is sometimes possible to transform the functional equation into a form like $f'(x) = g(x)$ (where $g(x)$ is integrable). Then integrate to get

$$f(x) = \int g(x)\mathrm{d}x + C,$$

and finally determine the expression of $f(x)$ based on the given conditions.

Example 2.11. Find the real coefficient polynomial $f(x)$ such that for any real coefficient polynomial $g(x)$, the equality $f(g(x)) = g(f(x))$ is true.

Solution. Let $g(x) = x + h$. Then $f(x+h) = f(x) + h$, implying

$$\frac{f(x+h) - f(x)}{h} = 1 \quad \text{for } h \neq 0.$$

Hence, $f'(x) = 1$, and $f(x) = x + C$, where C is a constant.

Let $g(x) \equiv 0$. Then $f(0) = 0$, so $f(x) = x$. Conversely, if $f(x) = x$, then for any real coefficient polynomial $g(x)$, one has $f(g(x)) = g(f(x))$.

Therefore, $f(x) = x$ is the unique solution to the problem.

(8) *Other elementary techniques*

(i) Substitution

The basic idea of the substitution method is to appropriately replace the independent variable in the functional equation with another variable. This results in a new functional equation, from which the unknown function can be derived.

Example 2.12 (Putnam Mathematical Competition 1971, B2).
Let $f(x)$ be a real valued function defined for all real x except for $x = 0$ and $x = 1$ satisfying the functional equation

$$f(x) + f\left(\frac{x-1}{x}\right) = 1 + x. \tag{I}$$

Find all functions $f(x)$ satisfying these conditions.

Solution. Replace x with $\frac{x-1}{x}$ and substitute into (I) to get

$$f\left(\frac{x-1}{x}\right) + f\left(\frac{1}{1-x}\right) = 1 + \frac{x-1}{x}. \tag{II}$$

Replace x with $\frac{1}{1-x}$ and substitute into (I) to get

$$f\left(\frac{1}{1-x}\right) + f(x) = 1 + \frac{1}{1-x}. \tag{III}$$

From (I) + (III) − (II), $f(x) = \frac{1+x^2-x^3}{2x(1-x)}$.

There is another similar problem:

- **(Korean Mathematical Olympiad 1999, Final Round, Problem 4).** Find all functions $f(x)$ such that

$$f\left(\frac{x-3}{x+1}\right) + f\left(\frac{3+x}{1-x}\right) = x \quad \text{for real } x \text{ except for } x = \pm 1.$$

(ii) Setting Values

Setting special values to the independent variable within the domain of the function to uncover hidden conditions in the problem. These new conditions are used to simplify the functional equation and subsequently determine the unknown function.

Example 2.13. Solve the functional equation: for any $x, y \in \mathbf{R}$,

$$f(x+y) + f(x-y) = 2f(x)\cos y.$$

Solution. Set $x = 0$ and $y = t$ to get

$$f(t) + f(-t) = 2f(0)\cos t. \tag{I}$$

Set $x = \frac{\pi}{2} + t$ and $y = \frac{\pi}{2}$ to get

$$f(\pi + t) + f(t) = 0. \tag{II}$$

Set $x = \frac{\pi}{2}$ and $y = \frac{\pi}{2} + t$ to get

$$f(\pi + t) + f(-t) = -2f\left(\frac{\pi}{2}\right)\sin t. \tag{III}$$

From (I) + (II) − (III), $f(t) = f(0)\cos t + f\left(\frac{\pi}{2}\right)\sin t$, i.e., $f(x) = a\cos x + b\sin x$, where $a = f(0)$ and $b = f\left(\frac{\pi}{2}\right)$ are constants. Upon verification, $f(x) = a\cos x + b\sin x$ satisfies the conditions.

(iii) Mathematical Induction

This method is suitable for functional equations whose domain is the set of positive integers. It begins with concrete calculations of $f(1), f(2), f(3), \ldots,$ leading to a conjecture about the expression of $f(n)$, and then use mathematical induction to prove the conjecture.

Example 2.14. Solve the functional equation: $f(n+1) = \cos\theta + f(n)\sin\theta$, where $f(1) = \cos\theta$ with $\theta \in [0, 2\pi)$ and $n \in \mathbf{N}_+$.

Solution. From the given conditions,

$$f(2) = \cos\theta + \cos\theta\sin\theta = \cos\theta(1 + \sin\theta),$$

$$f(3) = \cos\theta + \cos\theta(1 + \sin\theta)\sin\theta = \cos\theta(1 + \sin\theta + \sin^2\theta),$$

$$f(4) = \cos\theta + \cos\theta(1 + \sin\theta + \sin^2\theta)\sin\theta$$

$$= \cos\theta(1 + \sin\theta + \sin^2\theta + \sin^3\theta).$$

From this, we conjecture

$$f(n) = \cos\theta(1 + \sin\theta + \sin^2\theta + \cdots + \sin^{n-1}\theta) = \frac{\cos\theta(1 - \sin^n\theta)}{1 - \sin\theta},$$

which can be easily proven by mathematical induction.

Additionally, the properties of the function itself, such as monotonicity, continuity, boundedness, periodicity, etc., are quite important in solving functional equations. Whether explicitly stated as conditions in the problem or implicitly given from relationships the unknown function meets, the properties of the function cannot be overlooked.

Example 2.15. Find all strictly monotonic functions $f : \mathbf{R} \to \mathbf{R}$ satisfying

$$f(f(x) + y) = f(x + y) + f(0)$$

for all real numbers x and y.

Solution. Setting $y = -x$, we have $f(f(x) - x) = 2f(0)$. Since f is strictly monotonic, it must be injective. For any real numbers x and y,

$$f(f(x) - x) = 2f(0) = f(f(y) - y),$$

so $f(x) - x = f(y) - y$, implying $f(x) - x$ is a constant. Upon verification, all solutions are $f(x) = x + c$, where c is any given real number.

If the solution of the functional equation is restricted to polynomial functions, one can employ techniques specific to polynomials, such as analyzing degrees and coefficients, to solve polynomial functional equations. Such techniques will be presented in Chapter 5.

2.1.7 *Other important theorems*

Theorem 2.1 (Bolzano's Theorem). If a function $y = f(x)$ is continuous on a closed interval $[a, b]$ and $f(a) \cdot f(b) < 0$, then there exists a zero of $y = f(x)$ in the interval (a, b).

Theorem 2.2 (Boundedness Theorem). If a function $f(x)$ is continuous on a closed interval $[a, b]$, then $f(x)$ is bounded on $[a, b]$.

Theorem 2.3 (Rolle's Theorem). If a function $f(x)$ is continuous on a closed interval $[a, b]$, differentiable on the open interval (a, b), and $f(a) = f(b)$, then there exists at least one point ξ in (a, b) such that $f'(\xi) = 0$.

Theorem 2.4 (Lagrange's Mean Value Theorem). If a function $f(x)$ is continuous on a closed interval $[a, b]$ and differentiable on the open interval (a, b), then there exists at least one point ξ in (a, b) such that

$$f'(\xi) = \frac{f(b) - f(a)}{b - a}.$$

Rigorous proofs of these theorems belong to the realm of advanced mathematics and are not required to be mastered here. However, understanding them intuitively is relatively straightforward and helps deepen the understanding of functions. So they are provided for reference.

2.2   Problems and Solutions

2.2.1   *Proving properties*

Problem 2.1 (IMO 10-5, proposed by the German Democratic Republic). Let f be a real-valued function defined for all real numbers x such that, for some positive constant a, the equation

$$f(x + a) = \frac{1}{2} + \sqrt{f(x) - (f(x))^2} \qquad (*)$$

holds for all x.

(a) Prove that the function f is periodic (i.e., there exists a positive number b such that $f(x + b) = f(x)$ for all x).
(b) For $a = 1$, give an example of a non-constant function with the required properties.

Proof. (a) Since the equation (*) holds for any $x \in \mathbf{R}$, we have

$$f(x + 2a) = f((x + a) + a)$$

$$= \frac{1}{2} + \sqrt{f(x + a) - (f(x + a))^2}$$

$$= \frac{1}{2} + \sqrt{\begin{aligned}&\frac{1}{2} + \sqrt{f(x) - (f(x))^2} \\ &\quad - \left(\frac{1}{4} + \sqrt{f(x) - (f(x))^2} + f(x) - (f(x))^2\right)\end{aligned}}$$

$$= \frac{1}{2} + \sqrt{\frac{1}{4} - f(x) + (f(x))^2}$$

$$= \frac{1}{2} + \left|\frac{1}{2} - f(x)\right|.$$

Since $f(x+a) \geq \frac{1}{2}$ for any $x \in \mathbf{R}$, we see that $f(x) \geq \frac{1}{2}$. Therefore

$$f(x+2a) = \frac{1}{2} + f(x) - \frac{1}{2} = f(x).$$

Thus, $f(x)$ is a periodic function with a period of $2a$.

(b) When $a = 1$, the following is a periodic function with the prescribed property:

$$f(x) = \frac{1}{2} + \frac{1}{2} \left| \sin \frac{\pi x}{2} \right|.$$

Its period is 2.

Note. (i) If λ, μ are distinct constants and $f(x)$ is a function on $\mathbf{R}$ satisfying

$$f(x+\lambda) = \frac{a}{2} \pm \sqrt{b + af(x+\mu) - (f(x+\mu))^2},$$

where a, b are real numbers and $a \neq 0$, then $f(x)$ is a periodic function with a period of $2|\lambda - \mu|$.

In particular, when $a = 1$, $b = 0$, $\lambda > 0$, and $\mu = 0$, the function $f(x)$ above becomes the function in Problem 2.1.

(ii) Suppose $\{\lambda_n\}$ is an arithmetic sequence with a non-zero common difference and $f(x)$ is a function on $\mathbf{R}$ satisfying

$$f(x+\lambda_1) = \frac{a}{2} \pm \sqrt{b + \sum_{i=2}^{n} (af(x+\lambda_i) - (f(x+\lambda_i))^2)},$$

where a, b are real numbers and $a \neq 0$. Then $f(x)$ is a periodic function with a period of $n|d|$, where d is the common difference of the arithmetic sequence.

【Score Situation】 This particular problem saw the following distribution of scores among contestants: 38 contestants scored 7 points, 2 contestants scored 6 points, 20 contestants scored 5 points, 3 contestants scored 4 points, 8 contestants scored 3 points, 1 contestant scored 2 points, 2 contestants scored 1 point, and 17 contestants scored 0 point. The average score of this problem is 4.593, indicating that it was simple.

Among the top five teams in the team scores, the German Democratic Republic team achieved a total score of 304 points, the Soviet Union team achieved a total score of 298 points, the Hungary team achieved a total score of 291 points, the United Kingdom team achieved a total score of 263 points, and the Poland team achieved a total score of 262 points.

The gold medal cutoff for this IMO was set at 39 points (with 22 contestants earning gold medals), the silver medal cutoff was 33 points (with 22 contestants earning silver medals), and the bronze medal cutoff was 26 points (with 20 contestants earning bronze medals).

In this IMO, a total of 16 contestants achieved a perfect score of 40 points.

Problem 2.2 (IMO 11-2, proposed by Hungary). Let $a_1, a_2, \ldots, a_n$ be real constants, x a real variable, and

$$f(x) = \cos(a_1 + x) + \frac{1}{2}\cos(a_2 + x) + \frac{1}{4}\cos(a_3 + x)$$

$$+ \cdots + \frac{1}{2^{n-1}}\cos(a_n + x).$$

Given that $f(x_1) = f(x_2) = 0$, prove that $x_2 - x_1 = m\pi$ for some integer m.

Proof. For all real numbers x, obviously $\cos(a_i + x) \geq -1$. Therefore,

$$f(-a_1) = 1 + \frac{1}{2}\cos(a_2 - a_1) + \frac{1}{4}\cos(a_3 - a_1) + \cdots + \frac{1}{2^{n-1}}\cos(a_n - a_1)$$

$$\geq 1 - \frac{1}{2} - \frac{1}{4} - \cdots - \frac{1}{2^{n-1}}$$

$$= \frac{1}{2^{n-1}} > 0.$$

Hence, $f(x)$ is not identically zero. Furthermore,

$$f(x) = \sum_{k=1}^{n} \frac{1}{2^{k-1}}(\cos a_k \cdot \cos x - \sin a_k \cdot \sin x)$$

$$= \left(\sum_{k=1}^{n} \frac{1}{2^{k-1}}\cos a_k\right)\cos x - \left(\sum_{k=1}^{n} \frac{1}{2^{k-1}}\sin a_k\right)\sin x$$

$$= A\cos x - B\sin x,$$

where $A = \sum_{k=1}^{n} \frac{1}{2^{k-1}}\cos a_k$ and $B = \sum_{k=1}^{n} \frac{1}{2^{k-1}}\sin a_k$, and they cannot both be zero. Otherwise, $f(x)$ would be identically zero, leading to a contradiction.

If $A \neq 0$, then

$$f(x_1) = A\cos x_1 - B\sin x_1 = 0,$$

$$f(x_2) = A\cos x_2 - B\sin x_2 = 0.$$

In this case, if $\sin x_1 = 0$, then $\cos x_1 = 0$, which is a contradiction. Thus, $\sin x_1 \neq 0$, and similarly, $\sin x_2 \neq 0$. Therefore, $\cot x_1 = \cot x_2 = \frac{B}{A}$.

If $A = 0$, then $B \neq 0$. From $f(x_1) = f(x_2) = 0$, we have $\sin x_1 = \sin x_2 = 0$.

Both cases imply $x_2 - x_1 = m\pi$, where m is an integer.

Note. We can also solve this problem from the perspective of vectors. Suppose vectors $v_1, v_2, \ldots, v_n$ begin at the origin $(0,0)$, with lengths of $1, \frac{1}{2}, \frac{1}{4}, \ldots, \frac{1}{2^{n-1}}$ and angles $a_1, a_2, \ldots, a_n$ between themselves and the x-axis, respectively. Then the given expression in the problem is the sum of x-coordinates of these vectors rotated by angle x.

Let r be the length of $V = v_1 + v_2 + \cdots + v_n$. It is evident that $r > 0$, and thus the angle between V rorated by angle x and the x-axis should be $\frac{2k+1}{2}\pi$.

Furthermore, there are several similar problems:

- **(Estonia Team Selection Test 2002, Problem 5).** Let $0 < \alpha < \frac{\pi}{2}$ and $x_1, x_2, \ldots, x_n$ be real numbers such that

$$\sin x_1 + \sin x_2 + \cdots + \sin x_n \geq n \cdot \sin \alpha.$$

  Prove that

$$\sin(x_1 - \alpha) + \sin(x_2 - \alpha) + \cdots + \sin(x_n - \alpha) \geq 0.$$

- **(William Lowell Putnam Mathematical Competition 1967, A1).** Let $f(x) = a_1 \sin x + a_2 \sin 2x + \cdots + a_n \sin nx$, where $a_1, a_2, \ldots, a_n$ are real numbers. Given that $|f(x)| \leq |\sin x|$ for all real x, prove that $|a_1 + 2a_2 + \cdots + na_n| \leq 1$.

【Score Situation】This particular problem saw the following distribution of scores among contestants: 34 contestants scored 7 points, 5 contestants scored 6 points, 12 contestants scored 5 points, 32 contestants scored 4 points, 6 contestants scored 3 points, 5 contestants scored 2 points, 5 contestants scored 1 point, and 13 contestants scored 0 point. The average score of this problem is 4.366, indicating that it was simple.

Among the top five teams in the team scores, the scores of this problem are as follows: the Hungary team scored 53 points (with a total team score of 247 points), the German Democratic Republic team scored 45 points (with a total team score of 240 points), the Soviet Union team scored 52 points (with a total team score of 231 points), the Romania team scored 36 points (with a total team score of 219 points), and the United Kingdom team scored 43 points (with a total team score of 193 points).

The gold medal cutoff for this IMO was set at 40 points (with 3 contestants earning gold medals), the silver medal cutoff was 30 points (with 20 contestants earning silver medals), and the bronze medal cutoff was 24 points (with 21 contestants earning bronze medals).

In this IMO, only three contestants achieved a perfect score of 40 points, namely Tibor Fiala from Hungary, Vladimir Drinfeld from the Soviet Union, and Simon Phillips Norton from the United Kingdom.

Problem 2.3 (IMO 14-5, proposed by Bulgaria). Let f and g be real-valued functions defined for all real values of x and y and satisfying the equation

$$f(x+y) + f(x-y) = 2f(x)g(y)$$

for all $x, y \in \mathbf{R}$.

Prove that if $f(x)$ is not identically zero and $|f(x)| \leq 1$ for all $x \in \mathbf{R}$, then $|g(y)| \leq 1$ for all $y \in \mathbf{R}$.

Proof 1. Let M be the maximum value of $|f(x)|$. Suppose there exists a $y_0 \in \mathbf{R}$ such that $|g(y_0)| = 1 + r$, where $r > 0$. Then for any $x \in \mathbf{R}$,

$$2|f(x)||g(y_0)| = |f(x+y_0) + f(x-y_0)|$$
$$\leq |f(x+y_0)| + |f(x-y_0)|$$
$$\leq 2M.$$

Thus,

$$|f(x)| \leq \frac{M}{|g(y_0)|} = \frac{M}{1+r} = M - \delta < M, \quad \delta > 0,$$

which contradicts the definition of M. Therefore, the conclusion holds.

Proof 2. Since $f(x)$ is not identically zero, there exists a real number a such that $f(a) \neq 0$. Substituting $x = a + ny$ into the given functional equation and rearranging, we get

$$f(a+(n+1)y) - 2f(a+ny)g(y) + f(a+(n-1)y) = 0. \qquad (1)$$

With y fixed, let $f(a+ny)$ be denoted as f_n and $g(y)$ as g. The equation (1) can be written as a recurrence relation

$$f_{n+1} - 2gf_n + f_{n-1} = 0. \qquad (2)$$

If the conclusion does not hold, then there exists a real number y_0 such that $|g(y_0)| > 1$. For $y = y_0$, the equation $r^2 - 2g(y_0)r + 1 = 0$ has two distinct real roots r_1 and r_2, where $r_1 = g(y_0) + \sqrt{(g(y_0))^2 - 1}$ and $r_2 = g(y_0) - \sqrt{(g(y_0))^2 - 1}$.

Therefore, the solution to the recurrence relation (2) is $f_n = b_1 r_1^n + b_2 r_2^n$.

Clearly, $r_1 > 1$, and if $b_1 \neq 0$, then $\lim_{n \to +\infty} f_n = \infty$, which contradicts the assumption. If $b_1 = 0$, then $b_2 \neq 0$, otherwise f_n would be identically zero. Furthermore, $0 < r_2 < 1$, so $\lim_{n \to -\infty} f_n = \lim_{n \to -\infty} b_2 r_2^n = \infty$, which also contradicts the assumption.

In conclusion, $|g(y)| \leq 1$ for all $y \in \mathbf{R}$.

Note. It is evident that the recurrence relation $f_{n+1} - 2g f_n + f_{n-1} = 0$ is satisfied by the functions $f_n = \cos nx$ and $g = \cos x$. This is related to the Chebyshev polynomials

$$T_n(\cos \theta) = \cos n\theta,$$

which will be presented in Chapter 5.

Furthermore, there are several similar problems:

- **(All-Russian Mathematical Olympiad 2018, Regional Round, Grade 11, Problem 7).** A function $f(x)$ is defined over the entire real axis and for all real numbers x and y, it satisfies the condition

$$f(x) + f(y) = 2f\left(\frac{x+y}{2}\right) f\left(\frac{x-y}{2}\right).$$

  Is $f(x)$ necessarily an even function?

- **(William Lowell Putnam Mathematical Competition 1997, B2).** Let f be a twice-differentiable real-valued function satisfying

$$f(x) + f''(x) = -xg(x)f'(x),$$

  where $g(x) \geq 0$ for all real x. Prove that $|f(x)|$ is bounded.

- **(Asian Pacific Mathematics Olympiad 1989, Problem 5).** Let $g(x)$ be the composition inverse function to $f(x)$, determine all functions f from the reals to the reals such that:

  (i) $f(x)$ is strictly increasing;

  (ii) $f(x) + g(x) = 2x$ for all real x.

  (Note: f and g are said to be composition inverse if $f(g(x)) = x$ and $g(f(x)) = x$ for all real x).

【Score Situation】This particular problem saw the following distribution of scores among contestants: 9 contestants scored 7 points, 2 contestants scored 6 points, 2 contestants scored 5 points, no contestant scored 4 points, 1 contestant scored 3 points, 1 contestant scored 2 points, 3 contestants scored 1 point, and 15 contestants scored 0 point. The average score of this problem is 2.818, indicating that it had a certain level of difficulty.

Among the top five teams in the team scores, the scores of this problem are as follows: the Soviet Union team scored 54 points (with a total team score of 270 points), the Hungary team scored 40 points (with a total team score of 263 points), the German Democratic Republic team scored 46 points (with a total team score of 239 points), the Romania team scored 36 points (with a total team score of 208 points), and the United Kingdom team scored 31 points (with a total team score of 179 points).

The gold medal cutoff for this IMO was set at 40 points (with 8 contestants earning gold medals), the silver medal cutoff was 30 points (with 16 contestants earning silver medals), and the bronze medal cutoff was 19 points (with 30 contestants earning bronze medals).

In this IMO, a total of eight contestants achieved a perfect score of 40 points.

Problem 2.4 (IMO 15-5, proposed by Poland). Suppose G is a set of non-constant functions of a real variable x of the form

$$f(x) = ax + b, \quad \text{where } a \text{ and } b \text{ are real numbers},$$

and G has the following properties:

(a) if f and g are in G, then $g \circ f$ is in G; here $(g \circ f)(x) = g(f(x))$;
(b) if f is in G, then its inverse f^{-1} is in G; here the inverse of $f(x) = ax+b$ is $f^{-1}(x) = \frac{x-b}{a}$;
(c) for every f in G, there exists a real number x_f such that $f(x_f) = x_f$.

Prove that there exists a real number k such that $f(k) = k$ for all f in G.

Proof 1. First, we prove that

$$f(x) = x + b \in G \Rightarrow b = 0. \tag{1}$$

From (c), there exists an $x_f \in \mathbf{R}$ such that $f(x_f) = x_f + b = x_f$, so $b = 0$.

Next, we prove that if $f(x) = ax + b$, then b is uniquely determined by a.

If $g_1(x) = ax + b_1$ and $g_2(x) = ax + b_2$ are two functions in G, then from (a) and (b), $g_1^{-1}(g_2(x)) = \frac{(ax+b_2)-b_1}{a} = x + \frac{b_2-b_1}{a}$ is also a function in G. From (1), it is known that $b_1 = b_2$.

Finally, we prove that there exists a real number k such that $f(k) = k$ for all $f(x) = ax + b \in G$.

For $a = 1$, the function $f(x) = x$ in G obviously satisfies $f(k) = k$ for all real numbers k. Consider $m(x) = ax + b$ and $n(x) = cx + d$ in G, where $a \neq 1$ and $c \neq 1$. From (c), there exist x_m, x_n such that $ax_m + b = x_m$ and $cx_n + d = x_n$, i.e., $x_m = -\frac{b}{a-1}$ and $x_n = -\frac{d}{c-1}$.

From (a), $m(n(x)) = acx + ad + b$ and $n(m(x)) = acx + bc + d$ are both in G, and $ad + b = bc + d$. Hence

$$-\frac{b}{a-1} = -\frac{d}{c-1}.$$

Therefore, $k = x_m = x_n$.

Proof 2. From (a) and (b), it is known that G forms a group with respect to the composition operation "$\circ$." The function $i(x) = x$ serves as the identity element of G, as for any real number x, it holds that

$$(f \circ i)(x) = (i \circ f)(x) = f(x).$$

For each $f(x) = ax + b \in G$, it can be associated with its coefficient a, which is multiplicative, i.e., if $f(x) = ax + b$ and $g(x) = cx + d$, then $g \circ f$ is associated with ac.

It is evident that each linear function $f(x) = ax + b (a \neq 1)$ has a fixed point $x_f = \frac{b}{1-a}$, geometrically the intersection of the lines $y = ax + b$ and $y = x$.

For $a = 1$, the function $f(x) = ax + b = x + b$ has a fixed point if and only if $b = 0$. In this case, the function is $i(x) = x$, and every point is a fixed point.

Now consider the composite functions $m = f \circ g$ and $n = g \circ f$, where $f(x) = ax + b$ and $g(x) = cx + d$. It is easy to see that the slopes of m and n are both ac, and the slopes of m^{-1} and n^{-1} are both $\frac{1}{ac}$. Therefore, the slope of $m^{-1} \circ n$ is 1. Since $m^{-1} \circ n \in G$, it is the identity element of G, so

$$m^{-1} \circ n = i \Rightarrow m \circ m^{-1} \circ n = m \circ i \Rightarrow n = m \Rightarrow f \circ g = g \circ f.$$

Thus, the functions in G are commutative with respect to the composition operation "$\circ$." If G contains only the identity element, then there is no need to prove further. If G contains $f(x) = ax + b (a \neq 1)$ and $g(x) = cx + d (c \neq 1)$, then the fixed points of these functions are $\frac{b}{1-a}$ and $\frac{d}{1-c}$, respectively. Since $f \circ g = g \circ f$ and

$$(f \circ g)(x) = f(cx + d) = acx + ad + b,$$

$$(g \circ f)(x) = g(ax + b) = acx + bc + d,$$

we have $acx + ad + b = acx + bc + d$. Thus $\frac{b}{1-a} = \frac{d}{1-c}$, indicating $f(x)$ and $g(x)$ have a common fixed point.

Since $f(x)$ and $g(x)$ are arbitrary functions in G other than the identity element, by setting $k = \frac{b}{1-a}$, the number k is a common fixed point for all functions in G.

【Score Situation】 This particular problem saw the following distribution of scores among contestants: 62 contestants scored 6 points, 3 contestants scored 5 points, 2 contestants scored 4 points, 2 contestants scored 3 points, 2 contestants scored 2 points, 17 contestants scored 1 point, and 37 contestants scored 0 point. The average score of this problem is 3.376, indicating that it was relatively straightforward.

Among the top five teams in the team scores, the scores of this problem are as follows: the Soviet Union team scored 48 points (with a total team score of 254 points), the Hungary team scored 45 points (with a total team score of 215 points), the German Democratic Republic team scored 40 points (with a total team score of 188 points), the Poland team scored 42 points (with a total team score of 174 points), and the United Kingdom team scored 24 points (with a total team score of 164 points).

The gold medal cutoff for this IMO was set at 35 points (with 5 contestants earning gold medals), the silver medal cutoff was 27 points (with 15 contestants earning silver medals), and the bronze medal cutoff was 17 points (with 48 contestants earning bronze medals).

In this IMO, only one contestant achieved a perfect score of 40 points, namely Sergei Konyagin from the Soviet Union.

Problem 2.5 (IMO 19-6, proposed by Bulgaria). Let $f(n)$ be a function defined on the set of all positive integers and having all its values in the same set. Prove that if

$$f(n+1) > f(f(n))$$

for each positive integer n, then

$$f(n) = n \quad \text{for each } n.$$

Proof. We employ mathematical induction to prove that for any positive integer n, if $m \geq n$ is a positive integer, then $f(m) \geq n$.

For $n = 1$, the proposition is evidently true.

Suppose the proposition holds for $n = k - 1$, then for $m \geq k$, by the inductive hypothesis, $f(m-1) \geq k-1$. Thus,

$$f(m) > f(f(m-1)) \geq k-1,$$

which implies $f(m) \geq k$. Therefore, the proposition holds for $n = k$.

From this it follows that $f(n) \geq n$ and $f(n+1) > f(f(n)) \geq f(n)$, indicating that $f(n)$ is monotonically increasing. If there exists a positive integer n_0 such that $f(n_0) > n_0$, then $f(n_0) \geq n_0 + 1$ and $f(f(n_0)) \geq f(n_0 + 1)$, leading to a contradiction.

Hence, $f(n) = n$ for any positive integer n.

Note. In this problem, a proof by contradiction can also be used to show that if $m \geq n$, then $f(m) \geq n$.

Furthermore, there is a similar problem:

- **(Belarus Team Selection Test 2000, Problem 4).** Does there exist a function $f : \mathbf{N} \to \mathbf{N}$ such that

$$f(f(n-1)) = f(n+1) - f(n)$$

  for all $n \geq 2$?

【Score Situation】This particular problem saw the following distribution of scores among contestants: 10 contestants scored 8 points, 1 contestant scored 7 points, no contestant scored 6 points, no contestant scored 5 points, 3 contestants scored 4 points, 3 contestants scored 3 points, no contestant scored 2 points, 5 contestants scored 1 point, and 15 contestants scored 0 point. The average score for this problem is 3.054, indicating that it was relatively straightforward.

Among the top five teams in the team scores, the United States team achieved a total score of 202 points, the Soviet Union team achieved a total score of 192 points, the Hungary team achieved a total score of 190 points, the United Kingdom team achieved a total score of 190 points, and the Netherlands team achieved a total score of 185 points.

The gold medal cutoff for this IMO was set at 34 points (with 13 contestants earning gold medals), the silver medal cutoff was 24 points (with 29 contestants earning silver medals), and the bronze medal cutoff was 17 points (with 35 contestants earning bronze medals).

In this IMO, a total of five contestants achieved a perfect score of 40 points.

Problem 2.6 (IMO 28-4, proposed by Vietnam). Prove that there is no function f from the set of non-negative integers into itself such that $f(f(n)) = n + 1987$ for every n.

Proof 1. If such a function exists, then

$$f(n + 1987) = f(f(f(n))) = f(n) + 1987$$

for all $n \in \mathbf{N}$. Following this pattern,

$$
\begin{aligned}
f(n + 1987k) &= f((n + 1987(k - 1)) + 1987) \\
&= f(n + 1987(k - 1)) + 1987 \\
&= f(n) + 1987(k - 1) + 1987 \\
&= f(n) + 1987k.
\end{aligned}
$$

Let $r \in \mathbf{N}$ with $r \leq 1986$ and suppose

$$
f(r) = 1987k + l, \tag{1}
$$

where $k, l \in \mathbf{N}$ and $l \leq 1986$. By definition,

$$
f(f(r)) = r + 1987, \quad f(f(r)) = f(1987k + l) = f(l) + 1987k,
$$

implying

$$
r + 1987 = f(l) + 1987k. \tag{2}
$$

Since $r \leq 1986$, it follows that $k = 0$ or $k = 1$.

If $k = 1$, then from (1) and (2),

$$
f(r) = 1987 + l, \quad f(l) = r < 1987 + l = f(r),
$$

indicating $r \neq l$.

If $k = 0$, then from (1) and (2),

$$
f(r) = l, \quad f(l) = r + 1987 > l = f(r),
$$

indicating $r \neq l$.

Hence, integers in the set $\{0, 1, 2, \ldots, 1986\}$ always pair up such that

$$
f(l) = r, \quad f(r) = 1987 + l, \quad \text{or} \quad f(r) = l, \quad f(l) = 1987 + r,
$$

with $r \neq l$. However, the set $\{0, 1, 2, \ldots, 1986\}$ has 1987 integers and 1987 is odd, so they cannot be paired up.

Therefore, such a function f does not exist.

Proof 2. By contradiction, suppose such a function f exists.

It is evident that f is injective: If $f(n_1) = f(n_2)$, then $f(f(n_1)) = f(f(n_2))$, implying $n_1 + 1987 = n_2 + 1987$, thus $n_1 = n_2$. Therefore, $f(0)$, $f(1), \ldots, f(1986)$ are 1987 distinct integers. Let $M = \{f(0), f(1), \ldots, f(1986)\}$.

As shown in Proof 1, $f(n + 1987k) = f(n) + 1987k$. Hence, $f(n) \geq 1987k$ for $n \geq 1987k$. For any $m \in \{0, 1, \ldots, 1986\}$, since $f(f(m)) = m + 1987 < 2 \times 1987$, we get $f(m) < 2 \times 1987$. Let $I_1 = \{0, 1, \ldots, 1986\}$ and $I_2 = \{1987, 1988, \ldots, 3973\}$. Then $M \subseteq I_1 \cup I_2$. If $|M \cap I_1| = m_1$ and $|M \cap I_2| = m_2$, then $m_1 + m_2 = 1987$.

For any $f(a) \in M \cap I_1$, there exists a $b(0 \leq b < 1987)$ such that $f(a) = b$. Therefore,

$$f(b) = f(f(a)) = a + 1987,$$

implying $f(b) \in M \cap I_2$. Since different $f(a)$ correspond to different $f(b)$, it follows that $m_1 \leq m_2$.

Similarly, $m_1 \geq m_2$, thus $m_1 = m_2$. Consequently, $m_1 + m_2 = 2m_1 = 1987$, which is a contradiction.

Therefore, such a function f does not exist.

Note. There are several similar problems:

- **(MathPath Summer Program Qualifying Test 2023, Problem 6).** Let $f(n)$ be a function that takes positive integers as inputs and also gives positive integers as outputs. Is it possible to find a function such that

$$f(f(n)) = n + 5$$

for every positive integer n?

Either prove that this task is impossible, or else describe how to construct a function with the desired property.

- **(Vietnam Mathematical Olympiad 1993, Problem 3).** Find a function $f : \mathbf{N} \to \mathbf{N}$ such that

$$f(f(n)) = 1993n^{1945} \quad \text{for all } n \in \mathbf{N}.$$

- **(British Mathematical Olympiad 1992, 1st Round, Problem 5).** Let f be a function mapping positive integers into positive integers. Suppose that

$$f(n + 1) > f(n) \quad \text{and} \quad f(f(n)) = 3n \quad \text{for all positive integers } n.$$

Determine $f(1992)$.

- **(Japan Mathematical Olympiad 1991, Final Round, Problem 2).** The mappings p, q from $\mathbf{N}_+$ to $\mathbf{N}_+$ are defined as follows:

$$p(1) = 2, \quad p(2) = 3, \quad p(3) = 4, \quad p(4) = 1, \quad p(n) = n(n \geq 5),$$

$$q(1) = 3, \quad q(2) = 4, \quad q(3) = 2, \quad q(4) = 1, \quad q(n) = n(n \geq 5).$$

(a) Suppose f is a mapping from $\mathbf{N}_+$ to $\mathbf{N}_+$ such that $f(f(n)) = p(n)+2$. Give an example for f.

(b) Prove that $f(f(n)) = q(n)+2$ does not hold for any mapping f from $\mathbf{N}_+$ to $\mathbf{N}_+$.

【Score Situation】 This particular problem saw the following distribution of scores among contestants: 91 contestants scored 7 points, 7 contestants scored 6 points, 12 contestants scored 5 points, 1 contestant scored 4 points, 10 contestants scored 3 points, 21 contestants scored 2 points, 20 contestants scored 1 point, and 75 contestants scored 0 point. The average score for this problem is 3.523, indicating that it was relatively straightforward.

Among the top five teams in the team scores, the scores of this problem are as follows: the Romania team scored 42 points (with a total team score of 250 points), the Germany team scored 42 points (with a total team score of 248 points), the Soviet Union team scored 42 points (with a total team score of 235 points), the German Democratic Republic team scored 38 points (with a total team score of 231 points), and the United States team scored 36 points (with a total team score of 220 points).

The gold medal cutoff for this IMO was set at 42 points (with 22 contestants earning gold medals), the silver medal cutoff was 32 points (with 42 contestants earning silver medals), and the bronze medal cutoff was 18 points (with 56 contestants earning bronze medals).

In this IMO, a total of 22 contestants achieved a perfect score of 42 points.

Problem 2.7 (IMO 34-5, proposed by Germany). Does there exist a function $f : \mathbf{N}_+ \to \mathbf{N}_+$ such that:

(a) $f(1) = 2$;

(b) $f(f(n)) = f(n) + n$ for all $n \in \mathbf{N}_+$;

(c) $f(n) < f(n+1)$ for all $n \in \mathbf{N}_+$?

Solution. Such a function exists. Let $f(n) = \lfloor \alpha n + \beta \rfloor$ for $n \in \mathbf{N}_+$, where $\alpha = \frac{\sqrt{5}+1}{2}$ and $\beta = \frac{\sqrt{5}-1}{2}$, and $\lfloor x \rfloor$ denotes the greatest integer not exceeding x. Next, we prove that $f(n)$ satisfies the conditions:

$$f(1) = \lfloor \alpha + \beta \rfloor = \lfloor \sqrt{5} \rfloor = 2,$$

$$f(n+1) = \lfloor (n+1)\alpha + \beta \rfloor = \lfloor n\alpha + \alpha + \beta \rfloor$$
$$\geq \lfloor n\alpha + 1 + \beta \rfloor = f(n) + 1 > f(n).$$

Thus, $f(n)$ satisfies conditions (a) and (c), and is a function from $\mathbf{N}_+$ to $\mathbf{N}_+$.

$$f(f(n)) = \lfloor \alpha \lfloor \alpha n + \beta \rfloor + \beta \rfloor = \lfloor (\beta+1) \lfloor \alpha n + \beta \rfloor + \beta \rfloor$$
$$= \lfloor \alpha n + \beta \rfloor + \lfloor \beta \lfloor \alpha n + \beta \rfloor + \beta \rfloor$$
$$= f(n) + \lfloor \beta \lfloor \alpha n + \beta \rfloor + \beta \rfloor.$$

Since $\alpha n + \beta$ is not an integer,

$$\beta \lfloor \alpha n + \beta \rfloor + \beta < \beta(\alpha n + \beta) + \beta = n + \beta^2 + \beta = n + 1,$$
$$\beta \lfloor \alpha n + \beta \rfloor + \beta > \beta(\alpha n + \beta - 1) + \beta = n + \beta^2 > n.$$

Therefore, $\lfloor \beta \lfloor \alpha n + \beta \rfloor + \beta \rfloor = n$, and thus $f(f(n)) = f(n) + n$.
In conclusion, $f(n) = \lfloor \alpha n + \beta \rfloor$ satisfies all conditions.

Note. The constructed $f(n)$ is not unique; $f(n) = \left\lfloor \frac{\sqrt{5}+1}{2}n + \frac{1}{2} \right\rfloor$ also meets the conditions. In fact, from $f(n) > n$ and the given condition,

$$1 < \frac{f(f(n))}{f(n)} = 1 + \left(\frac{f(n)}{n}\right)^{-1} < 2,$$

suggesting that the ratio $\frac{f(n)}{n}$ approaches a limit value α as n tends to infinity, satisfying $\alpha = 1 + \alpha^{-1}(\alpha \geq 1)$. Hence $\alpha = \frac{\sqrt{5}+1}{2}$.

This leads to the conjecture that $f(n)$ is an integer close to αn, and thus $f(n) = \lfloor \alpha n + \beta \rfloor$ is constructed, and upon verification, it is sufficient that $1 - \alpha^{-1} \leq \beta \leq \alpha^{-1}$.

Furthermore, there are several similar problems:

- **(USA TST Selection Test 2017, Problem 6).** A sequence of positive integers $(a_n)_{n \geq 1}$ is of *Fibonacci type* if it satisfies the recursive relation $a_{n+2} = a_{n+1} + a_n$ for all $n \geq 1$. Is it possible to partition the set of positive integers into an infinite number of Fibonacci type sequences?
- **(From the "Problems in Mathematics" in the Hungarian journal *KöMaL*, January 2001, A.244).** A sequence of numbers is called of Fibonacci-type if each term, after the first two, is the sum of the previous two.

  Prove that the set of positive integers can be partitioned into the disjoint union of infinite Fibonacci-type sequences.
- **(From the "Problems in Mathematics" in the Hungarian journal *KöMaL*, September 2000, B.3429).** Let $q = \frac{1+\sqrt{5}}{2}$ and $f : \mathbf{N} \to \mathbf{N}$ be a function satisfying

$$|f(n) - qn| < \frac{1}{q}$$

for every $n \in \mathbf{N}$. Prove that $f(f(n)) = f(n) + n$.

- **(Estonia Team Selection Test 2000, Problem 4).** Find all functions $f : \mathbf{N} \to \mathbf{N}$ such that

$$f(f(f(n))) + f(f(n)) + f(n) = 3n$$

  for all $n \in \mathbf{N}$.

- **(Estonia Team Selection Test 2000, Problem 6).** We call an infinite sequence of positive integers an F-sequence if every term of this sequence (starting from the third term) equals the sum of the two preceding terms. Is it possible to decompose the set of all positive integers into

  (a) a finite;
  (b) an infinite

  number of F-sequences having no common members?

- **(William Lowell Putnam Mathematical Competition 1988, A5).** Prove that there exists a unique function f from the set $\mathbf{R}_+$ of positive real numbers to $\mathbf{R}_+$ such that

$$f(f(x)) = 6x - f(x) \quad \text{and} \quad f(x) > 0 \quad \text{for all } x > 0.$$

- Problem 2.10 (IMO 20-3) in this chapter.

【Score Situation】 This particular problem saw the following distribution of scores among contestants: 91 contestants scored 7 points, 27 contestants scored 6 points, 19 contestants scored 5 points, 32 contestants scored 4 points, 65 contestants scored 3 points, 44 contestants scored 2 points, 92 contestants scored 1 point, and 43 contestants scored 0 point. The average score for this problem is 3.383, indicating that it was relatively straightforward.

Among the top five teams in the team scores, the scores of this problem are as follows: the China team scored 39 points (with a total team score of 215 points), the Germany team scored 40 points (with a total team score of 189 points), the Bulgaria team scored 42 points (with a total team score of 178 points), the Russia team scored 38 points (with a total team score of 177 points), and the Chinese Taiwan team scored 33 points (with a total team score of 162 points).

The gold medal cutoff for this IMO was set at 30 points (with 35 contestants earning gold medals), the silver medal cutoff was 20 points (with 66 contestants earning silver medals), and the bronze medal cutoff was 11 points (with 97 contestants earning bronze medals).

In this IMO, only two contestants achieved a perfect score of 42 points, namely Hong Zhou from China and Hung-Wu Wu from Chinese Taiwan.

Problem 2.8 (IMO 52-3, proposed by Belarus). Let $f : \mathbf{R} \to \mathbf{R}$ be a real-valued function defined on the set of real numbers that satisfies

$$f(x+y) \leq yf(x) + f(f(x))$$

for all $x, y \in \mathbf{R}$. Prove that $f(x) = 0$ for all $x \leq 0$.

Proof 1. By setting $y = f(x) - x$ in the given inequality, we obtain

$$f(f(x)) \leq (f(x) - x)f(x) + f(f(x)),$$

from which it follows that $(f(x)-x)f(x) \geq 0$. Consequently, for any $x \in \mathbf{R}$,

$$(f(f(x)) - f(x))f(f(x)) \geq 0.$$

By setting $y = 0$ in the given inequality, we have $f(x) \leq f(f(x))$. Thus

$$f(f(x)) \geq 0 \quad \text{or} \quad f(x) = f(f(x)) < 0. \tag{1}$$

Case 1: There exists an $x_0 \in \mathbf{R}$ such that $f(x_0) > 0$.
For any $y \in \mathbf{R}$,

$$f(x_0 + y) \leq yf(x_0) + f(f(x_0)).$$

Hence, $f(x_0 + y) < 0$ for any $y < -\frac{f(f(x_0))}{f(x_0)}$. Consequently, $f(z) < 0$ for any real number $z < x_0 - \frac{f(f(x_0))}{f(x_0)}$.

Therefore, when $z < \min\left\{0, x_0 - \frac{f(f(x_0))}{f(x_0)}\right\}$, both $z < 0$ and $f(z) < 0$ are satisfied. Combining them with $(f(z) - z)f(z) \geq 0$ leads to

$$f(z) \leq z < \min\left\{0, x_0 - \frac{f(f(x_0))}{f(x_0)}\right\} \leq x_0 - \frac{f(f(x_0))}{f(x_0)},$$

implying $f(f(z)) = f(x_0 + (f(z) - x_0)) < 0$.

Thus, $f(f(z)) = f(z) < 0$ for any $z < \min\left\{0, x_0 - \frac{f(f(x_0))}{f(x_0)}\right\}$. As a result, for any $y \in \mathbf{R}$,

$$f(z+y) \leq yf(z) + f(f(z)) = (y+1)f(z).$$

Setting $y = x_0 - z$ yields $f(x_0) \leq (1 + x_0 - z)f(z)$. When z is sufficiently small, $1 + x_0 - z > 0$ and $f(z) < 0$, leading to $f(x_0) < 0$, a contradiction.

Case 2: $f(x) \leq 0$ for any $x \in \mathbf{R}$.
In this case,

$$f(f(x)) \leq 0. \tag{2}$$

If $f(f(x)) \geq 0$ for any $x \in \mathbf{R}$, then $f(f(x)) = 0$ for any $x \in \mathbf{R}$. Thus, $f(0) = f(f(f(x))) = 0$. From the conditions, $f(x+y) \leq yf(x)$ for any $x, y \in \mathbf{R}$, which implies $f(0) \leq -xf(x)$. For $x < 0$, it follows that $f(x) \geq 0$, leading to $f(x) = 0$ for $x \leq 0$.

If there exists a z_0 such that $f(f(z_0)) < 0$, then from (1) it is known that $f(z_0) = f(f(z_0)) < 0$. Therefore, there exists a $t = f(z_0) < 0$ such that $f(t) = t$. And if there exist t_1, t_2 both satisfying $f(t) = t$, where $t_1 < t_2 < 0$, then

$$t_2 = f(t_2) \leq (t_2 - t_1)f(t_1) + f(f(t_1)) = (t_2 - t_1)t_1 + t_1,$$

leading to $(t_2 - t_1)(1 - t_1) \leq 0$, a contradiction. Thus, such t is unique, and further, there exists an $x_1 \in \mathbf{R}$ such that $f(x_1) \neq t$.

(Otherwise, $f(x) = t$ for all $x \in \mathbf{R}$, implying $t \leq yt + t$ for any $y \in \mathbf{R}$, but it is not true for $y > 0$.)

For this x_1, we have $f(f(x_1)) \neq f(x_1)$, and from (1), $f(f(x_1)) \geq 0 > t$. Thus $f(f(f(x_1))) \geq 0$.

(Otherwise, from (1), $0 \leq f(f(x_1)) = f(f(f(x_1))) < 0$, a contradiction.) Combining with (2) implies

$$0 = f(f(x_1)) = f(f(f(x_1))) = f(0).$$

Thus, it still holds that $f(x + y) \leq yf(x)$, leading to $f(0) = 0 \leq -xf(x)$. Therefore, $f(x) \geq 0$ for $x < 0$, which implies $f(x) = 0$ for $x \leq 0$.

In conclusion, $f(x) = 0$ for any $x \leq 0$.

Proof 2. Set $y = t - x$, and then the given inequality can be rewritten as

$$f(t) \leq tf(x) - xf(x) + f(f(x)). \tag{3}$$

In (3), by setting $t = f(a)$, $x = b$ and $t = f(b)$, $x = a$, we have

$$f(f(a)) - f(f(b)) \leq f(a)f(b) - bf(b),$$

$$f(f(b)) - f(f(a)) \leq f(a)f(b) - af(a).$$

Adding the above two inequalities yields $2f(a)f(b) \geq af(a) + bf(b)$.

Setting $b = 2f(a)$ gives $bf(b) \geq af(a) + bf(b)$, so $af(a) \leq 0$. Therefore, $f(a) \leq 0$ when $a > 0$, and $f(a) \geq 0$ when $a < 0$.

Suppose there exists a real number x_0 such that $f(x_0) > 0$. Then from (3), it is known that $f(t) < 0$ for every $t < \min\left\{\frac{x_0 f(x_0) - f(f(x_0))}{f(x_0)}, 0\right\}$, which contradicts that $f(t) \geq 0$ for $t < 0$.

Therefore, $f(x) \leq 0$ for all real numbers x. Combined with $f(x) \geq 0$ for $x < 0$, it follows that $f(x) = 0$ for any real number $x < 0$. Setting $t = x$ in (3), we have $f(x) \leq f(f(x))$, and by taking $x = -1$, we obtain $0 \leq f(0)$. Since $f(x) \leq 0$ for any real number x, it follows that $f(0) = 0$.

In conclusion, $f(x) = 0$ for any $x \leq 0$.

Note. There are several similar problems:

- **(Asian Pacific Mathematics Olympiad 2002, Problem 5).** Let **R** denote the set of all real numbers. Find all functions f from **R** to **R** satisfying:

  (i) there are only finitely many s in **R** such that $f(s) = 0$;
  (ii) $f(x^4 + y) = x^3 f(x) + f(f(y))$ for all x, y in **R**.

- **(Romania Team Selection Test 2001, Problem 11).** Prove that there is no function $f : (0, +\infty) \to (0, +\infty)$ such that

  $$f(x + y) \geq f(x) + yf(f(x))$$

  for all $x, y \in (0, +\infty)$.

【Score Situation】 This particular problem saw the following distribution of scores among contestants: 51 contestants scored 7 points, 5 contestants scored 6 points, 3 contestants scored 5 points, 7 contestants scored 4 points, 13 contestants scored 3 points, 34 contestants scored 2 points, 57 contestants scored 1 point, and 393 contestants scored 0 point. The average score for this problem is 1.055, indicating that it was relatively challenging.

Among the top five teams in the team scores, the scores of this problem are as follows: the China team scored 42 points (with a total team score of 189 points), the United States team scored 35 points (with a total team score of 184 points), the Singapore team scored 23 points (with a total team score of 179 points), the Russia team scored 23 points (with a total team score of 161 points), and the Thailand team scored 23 points (with a total team score of 160 points).

The gold medal cutoff for this IMO was set at 28 points (with 54 contestants earning gold medals), the silver medal cutoff was 22 points (with 90 contestants earning silver medals), and the bronze medal cutoff was 16 points (with 137 contestants earning bronze medals).

In this IMO, only one contestant achieved a perfect score of 42 points, namely Lisa Sauermann from Germany.

Problem 2.9 (IMO 54-5, proposed by Bulgaria). Let $\mathbf{Q}_+$ be the set of positive rational numbers. Let $f : \mathbf{Q}_+ \to \mathbf{R}$ be a function satisfying the following three conditions:

(i) for all $x, y \in \mathbf{Q}_+$, we have $f(x)f(y) \geq f(xy)$;
(ii) for all $x, y \in \mathbf{Q}_+$, we have $f(x + y) \geq f(x) + f(y)$;
(iii) there exists a rational number $a > 1$ such that $f(a) = a$.

Prove that $f(x) = x$ for all $x \in \mathbf{Q}_+$.

Proof. Let $\mathbf{N}_+$ denote the set of all positive integers. In the inequality (i)

$$f(x)f(y) \geq f(xy), \tag{1}$$

setting $x = 1$ and $y = a$ gives $f(1) \geq 1$. Then, from

$$f(x+y) \geq f(x) + f(y), \tag{2}$$

applying mathematical induction on n, we can deduce for any $n \in \mathbf{N}_+$ and $x \in \mathbf{Q}_+$,

$$f(nx) \geq nf(x). \tag{3}$$

In particular,

$$f(n) \geq nf(1) \geq n. \tag{4}$$

For any $x = \frac{p}{q} \in \mathbf{Q}_+$ and $p, q \in \mathbf{N}_+$, $f(q) \cdot f(\frac{p}{q}) \geq f(p) \geq p$, thus $f(x) > 0$ for $x \in \mathbf{Q}_+$.

From (2), f is strictly increasing, and for any rational number $x \geq 1$,

$$f(x) \geq f(\lfloor x \rfloor) \geq \lfloor x \rfloor > x - 1,$$

where $\lfloor x \rfloor$ denotes the greatest integer not exceeding x.

From (1), using mathematical induction gives $(f(x))^n \geq f(x^n)$. Therefore,

$$(f(x))^n \geq f(x^n) > x^n - 1,$$

which implies $f(x) > \sqrt[n]{x^n - 1}$ for any rational number $x > 1$ and $n \in \mathbf{N}_+$. Thus, for any ratinoal number $x > 1$,

$$f(x) \geq x. \tag{5}$$

(If there exists a rational number $x_0 > 1$ such that $1 < f(x_0) = y < x_0$, then for sufficiently large n,

$$x_0^n - y^n = (x_0 - y)(x_0^{n-1} + x_0^{n-2}y + \cdots + y^{n-1}) > n(x_0 - y),$$

from which $x_0^n - y^n > 1$. Therefore, $y = f(x_0) > \sqrt[n]{x_0^n - 1} > y$, a contradiction.)

From (1) and (5), $a^n = (f(a))^n \geq f(a^n) \geq a^n$, from which $f(a^n) = a^n$. Then, for any rational number $x > 1$, we can choose an $n_0 \in \mathbf{N}_+$ such that $a^{n_0} - x > 1$. In this case, from (2) and (5),

$$a^{n_0} = f(a^{n_0}) \geq f(x) + f(a^{n_0} - x) \geq x + (a^{n_0} - x) = a^{n_0}.$$

Thus $f(x) = x$ for any rational number $x > 1$. Combined with $f(1) \geq 1$, we conclude $f(1) = 1$.

(Otherwise, if $f(1) = t > 1$, then

$$t = f(t) \geq f(1) + f(t-1) = t + f(t-1) > t,$$

a contradiction.)

Finally, for any $x \in \mathbf{Q}_+$ and any $n \in \mathbf{N}_+$, from (1) and (3) it follows that

$$nf(x) = f(n)f(x) \geq f(nx) \geq nf(x),$$

so $f(nx) = nf(x)$. Consequently, $f\left(\frac{m}{n}\right) = \frac{f(m)}{n} = \frac{m}{n}$ for any $m, n \in \mathbf{N}_+$.

Note. The condition $f(a) = a > 1$ is essential. Indeed, for $b \geq 1$, the function $f(x) = bx^2$ satisfies (1) and (2) for any $x, y \in \mathbf{Q}_+$, with a unique fixed point $\frac{1}{b} \leq 1$.

【Score Situation】This particular problem saw the following distribution of scores among contestants: 135 contestants scored 7 points, 19 contestants scored 6 points, 10 contestants scored 5 points, no contestant scored 4 points, 11 contestants scored 3 points, 33 contestants scored 2 points, 84 contestants scored 1 point, and 235 contestants scored 0 point. The average score for this problem is 2.452, indicating that it had a certain level of difficulty.

Among the top five teams in the team scores, the scores of this problem are as follows: the China team scored 42 points (with a total team score of 208 points), the South Korea team scored 36 points (with a total team score of 204 points), the United States team scored 40 points (with a total team score of 190 points), the Russia team scored 40 points (with a total team score of 187 points), and the North Korea team scored 41 points (with a total team score of 184 points).

The gold medal cutoff for this IMO was set at 31 points (with 45 contestants earning gold medals), the silver medal cutoff was 24 points (with 92 contestants earning silver medals), and the bronze medal cutoff was 15 points (with 141 contestants earning bronze medals).

In this IMO, no contestant achieved a perfect score of 42 points.

2.2.2 *Determining values*

Problem 2.10 (IMO 20-3, proposed by the United Kingdom). The set of all positive integers is the union of two disjoint subsets $\{f(1), f(2), \ldots, f(n), \ldots\}$ and $\{g(1), g(2), \ldots, g(n), \ldots\}$, where

$$f(1) < f(2) < \cdots < f(n) < \cdots,$$
$$g(1) < g(2) < \cdots < g(n) < \cdots,$$

and $g(n) = f(f(n)) + 1$ for all $n \geq 1$.
Determine $f(240)$.

Solution 1. Let $A = \{f(k)\}$ and $B = \{g(k)\}$, where $k \in \mathbf{N}_+$. Each positive integer is either in A or B, but not in both. For $k \neq 1$, we see that $f(1)$ is less than both $f(k)$ and $g(k)$, so $f(1) = 1$.

From the conditions, the nth positive integer not in A is $g(n) = f(f(n)) + 1$, implying that in the positive integers $1, 2, \ldots, g(n)$, there are n integers in B and $f(n)$ integers in A. Thus,

$$g(n) = f(n) + n, \quad f(f(n)) = f(n) + n - 1.$$

Since $g(n+1) - g(n) = f(n+1) + (n+1) - f(n) - n \geq 2$, we have $g(n) + 1 \in A$. Hence,

$$f(1) = 1,$$
$$g(1) = 2,$$
$$f(2) = 3,$$
$$f(f(2)) = f(3) = f(2) + 2 - 1 = 4,$$
$$f(f(3)) = f(4) = f(3) + 3 - 1 = 6,$$
$$f(f(4)) = f(6) = f(4) + 4 - 1 = 9,$$
$$f(f(6)) = f(9) = f(6) + 6 - 1 = 14,$$
$$f(f(9)) = f(14) = f(9) + 9 - 1 = 22,$$
$$f(f(14)) = f(22) = f(14) + 14 - 1 = 35,$$
$$f(f(22)) = f(35) = f(22) + 22 - 1 = 56,$$
$$f(f(35)) = f(56) = f(35) + 35 - 1 = 90,$$
$$g(35) = f(f(35)) + 1 = 91,$$
$$f(57) = 92,$$
$$f(f(57)) = f(92) = f(57) + 57 - 1 = 148,$$
$$f(f(92)) = f(148) = f(92) + 92 - 1 = 239,$$
$$f(f(148)) = f(239) = f(148) + 148 - 1 = 386,$$
$$g(148) = f(f(148)) + 1 = f(239) + 1 = 387,$$
$$f(240) = 388.$$

Therefore, $f(240) = 388$.

Note. The expressions of these two functions are actually

$$f(n) = \left\lfloor \frac{\sqrt{5}+1}{2} n \right\rfloor \quad \text{and} \quad g(n) = \left\lfloor \frac{\sqrt{5}+3}{2} n \right\rfloor.$$

This is a special case of Beatty's theorem.

◎ **Beatty's Theorem.** If two irrational numbers α, β are both greater than 1 and $\frac{1}{\alpha} + \frac{1}{\beta} = 1$, then the functions $f(n) = \lfloor n\alpha \rfloor$ and $g(n) = \lfloor n\beta \rfloor$ satisfy:

(i) $f(n), g(n)$ are increasing functions from $\mathbf{N}_+$ to $\mathbf{N}_+$;
(ii) $f(\mathbf{N}_+) \cup g(\mathbf{N}_+) = \mathbf{N}_+$;
(iii) $f(\mathbf{N}_+) \cap g(\mathbf{N}_+) = \varnothing$.

In this problem, $\alpha = \frac{\sqrt{5}+1}{2}$ and $\beta = \frac{\sqrt{5}+3}{2}$.

Solution 2. By Beatty's theorem, one can choose α, β such that $\{f(n)\}$ and $\{g(n)\}$ satisfy

$$f(\mathbf{N}_+) \cup g(\mathbf{N}_+) = \mathbf{N}_+ \quad \text{and} \quad f(\mathbf{N}_+) \cap g(\mathbf{N}_+) = \varnothing.$$

The key is to derive $g(n) = f(f(n)) + 1$, leading to $g(n) = f(n) + n$, i.e.,

$$\lfloor n\beta \rfloor = \lfloor n\alpha \rfloor + n = \lfloor n\alpha + n \rfloor$$

for all $n \in \mathbf{N}_+$. If $n\beta = n\alpha + n$, i.e., $\beta = \alpha + 1$, then $\frac{1}{\alpha} + \frac{1}{\alpha+1} = 1$, yielding $\alpha = \frac{\sqrt{5}+1}{2}$, and thus $\beta = \frac{\sqrt{5}+3}{2}$. Therefore,

$$f(240) = \lfloor 240\alpha \rfloor = \lfloor 120 \times (1 + \sqrt{5}) \rfloor = 120 + \lfloor 120\sqrt{5} \rfloor$$

$$= 120 + \lfloor 120 \times 2.236 \rfloor = 388.$$

Note. There is a similar problem:

- (From the "**Problems in Mathematics**" in the **Hungarian journal *KöMaL*, January 2001, A.255**). Let $q = \frac{1+\sqrt{5}}{2}$ and define a function f on the set of positive integers as follows:

  (a) start with $f(1) = 2$ and let $n > 1$;
  (b) if the values $f(1), f(2), \ldots, f(n-1)$ have already been defined and n is not among these values, then set $f(n) = f(n-1) + 1$;
  (c) otherwise consider k, the smallest positive integer, for which $f(k) = n$, and let $f(n) = n + k$.

  Prove that $qn - \frac{1}{q} < f(n) \leq qn + \frac{1}{q^2}$ for any positive integer n.

【Score Situation】 This particular problem saw the following distribution of scores among contestants: 8 contestants scored 8 points, 2 contestants scored 7 points, 1 contestant scored 6 points, 3 contestants scored 5 points, 7 contestants scored 4 points, 4 contestants scored 3 points, 6 contestants scored 2 points, 8 contestants scored 1 point, and 9 contestants scored 0 point. The average score for this problem is 3.313, indicating that it was relatively straightforward.

Among the top five teams in the team scores, the scores of this problem are as follows: the Romania team scored 40 points (with a total team score of 237 points), the United States team scored 47 points (with a total team score of 225 points), the United Kingdom team scored 45 points (with a total team score of 201 points), the Vietnam team scored 45 points (with a total team score of 200 points), and the Czechoslovakia team scored 24 points (with a total team score of 195 points).

The gold medal cutoff for this IMO was set at 35 points (with 5 contestants earning gold medals), the silver medal cutoff was 27 points (with 20 contestants earning silver medals), and the bronze medal cutoff was 22 points (with 38 contestants earning bronze medals).

In this IMO, only one contestant achieved a perfect score of 40 points, namely Mark Kleiman from the United States.

Problem 2.11 (IMO 22-6, proposed by Finland).

For all non-negative integers x and y, let $f(x, y)$ be a function satisfying:

(a) $f(0, y) = y + 1$;
(b) $f(x + 1, 0) = f(x, 1)$;
(c) $f(x + 1, y + 1) = f(x, f(x + 1, y))$.

Determine $f(4, 1981)$.

Solution. Set $x = 0$, and from (a) and (b), it is known that $f(1, 0) = f(0, 1) = 2$. Substituting $x = 0$ and $y = n - 1$ into (c) and utilizing (a), we have

$$f(1, n) = f(0, f(1, n - 1))$$
$$= f(1, n - 1) + 1$$
$$= n + f(1, 0)$$
$$= n + 2. \tag{1}$$

From (c) and (1),

$$f(2, n) = f(1, f(2, n - 1))$$
$$= f(2, n - 1) + 2$$
$$= 2n + f(2, 0),$$

where $f(2,0) = f(1,1) = 1 + 2 = 3$. Thus

$$f(2,n) = 2n + 3. \tag{2}$$

From (c) and (2),

$$\begin{aligned}
f(3,n) + 3 &= f(2, f(3, n-1)) + 3 \\
&= 2f(3, n-1) + 6 \\
&= 2(f(3, n-1) + 3) \\
&= \cdots\cdots\cdots\cdots \\
&= 2^n(f(3,0) + 3),
\end{aligned}$$

where $f(3,0) = f(2,1) = 5$. Hence

$$f(3,n) = 2^{n+3} - 3. \tag{3}$$

From (c) and (3),

$$\begin{aligned}
f(4,n) + 3 &= f(3, f(4, n-1)) + 3 \\
&= 2^{f(4,n-1)+3} \\
&= \cdots\cdots\cdots\cdots \\
&= 2^{2^{\cdot^{\cdot^{\cdot 2^{f(4,0)+3}}}}},
\end{aligned}$$

with 2 repeating n times. Since $f(4,0) + 3 = f(3,1) + 3 = 2^4$,

$$f(4,n) = -3 + 2^{2^{\cdot^{\cdot^{\cdot 2}}}},$$

with 2 repeating $n+3$ times.

Therefore, $f(4,1981) = -3 + 2^{2^{\cdot^{\cdot^{\cdot 2}}}}$, with 2 repeating 1984 times.

Note. This function, defined by a double recursion, is known as the Ackermann function. Wilhelm Ackermann, a student of Hilbert, proved that this function grows faster than any single-variable recursive function. Computer scientists are also interested in it as it outpaces the growth of any simple loop program.

【Score Situation】 This particular problem saw the following distribution of scores among contestants: 44 contestants scored 7 points, 3 contestants scored 6 points, no contestant scored 5 points, 1 contestant scored 4 points, no contestant scored 3 points, 1 contestant scored 2 points, no contestant scored 1 point, and 2 contestants scored 0 point. The average score for this problem is 6.510, indicating that it was simple.

Among the top five teams in the team scores, the scores of this problem are as follows: the United States team scored 55 points (with a total team score of 314 points), the Germany team scored 56 points (with a total team score of 312 points), the United Kingdom team scored 55 points (with a total team score of 301 points), the Austria team scored 55 points (with a total team score of 290 points), and the Bulgaria team scored 55 points (with a total team score of 287 points).

The gold medal cutoff for this IMO was set at 41 points (with 36 contestants earning gold medals), the silver medal cutoff was 34 points (with 37 contestants earning silver medals), and the bronze medal cutoff was 26 points (with 30 contestants earning bronze medals).

In this IMO, a total of 26 contestants achieved a perfect score of 42 points.

Problem 2.12 (IMO 23-1, proposed by the United Kingdom). A function $f(n)$ is defined for all positive integers n and takes on non-negative integer values. Also, for all m and n,

$$f(m+n) - f(m) - f(n) = 0 \quad \text{or} \quad 1,$$

$$f(2) = 0, \quad f(3) > 0, \quad \text{and} \quad f(9999) = 3333.$$

Determine $f(1982)$.

Solution. From the given condition,

$$f(m+n) \geq f(m) + f(n). \tag{1}$$

Setting $m = n = 1$, we obtain $f(2) \geq 2f(1)$. However, since $f(2) = 0$ and $f(1)$ is a non-negative integer, it follows that $f(1) = 0$. Setting $m = 2$ and $n = 1$, we get

$$f(3) - f(2) + f(1) = 0 \quad \text{or} \quad 1.$$

Given that $f(3) > 0$, it implies $f(3) = 1$. Now, we prove that $f(3k) = k$ for $k \leq 3333$. From (1),

$$f(3k) \geq f(3k-3) + f(3)$$

$$\geq f(3k-6) + 2f(3)$$

$$\geq \cdots\cdots\cdots$$

$$\geq kf(3)$$

$$= k.$$

If $f(3k) > k$, then $f(3k) \geq k+1$, leading to

$$f(9999) \geq f(9999 - 3k) + f(3k)$$

$$\geq 3333 - k + k + 1$$

$$> 3333,$$

which contradicts the given conditions. Therefore, $f(3k) = k$.

Hence, $1982 = f(3 \times 1982) \geq 3f(1982)$, i.e., $f(1982) \leq \frac{1982}{3} < 661$. However, $f(1982) \geq f(3 \times 660) + f(2) = 660$, and thus $f(1982) = 660$.

Note. There exists a function that meets the conditions, such as $f(n) = \lfloor \frac{n}{3} \rfloor$.

Furthermore, there are several similar problems:

- **(Korea Winter Program Test 2018, Problem 1).** Find all functions $f : \mathbf{R} \to \mathbf{R}$ satisfying the following conditions:

  (i) $f(x+y) - f(x) - f(y) \in \{0, 1\}$ for all $x, y \in \mathbf{R}$;

  (ii) $\lfloor f(x) \rfloor = \lfloor x \rfloor$ for all $x \in \mathbf{R}$.

- **(Korean Mathematical Olympiad 2007, Final Round, Problem 6).** Let $f : \mathbf{N} \to \mathbf{N}$ be a function satisfying $kf(n) \leq f(kn) \leq kf(n)+k-1$ for all $k, n \in \mathbf{N}$.

  (a) Prove that $f(a) + f(b) \leq f(a+b) \leq f(a) + f(b) + 1$ for all $a, b \in \mathbf{N}$.

  (b) If f satisfies $f(2007n) \leq 2007f(n) + 200$ for every $n \in \mathbf{N}$, show that there exists $c \in \mathbf{N}$ such that $f(2007c) = 2007f(c)$.

【Score Situation】 This particular problem saw the following distribution of scores among contestants: 54 contestants scored 7 points, 4 contestants scored 6 points, no contestant scored 5 points, 20 contestants scored 4 points, 11 contestants scored 3 points, 8 contestants scored 2 points, 18 contestants scored 1 point, and 4 contestants scored 0 point. The average score for this problem is 4.613, indicating that it was simple.

Among the top five teams in the team scores, the scores of this problem are as follows: the Germany team scored 28 points (with a total team score of 145 points), the Soviet Union team scored 28 points (with a total team score of 137 points), the German Democratic Republic team scored 28 points (with a total team score of 136 points), the United States team scored 25 points (with a total team score of 136 points), and the Vietnam team scored 25 points (with a total team score of 133 points).

The gold medal cutoff for this IMO was set at 37 points (with 10 contestants earning gold medals), the silver medal cutoff was 30 points (with 20 contestants earning silver medals), and the bronze medal cutoff was 21 points (with 31 contestants earning bronze medals).

In this IMO, only three contestants achieved a perfect score of 42 points, namely Bruno Haible from Germany, Grigori Perelman from the Soviet Union, and Lê Tự Quốc Thắng from Vietnam.

Problem 2.13 (IMO 29-3, proposed by the United Kingdom). A function f is defined on the positive integers by

$$f(1) = 1, \quad f(3) = 3, \quad f(2n) = f(n),$$

$$f(4n+1) = 2f(2n+1) - f(n), \quad f(4n+3) = 3f(2n+1) - 2f(n),$$

for all positive integers n.

Determine the number of positive integers n, less than or equal to 1988, for which $f(n) = n$.

Solution. Let the binary representation of n be $(a_s a_{s-1} \cdots a_1)_2$. We prove that

$$f(n) = (a_1 a_2 \cdots a_s)_2. \tag{1}$$

For $n \in \{1, 2, 3\}$, it is evident that (1) holds. Since $f(2n) = f(n)$, it suffices to consider the case where n is odd. Assume (1) holds for $n < k$. For $n = k$, we consider $f(k)$.

(i) Suppose $k = 4m + 1$. Let $4m + 1 = (a_t a_{t-1} \cdots a_0)_2$, where $a_0 = 1$ and $a_1 = 0$. Then

$$m = (a_t a_{t-1} \cdots a_2)_2, \quad 2m + 1 = (a_t a_{t-1} \cdots a_2 1)_2.$$

From the inductive hypothesis,

$$f(2m+1) = (1 a_2 a_3 \cdots a_t)_2, \quad f(m) = (a_2 a_3 \cdots a_t)_2,$$

leading to

$$
\begin{aligned}
f(4m+1) &= 2f(2m+1) - f(m) \\
&= (1 a_2 a_3 \cdots a_t)_2 + (1 a_2 a_3 \cdots a_t)_2 - (a_2 a_3 \cdots a_t)_2 \\
&= (10 a_2 a_3 \cdots a_t)_2 + (a_2 a_3 \cdots a_t)_2 - (a_2 a_3 \cdots a_t)_2 \\
&= (10 a_2 a_3 \cdots a_t)_2 = (a_0 a_1 a_2 a_3 \cdots a_t)_2,
\end{aligned}
$$

confirming (1).

(ii) Let $k = 4m + 3$ with $4m + 3 = (a_t a_{t-1} \cdots a_0)_2$, where $a_0 = a_1 = 1$. Then

$$4m = (a_t a_{t-1} \cdots a_2 00)_2,$$

$$m = (a_t a_{t-1} \cdots a_2)_2,$$

$$2m + 1 = (a_t a_{t-1} \cdots a_2 1)_2.$$

From the inductive hypothesis and given conditions,

$$f(4m + 3) = 3f(2m + 1) - 2f(m)$$

$$= 3(1a_2 \cdots a_t)_2 - 2(a_2 a_3 \cdots a_t)_2$$

$$= (11a_2 \cdots a_t)_2 + 2(a_2 a_3 \cdots a_t)_2 - 2(a_2 a_3 \cdots a_t)_2$$

$$= (11a_2 a_3 \cdots a_t)_2 = (a_0 a_1 a_2 \cdots a_t)_2,$$

thereby confirming (1).

Since $f(n) = n$ is equivalent to $(a_1 a_2 \cdots a_t)_2 = (a_t a_{t-1} \cdots a_1)_2$, where $n = (a_t a_{t-1} \cdots a_1)_2$, the binary representation of n satisfying $f(n) = n$ is "symmetric." Thus, if the binary representation of n has $2m$ digits, then it is entirely determined by the first m digits.

Given the first digit 1, followed by $m - 1$ digits each chosen from 0 or 1, there are 2^{m-1} symmetric binary numbers with $2m$ digits. Similarly, there are 2^{m-1} symmetric binary numbers with $2m - 1$ digits.

Since $2^{10} < 1988 < 2^{11}$, there are $1 + 1 + 2 + 2 + 4 + 4 + 8 + 8 + 16 + 16 + 32 = 94$ symmetric binary numbers less than 2048 that satisfy $f(n) = n$. Since $1988 = (11111000100)_2$, there are only two symmetric binary numbers exceeding 1988, namely $(11111011111)_2$ and $(11111111111)_2$.

Hence, there are 92 positive integers not exceeding 1988 such that $f(n) = n$.

Note. There is a similar problem:

- **(Asian Pacific Mathematics Olympiad 2008, Problem 4).** Consider a function $f : \mathbf{N}_0 \to \mathbf{N}_0$, where $\mathbf{N}_0$ is the set of all non-negative integers, defined by the following conditions:
    (i) $f(0) = 0$;
    (ii) $f(2n) = 2f(n)$;
    (iii) $f(2n + 1) = n + 2f(n)$ for all $n \geq 0$.
    (a) Determine the three sets $L := \{n | f(n) < f(n + 1)\}$, $E := \{n | f(n) = f(n + 1)\}$, and $G := \{n | f(n) > f(n + 1)\}$.
    (b) For each $k \geq 0$, find a formula for $a_k := \max\{f(n) : 0 \leq n \leq 2^k\}$ in terms of k.

- **(British Mathematical Olympiad 2007, 1st Round, Problem 6).** A function f is defined on the set of positive integers by

$$f(1) = 1, \quad f(2n) = 2f(n),$$

$$\text{and} \quad nf(2n + 1) = (2n + 1)(f(n) + n) \quad \text{for all } n \geq 1.$$

(i) Prove that $f(n)$ is always an integer.

(ii) For how many positive integers less than 2007 is $f(n) = 2n$?

【Score Situation】This particular problem saw the following distribution of scores among contestants: 31 contestants scored 7 points, 9 contestants scored 6 points, 4 contestants scored 5 points, 3 contestants scored 4 points, 1 contestant scored 3 points, 13 contestants scored 2 points, 132 contestants scored 1 point, and 75 contestants scored 0 point. The average score for this problem is 1.731, indicating that it was relatively challenging.

Among the top five teams in the team scores, the scores of this problem are as follows: the Soviet Union team scored 36 points (with a total team score of 217 points), the Romania team scored 35 points (with a total team score of 201 points), the China team scored 17 points (with a total team score of 201 points), the Germany team scored 25 points (with a total team score of 174 points), and the Vietnam team scored 12 points (with a total team score of 166 points).

The gold medal cutoff for this IMO was set at 32 points (with 17 contestants earning gold medals), the silver medal cutoff was 23 points (with 48 contestants earning silver medals), and the bronze medal cutoff was 14 points (with 65 contestants earning bronze medals).

In this IMO, a total of five contestants achieved a perfect score of 42 points.

2.2.3   *Deriving expressions*

Problem 2.14 (IMO 24-1, proposed by the United Kingdom). Find all functions f defined on the set of positive real numbers which take positive real values and satisfy the conditions:

(a) $f(xf(y)) = yf(x)$ for all positive x and y;

(b) $f(x) \to 0$ as $x \to +\infty$.

Solution. First, we prove that 1 is in the range of f. For any $x_0 > 1$, set $y_0 = \frac{1}{f(x_0)}$ and substitute it into (a) to obtain

$$f(x_0 f(y_0)) = y_0 f(x_0) = 1,$$

implying that 1 is within the range of f.

Next, suppose $y_0 \in \mathbf{R}_+$ satisfies $f(y_0) = 1$. Then, from (a),

$$f(x) = y_0 f(x),$$

implying $y_0 = 1$, i.e., $f(1) = 1$.

For any positive real number x, let $a = xf(x)$ and substitute $y = x$ into (a) to yield $f(a) = a$. If $a > 1$, then

$$f(a^2) = f(af(a)) = af(a) = a^2.$$

By mathematical induction, as $n \to +\infty$, $f(a^{2^n}) = a^{2^n} \to +\infty$, contradicting (b).

If $a < 1$, then

$$1 = f(1) = f\left(\frac{1}{a}f(a)\right) = af\left(\frac{1}{a}\right),$$

implying $f\left(\frac{1}{a}\right) = \frac{1}{a}$. However, since $\frac{1}{a} > 1$, following the case for $a > 1$ by setting $a' = \frac{1}{a}$, we also come to a contradiction.

Therefore, $a = 1$, implying $f(x) = \frac{1}{x}$. It is easy to verify that $f(x) = \frac{1}{x}$ meets the conditions.

Note. Another approach to deriving $f(1) = 1$ involves substituting values into the condition. Setting $x = y = 1$ gives $f(f(1)) = f(1)$, and setting $x = 1$ and $y = f(1)$ yields $f(f(f(1))) = f(1)f(1)$. Therefore, $f(1) = f(1)f(1)$. Since $f(1) > 0$, we have $f(1) = 1$.

Furthermore, there are several similar problems:

- **(Japan Team Selection Test 2019, Problem 4)** Let $\mathbf{Q}_+$ denote the set of all positive rational numbers. Determine all functions $f : \mathbf{Q}_+ \to \mathbf{Q}_+$ satisfying

$$f(x^2(f(y))^2) = f(y)(f(x))^2 \quad \text{for all } x, y \in \mathbf{Q}_+.$$

- Problem 2.16 (IMO 31-4) in this chapter.

【Score Situation】 This particular problem saw the following distribution of scores among contestants: 12 contestants scored 7 points, 1 contestant scored 6 points, 1 contestant scored 5 points, no contestant scored 4 points, 9 contestants scored 3 points, 4 contestants scored 2 points, 13 contestants scored 1 point, and no contestant scored 0 point. The average score of this problem is 3.575, indicating that it was relatively straightforward.

Among the top five teams in the team scores, the Germany team achieved a total score of 212 points, the United States team achieved a total score of 171 points, the Hungary team achieved a total score of 170 points, the Soviet Union team achieved a total score of 169 points, and the Romania team achieved a total score of 161 points.

The gold medal cutoff for this IMO was set at 38 points (with 9 contestants earning gold medals), the silver medal cutoff was 26 points (with 27 contestants earning silver medals), and the bronze medal cutoff was 15 points (with 57 contestants earning bronze medals).

In this IMO, a total of four contestants achieved a perfect score of 42 points.

Problem 2.15 (IMO 27-5, proposed by the United Kingdom). Find all functions f, defined on the non-negative real numbers and taking non-negative real values, such that:

(a) $f(xf(y))f(y) = f(x+y)$ for all $x, y \geq 0$;
(b) $f(2) = 0$;
(c) $f(x) \neq 0$ for $0 \leq x < 2$.

Solution. Setting $y = 2$ and substituting it into (a), we obtain

$$f(x+2) = 0. \tag{1}$$

Since $x \geq 0$, from (1), $f(x) = 0$ for $x \geq 2$. Furthermore, from (c), $x \geq 2$ if $f(x) = 0$.

Hence, for $x + y \geq 2$,

$$f(x+y) = 0 = f(xf(y))f(y),$$

implying either $y \geq 2$ or $xf(y) \geq 2$. For $0 \leq y < 2$, the inequality $xf(y) \geq 2$ is equivalent to $x + y \geq 2$, i.e.,

$$x \geq \frac{2}{f(y)} \Leftrightarrow x \geq 2 - y,$$

leading to $\frac{2}{f(y)} = 2 - y$, and therefore $f(y) = \frac{2}{2-y}$.
In conclusion, the function is

$$f(x) = \begin{cases} \frac{2}{2-x}, & 0 \leq x < 2, \\ 0, & x \geq 2. \end{cases}$$

Note. This problem is the same as the 20th problem in 2004 Germany Team Selection Test. Furthermore, there is a similar problem:

- **(International Mathematics Competition for University Students 2000, Problem 11).** Let $\mathbf{R}_+$ be the set of positive real numbers. Find all functions $f : \mathbf{R}_+ \to \mathbf{R}_+$ such that for all $x, y \in \mathbf{R}_+$,

$$f(x)f(yf(x)) = f(x+y).$$

【Score Situation】 This particular problem saw the following distribution of scores among contestants: 57 contestants scored 7 points, 41 contestants scored 6 points, 8 contestants scored 5 points, 13 contestants scored 4 points, 30 contestants scored 3 points, 14 contestants scored 2 points, 17 contestants scored 1 point, and 30 contestants scored 0 point. The average score for this problem is 4.152, indicating that it was simple.

Among the top five teams in the team scores, the scores of this problem are as follows: the Soviet Union team scored 41 points (with a total team score of 203 points), the United States team scored 39 points (with a total team score of 203 points), the Germany team scored 40 points (with a total team score of 196 points), the China team scored 30 points (with a total team score of 177 points), and the German Democratic Republic team scored 42 points (with a total team score of 172 points).

The gold medal cutoff for this IMO was set at 34 points (with 18 contestants earning gold medals), the silver medal cutoff was 26 points (with 41 contestants earning silver medals), and the bronze medal cutoff was 17 points (with 48 contestants earning bronze medals).

In this IMO, only three contestants achieved a perfect score of 42 points, namely Vladimir Roganov and Stanislav Smirnov from the Soviet Union, and Géza Kós from Hungary.

Problem 2.16 (IMO 31-4, proposed by Turkey). Let $\mathbf{Q}_+$ be the set of positive rational numbers. Construct a function $f : \mathbf{Q}_+ \to \mathbf{Q}_+$ such that

$$f(xf(y)) = \frac{f(x)}{y}$$

for all x and y in $\mathbf{Q}_+$.

Solution. Suppose a function f satisfies

$$f(xf(y)) = \frac{f(x)}{y}, \quad \text{where } x, y \in \mathbf{Q}_+, \tag{1}$$

and $f(x) \in \mathbf{Q}_+$ for any $x \in \mathbf{Q}_+$. If $y_1, y_2 \in \mathbf{Q}_+$ such that $f(y_1) = f(y_2)$, then from (1),

$$\frac{f(x)}{y_1} = f(xf(y_1)) = f(xf(y_2)) = \frac{f(x)}{y_2},$$

and since $f(x) > 0$, it follows that $y_1 = y_2$. Hence f is injective from $\mathbf{Q}_+$ to $\mathbf{Q}_+$.

Setting $x = y = 1$ in (1) gives $f(f(1)) = f(1)$, and as f is injective, $f(1) = 1$. Setting $x = 1$ in (1) gives

$$f(f(y)) = \frac{f(1)}{y} = \frac{1}{y}. \tag{2}$$

Setting $y = f(t)$ (where $t \in \mathbf{Q}_+$) in (1) gives

$$f(xf(f(t))) = \frac{f(x)}{f(t)},$$

and from (2), $f(xf(f(t))) = f\left(\frac{x}{t}\right) = \frac{f(x)}{f(t)}$. Thus

$$f(xy) = f(x)f(y), \quad \text{where } x, y \in \mathbf{Q}_+. \tag{3}$$

Conversely, if $f : \mathbf{Q}_+ \to \mathbf{Q}_+$ satisfies (2) and (3), then f must satisfy (1).

Let P be the set of all prime numbers, with $p_1 = 2$, $p_2 = 3, \ldots$, and generally, p_k being the kth prime number in ascending order. For a given function $g : P \to \mathbf{Q}_+$, construct a function $f : \mathbf{Q}_+ \to \mathbf{Q}_+$ as follows:

If $x = p_1^{\alpha_1} p_2^{\alpha_2} \cdots p_s^{\alpha_s}$, where $\alpha_1, \alpha_2, \ldots, \alpha_s$ are integers, then define

$$f(x) = (g(p_1))^{\alpha_1} (g(p_2))^{\alpha_2} \cdots (g(p_s))^{\alpha_s}. \tag{4}$$

Since $\alpha_1, \alpha_2, \ldots, \alpha_s$ can be 0 or negative integers, (4) defines f for any positive rational number. Such a function obviously satisfies (3). To ensure f satisfies (2), choose $g : P \to \mathbf{Q}_+$ as follows:

$$g(p_k) = \begin{cases} p_{k+1}, & k = 1, 3, 5, 7, \ldots, \\ \frac{1}{p_{k-1}}, & k = 2, 4, 6, 8, \ldots. \end{cases} \tag{5}$$

It is easy to verify that the function f defined by (4) and (5) satisfies (2), and therefore such a defined f satisfies (1).

Note. The function f satisfying (1) is not unique.

【Score Situation】 This particular problem saw the following distribution of scores among contestants: 69 contestants scored 7 points, 3 contestants scored 6 points, 3 contestants scored 5 points, 10 contestants scored 4 points, 46 contestants scored 3 points, 70 contestants scored 2 points, 76 contestants scored 1 point, and 31 contestants scored 0 point. The average score for this problem is 2.955, indicating that it had a certain level of difficulty.

Among the top five teams in the team scores, the scores of this problem are as follows: the China team scored 42 points (with a total team score of 230 points), the Soviet Union team scored 35 points (with a total team score of 193 points), the United States team scored 37 points (with a total team score of 174 points), the Romania team scored 30 points (with a total team score of 171 points), and the France team scored 38 points (with a total team score of 168 points).

The gold medal cutoff for this IMO was set at 34 points (with 23 contestants earning gold medals), the silver medal cutoff was 23 points (with 56 contestants earning silver medals), and the bronze medal cutoff was 16 points (with 76 contestants earning bronze medals).

In this IMO, a total of four contestants achieved a perfect score of 42 points.

Problem 2.17 (IMO 33-2, proposed by India). Let $\mathbf{R}$ denote the set of all real numbers. Find all functions $f : \mathbf{R} \to \mathbf{R}$ such that

$$f(x^2 + f(y)) = y + (f(x))^2 \quad \text{for all } x, y \in \mathbf{R}.$$

Solution 1. Suppose

$$f(x^2 + f(y)) = y + (f(x))^2. \tag{1}$$

First, we prove that $f(0) = 0$. Set $x = 0$ and $t = (f(0))^2$. Then

$$f(f(y)) = y + t. \tag{2}$$

Replacing y with $x^2 + f(f(y))$ in (2), we get

$$f(f(x^2 + f(f(y)))) = x^2 + f(f(y)) + t. \tag{3}$$

From (1),

$$f(f(x^2 + f(f(y)))) = f((f(x))^2 + f(y))$$
$$= y + (f(f(x)))^2. \tag{4}$$

Combining (2), (3), and (4), we see that

$$x^2 + y + 2t = y + (x + t)^2 \quad \text{and} \quad 2t = t^2 + 2tx$$

for any real number x. It follows that $t = 0$, i.e., $f(0) = 0$. Thus, (2) becomes

$$f(f(y)) = y. \tag{5}$$

For $x \geq 0$, from (1) and (5),

$$f(x + y) = f(x + f(f(y))) = f(y) + (f(\sqrt{x}))^2 \geq f(y),$$

indicating that $f(x)$ is an increasing function on $\mathbf{R}$, i.e., $f(x) \geq f(y)$ for $x \geq y$.

If there exists an x_0 such that $f(x_0) > x_0$, then $f(f(x_0)) \geq f(x_0) > x_0$, which is a contradiction. Similarly, if there exists an x_0 such that $f(x_0) < x_0$, then $f(f(x_0)) \leq f(x_0) < x_0$, which is also a contradiction.

Therefore, $f(x) = x$. Clearly, $f(x) = x$ satisfies the given conditions.

Solution 2. From the conditions and the fact that y can take any real number, f is surjective. Moreover, if $f(y_1) = f(y_2)$, then

$$y_1 + (f(x))^2 = f(x^2 + f(y_1)) = f(x^2 + f(y_2)) = y_2 + (f(x))^2,$$

implying $y_1 = y_2$, and thus f is injective. Therefore, f is bijective.

From the conditions,

$$y + (f(x))^2 = f(x^2 + f(y)) = f((-x)^2 + f(y)) = y + (f(-x))^2,$$

For $x \neq 0$, we have $x \neq -x$ and $f(x) \neq f(-x)$. It follows that $f(x) = -f(-x) \neq 0$. Since f is surjective, there must be $f(0) = 0$, and if we set $x = 0$ in (1), then $f(f(y)) = y$.

The rest of the proof is similar to Solution 1.

Note. (i) The solution to the functional equation $f(xf(y) + x^2) = xy + (f(x))^2$ for all $x, y \in \mathbf{R}$ is $f(x) = x$.

(ii) The solutions to the functional equation $f(x^l f^{(m)}(y) + x^n) = x^l y + (f(x))^n$ for all $x, y \in \mathbf{R}$ are as follows:

(a) except for $n = 1$ and $l = 0$, if and only if both m, n are odd, then there are two solutions $f(x) = x$ and $f(x) = -x$;
(b) except for $n = 1$ and $l = 0$, if m, n are not both odd, then there is only one solution $f(x) = x$;
(c) for $n = 1$ and $l = 0$, the solutions are not yet fully determined.

Here, $l \in \{0, 1, 2, 3, \ldots\}$, $m, n \in \{1, 2, 3, \ldots\}$, and $f^{(m)}(x)$ represents the mth iteration of f.

Furthermore, there are several similar problems:

- **(Japan Team Selection Test 2023, Problem 5).** Let $\mathbf{R}$ be the set of real numbers and $\mathcal{F}$ be the set of all functions $f : \mathbf{R} \to \mathbf{R}$ such that

$$f(x + f(y)) = f(x) + f(y) \quad \text{for all } x, y \in \mathbf{R}.$$

  Find all rational numbers q such that for every function $f \in \mathcal{F}$, there exists some $z \in \mathbf{R}$ satisfying $f(z) = qz$.
- **(Chinese Team Selection Test 2021, Problem 17).** Determine all $f : \mathbf{R} \to \mathbf{R}$ such that

$$f(xf(y) + y^{2021}) = yf(x) + (f(y))^{2021} \quad \text{for all } x, y \in \mathbf{R}.$$

- **(Korean Mathematical Olympiad 2021, Final Round, Problem 6).** Find all functions $f : \mathbf{R} \to \mathbf{R}$ such that

$$f(x^2 - g(y)) = (g(x))^2 - y \quad \text{for all } x, y \in \mathbf{R}.$$

- **(European Girls' Mathematical Olympiad 2021, Problem 2).** Find all functions $f : \mathbf{Q} \to \mathbf{Q}$ such that

$$f(xf(x) + y) = f(y) + x^2$$

for all $x, y \in \mathbf{Q}$. Here, $\mathbf{Q}$ denotes the set of rational numbers.

- **(Korean Mathematical Olympiad 2015, Final Round, Problem 1).** Find all functions $f : \mathbf{R} \to \mathbf{R}$ such that

$$f(x^{2015} + (f(y))^{2015}) = y^{2015} + (f(x))^{2015} \quad \text{for all } x, y \in \mathbf{R}.$$

- **(Kyrgyzstan Mathematical Olympiad 2012, Problem 4).** Find all functions $f : \mathbf{R} \to \mathbf{R}$ such that

$$f((f(x))^2 + f(y)) = y + xf(x) \quad \text{for all } x, y \in \mathbf{R}.$$

- **(USA Mathematical Olympiad Program 2006, Homework).** Find all functions $f : \mathbf{R} \to \mathbf{R}$ satisfying

$$f(x + f(y)) = x + f(f(y))$$

for all real numbers x and y, with the additional constraint $f(2004) = 2005$.

- **(Japan Mathematical Olympiad 2004, Final Round, Problem 2).** Find all functions $f : \mathbf{R} \to \mathbf{R}$ such that

$$f(xf(x) + f(y)) = y + (f(x))^2 \quad \text{for all } x, y \in \mathbf{R}.$$

- **(Estonia Team Selection Test 2001, Problem 3).** Let k be a fixed real number. Find all functions $f : \mathbf{R} \to \mathbf{R}$ such that

$$kf(x^2 + y) = f(y) + (f(x))^2 \quad \text{for all } x, y \in \mathbf{R}.$$

【Score Situation】This particular problem saw the following distribution of scores among contestants: 67 contestants scored 7 points, 3 contestants scored 6 points, 18 contestants scored 5 points, 34 contestants scored 4 points, 43 contestants scored 3 points, 41 contestants scored 2 points, 113 contestants scored 1 point, and 31 contestants scored 0 point. The average score for this problem is 2.963, indicating that it had a certain level of difficulty.

Among the top five teams in the team scores, the scores of this problem are as follows: the China team scored 42 points (with a total team score of 240 points), the United States team scored 32 points (with a total team score of 181 points), the Romania team scored 42 points (with a total team score of 177 points), the Commonwealth of Independent States team scored 30 points (with a total team score of 176 points), and the United Kingdom team scored 24 points (with a total team score of 168 points).

The gold medal cutoff for this IMO was set at 32 points (with 26 contestants earning gold medals), the silver medal cutoff was 24 points (with 55 contestants earning silver medals), and the bronze medal cutoff was 14 points (with 74 contestants earning bronze medals).

In this IMO, a total of four contestants achieved a perfect score of 42 points.

Problem 2.18 (IMO 35-5, proposed by the United Kingdom). Let S be the set of real numbers strictly greater than -1. Find all functions $f : S \to S$ satisfying the two conditions:

(a) $f(x + f(y) + xf(y)) = y + f(x) + yf(x)$ for all x and y in S;
(b) $\frac{f(x)}{x}$ is strictly increasing on each of the intervals $(-1, 0)$ and $(0, +\infty)$.

Solution. Since $\frac{f(x)}{x}$ is strictly increasing on the interval $(-1, 0)$, the equation $\frac{f(x)}{x} = 1$ has at most one solution $x = u \in (-1, 0)$, i.e., there exists at most one $u \in (-1, 0)$ such that $f(u) = u$. If such u exists, then setting $x = y = u$ in (a) yields

$$f(u^2 + 2u) = u^2 + 2u.$$

Since $u \in (-1, 0)$ and $u^2 + 2u = (u+1)^2 - 1$, it follows that $u^2 + 2u \in (-1, 0)$. By the uniqueness,

$$u^2 + 2u = u,$$

implying $u = -1$ or $u = 0$, which contradicts $u \in (-1, 0)$.

Therefore, $f(u) \neq u$ for any $u \in (-1, 0)$. Similarly, $f(u) \neq u$ for any $u \in (0, +\infty)$. Thus, $f(u) = u$ can only occur at $u = 0$.

Substituting $x = y$ in (a), we obtain

$$f(x + (1+x)f(x)) = x + (1+x)f(x),$$

leading to $x + (1+x)f(x) = 0$, i.e., $f(x) = -\frac{x}{1+x}$.

It is easy to verify that $f(x) = -\frac{x}{1+x}$ satisfies all the given conditions. Thus, the only function that satisfies the conditions is $f(x) = -\frac{x}{1+x}$.

【Score Situation】This particular problem saw the following distribution of scores among contestants: 84 contestants scored 7 points, 38 contestants scored 6 points, 23 contestants scored 5 points, 22 contestants scored 4 points, 29 contestants scored 3 points, 40 contestants scored 2 points, 54 contestants scored 1 point, and 95 contestants scored 0 point. The average score for this problem is 3.221, indicating that it was relatively straightforward.

Among the top five teams in the team scores, the scores of this problem are as follows: the United States team scored 42 points (with a total team score of 252 points), the China team scored 40 points (with a total team score of 229 points), the Russia team scored

40 points (with a total team score of 224 points), the Bulgaria team scored 41 points (with a total team score of 223 points), and the Hungary team scored 31 points (with a total team score of 221 points).

The gold medal cutoff for this IMO was set at 40 points (with 30 contestants earning gold medals), the silver medal cutoff was 30 points (with 64 contestants earning silver medals), and the bronze medal cutoff was 19 points (with 98 contestants earning bronze medals).

In this IMO, a total of 22 contestants achieved a perfect score of 42 points.

Problem 2.19 (IMO 40-6, proposed by Japan). Determine all functions $f : \mathbf{R} \to \mathbf{R}$ such that

$$f(x - f(y)) = f(f(y)) + xf(y) + f(x) - 1 \tag{1}$$

for all real numbers x and y.

Solution. Setting $x = f(y)$ in (1), we get

$$f(0) = f(f(y)) + (f(y))^2 + f(f(y)) - 1,$$

which leads to $f(f(y)) = \frac{1 + f(0) - (f(y))^2}{2}$. Furthermore, replacing x with $f(x)$ in (1), we get

$$f(f(x) - f(y)) = f(f(y)) + f(x)f(y) + f(f(x)) - 1$$

$$= \frac{1 + f(0) - (f(y))^2}{2} + f(x)f(y) + \frac{1 + f(0) - (f(x))^2}{2} - 1$$

$$= f(0) - \frac{1}{2}(f(x) - f(y))^2. \tag{2}$$

Since the zero function does not satisfy (1), there exists a $y_0 \in \mathbf{R}$ such that $f(y_0) \neq 0$. From (1),

$$f(x - f(y_0)) - f(x) = xf(y_0) + f(f(y_0)) - 1.$$

Therefore, for any real number z, there exists an $x_0 \in \mathbf{R}$ such that

$$f(x_0 - f(y_0)) - f(x_0) = z.$$

Setting $x = x_0 - f(y_0)$ and $y = x_0$ in (2) gives

$$f(z) = f(0) - \frac{1}{2}z^2. \tag{3}$$

Substituting (3) into (1) yields $f(0) = 1$. Hence $f(x) = 1 - \frac{1}{2}x^2$.

Upon verification, $f(x) = 1 - \frac{1}{2}x^2$ satisfies (1). Therefore, the only function that meets the conditions is $f(x) = 1 - \frac{1}{2}x^2$.

【Score Situation】This particular problem saw the following distribution of scores among contestants: 11 contestants scored 7 points, 5 contestants scored 6 points, 7 contestants scored 5 points, 8 contestants scored 4 points, 21 contestants scored 3 points, 28 contestants scored 2 points, 225 contestants scored 1 point, and 145 contestants scored 0 point. The average score for this problem is 1.151, indicating that it was relatively challenging.

Among the top five teams in the team scores, the scores of this problem are as follows: the China team scored 15 points (with a total team score of 182 points), the Russia team scored 11 points (with a total team score of 182 points), the Vietnam team scored 18 points (with a total team score of 177 points), the Romania team scored 23 points (with a total team score of 173 points), and the Bulgaria team scored 29 points (with a total team score of 170 points).

The gold medal cutoff for this IMO was set at 28 points (with 38 contestants earning gold medals), the silver medal cutoff was 19 points (with 70 contestants earning silver medals), and the bronze medal cutoff was 12 points (with 118 contestants earning bronze medals).

In this IMO, no contestant achieved a perfect score of 42 points.

Problem 2.20 (IMO 43-5, proposed by India). Find all functions $f : \mathbf{R} \to \mathbf{R}$ such that

$$(f(x) + f(z))(f(y) + f(t)) = f(xy - zt) + f(xt + yz) \qquad (1)$$

for all $x, y, z, t \in \mathbf{R}$.

Solution 1. Setting $x = y = z = 0$ in (1) gives

$$2f(0)(f(0) + f(t)) = 2f(0). \qquad (2)$$

Substituting $t = 0$ into (2), we get $4(f(0))^2 = 2f(0)$, which implies

$$f(0) = 0 \quad \text{or} \quad f(0) = \frac{1}{2}.$$

If $f(0) = \frac{1}{2}$, then substituting it into (2) gives $f(t) = \frac{1}{2}$, i.e., $f(x) = \frac{1}{2}$ for all $x \in \mathbf{R}$. If $f(0) = 0$, then setting $z = t = 0$ in (1) gives

$$f(x)f(y) = f(xy). \qquad (3)$$

By setting $x = y = 1$ in (3), we get $(f(1))^2 = f(1)$, so $f(1) = 0$ or $f(1) = 1$.
If $f(1) = 0$, then (3) implies $f(x) = f(x)f(1) = 0$ for all $x \in \mathbf{R}$.
If $f(1) = 1$, setting $x = 0$ and $y = t = 1$ in (1), we get

$$2f(z) = f(-z) + f(z),$$
$$f(z) = f(-z),$$

indicating $f(x)$ is an even function. Since setting $x = y$ and $z = t = 0$ in (1) gives

$$f(x^2) = (f(x))^2 \geq 0,$$

and since $f(x)$ is even, $f(x) \geq 0$ for all $x \in \mathbf{R}$.

Setting $x = t$ and $y = z$ in (1), we get

$$(f(x) + f(y))^2 = f(x^2 + y^2),$$

which implies

$$f(x^2 + y^2) = (f(x))^2 + 2f(x)f(y) + (f(y))^2$$
$$\geq (f(x))^2 = f(x^2).$$

Hence, $u \geq v \geq 0$ implies $f(u) \geq f(v)$, i.e., $f(x)$ is increasing for $x > 0$. Setting $y = z = t = 1$ in (1), we find

$$2(f(x) + 1) = f(x - 1) + f(x + 1), \tag{4}$$

and using (4) and mathematical induction, we see that $f(n) = n^2$ for non-negative integers n. From $f(xy) = f(x)f(y)$, it follows that $f(a) = a^2$ for all positive rational numbers a.

Next, we prove that $f(x) = x^2$ for $x > 0$. Otherwise, there exists $x_0 \in \mathbf{R}$ such that $f(x_0) \neq x_0^2$.

If $f(x_0) < x_0^2$, by choosing a rational number a such that $\sqrt{f(x_0)} < a < x_0$, then $f(x_0) \geq f(a) = a^2 > f(x_0)$, which is a contradiction. If $f(x_0) > x_0^2$, then there is a similar contradiction. Therefore, $f(x) = x^2$ for $x > 0$. Since $f(x)$ is even, $f(x) = x^2$ for $x \in \mathbf{R}$.

Upon verification, the solutions satisfying the conditions are $f(x) = 0$, $f(x) = \frac{1}{2}$, and $f(x) = x^2$.

Solution 2. Setting $y = t = 0$ and $z = x$ in (1), we obtain $4f(0)f(x) = 2f(0)$.

(i) If $f(0) \neq 0$, then $f(x) = \frac{1}{2}$ for any $x \in \mathbf{R}$, which is clearly a solution.
(ii) If $f(0) = 0$, then setting $z = t = 0$ gives $f(xy) = f(x)f(y)$. If there exists $x_0 \neq 0$ such that $f(x_0) = 0$, then for any $x \in \mathbf{R}$,

$$f(x) = f\left(x_0 \cdot \frac{x}{x_0}\right) = f(x_0)f\left(\frac{x}{x_0}\right) = 0,$$

which is also a solution.

(iii) If $f(0) = 0$ and $f(x) \neq 0$ for $x \neq 0$, then setting $x = y = 0$ gives

$$f(-zt) = f(z)f(t) = f(zt), \qquad (5)$$

indicating $f(x)$ is even. Setting $z = t = 1$ in (5), we find $f(1) = (f(1))^2$, and since $f(1) \neq 0$, it follows that $f(1) = 1$.

For $x > 0$, clearly $f(x) = f(\sqrt{x} \cdot \sqrt{x}) = f(\sqrt{x})f(\sqrt{x}) > 0$. Setting $z = y$ and $t = x$ in (1) gives

$$f(x^2 + y^2) = (f(x) + f(y))^2.$$

Let $g(x) = \sqrt{f(x)}$. Then

$$g(0) = 0, \quad g(1) = 1, \quad g(-x) = g(x), \quad g(xy) = g(x)g(y).$$

It is evident that $g(x) > 0$ for $x \neq 0$, and

$$g(x^2 + y^2) = (g(x))^2 + (g(y))^2 = g(x^2) + g(y^2),$$

implying $g(x)$ satisfies Cauchy's functional equation. Thus,

$$\sqrt{f(x)} = g(x) = g(1) \cdot x = x, \quad \text{and so } f(x) = x^2.$$

Upon verification, the solutions satisfying the conditions are $f(x) = 0$, $f(x) = \frac{1}{2}$, and $f(x) = x^2$.

Note. There are several similar problems:

- **(Bolivia Team Selection Test 2021, Problem 1).** Find all functions $f : \mathbf{Q} \to \mathbf{Q}$ such that

$$f(x + y) + f(x - y) = 2f(x) + 2f(y)$$

  for all $x, y \in \mathbf{Q}$.

- **(Korean Mathematical Olympiad 2015, 2nd Round, Problem 5).** Find all functions $f : \mathbf{R} \to \mathbf{R}$ such that for all real numbers x, y, z,

$$(f(x) + 1)(f(y) + f(z)) = f(xy + z) + f(xz - y).$$

- **(United States of America Junior Mathematical Olympiad 2015, Problem 4).** Find all functions $f : \mathbf{Q} \to \mathbf{Q}$ such that

$$f(x) + f(t) = f(y) + f(z)$$

  for all rational numbers $x < y < z < t$ that form an arithmetic progression.

- **(ELMO 2011, called "English Language Master's Open" that time, Problem 4).** Find all functions $f : \mathbf{R}_+ \to \mathbf{R}_+$, where $\mathbf{R}_+$ denotes the positive reals, such that whenever $a > b > c > d > 0$ are real numbers with $ad = bc$,
$$f(a+d) + f(b-c) = f(a-d) + f(b+c).$$

【Score Situation】 This particular problem saw the following distribution of scores among contestants: 66 contestants scored 7 points, 25 contestants scored 6 points, 10 contestants scored 5 points, no contestant scored 4 points, 21 contestants scored 3 points, 101 contestants scored 2 points, 159 contestants scored 1 point, and 97 contestants scored 0 point. The average score for this problem is 2.267, indicating that it had a certain level of difficulty.

Among the top five teams in the team scores, the scores of this problem are as follows: the China team scored 42 points (with a total team score of 212 points), the Russia team scored 42 points (with a total team score of 204 points), the United States team scored 34 points (with a total team score of 171 points), the Bulgaria team scored 36 points (with a total team score of 167 points), and the Vietnam team scored 40 points (with a total team score of 166 points).

The gold medal cutoff for this IMO was set at 29 points (with 39 contestants earning gold medals), the silver medal cutoff was 23 points (with 73 contestants earning silver medals), and the bronze medal cutoff was 14 points (with 120 contestants earning bronze medals).

In this IMO, only three contestants achieved a perfect score of 42 points, namely Yunhao Fu and Botong Wang from China, and Andrei Khaliavine from Russia.

Problem 2.21 (IMO 49-4, proposed by South Korea). Find all functions $f : (0, +\infty) \to (0, +\infty)$ such that
$$\frac{(f(w))^2 + (f(x))^2}{f(y^2) + f(z^2)} = \frac{w^2 + x^2}{y^2 + z^2}$$
for all positive real numbers w, x, y, z satisfying $wx = yz$.

Solution. Setting $w = x = y = z = 1$ gives $(f(1))^2 = f(1)$. Hence $f(1) = 1$.

For any $t > 0$, setting $w = t$, $x = 1$, and $y = z = \sqrt{t}$ yields
$$\frac{(f(t))^2 + 1}{2f(t)} = \frac{t^2 + 1}{2t},$$
and after simplifying, we get $(tf(t) - 1)(f(t) - t) = 0$. Therefore, for each $t > 0$,
$$f(t) = t \quad \text{or} \quad f(t) = \frac{1}{t}. \tag{1}$$

If there exist $b, c \in (0, +\infty)$ such that $f(b) \neq b$ and $f(c) \neq \frac{1}{c}$, then from (1) we know b and c are both not equal to 1, and $f(b) = \frac{1}{b}$ and $f(c) = c$.

Setting $w = b$, $x = c$, and $y = z = \sqrt{bc}$, we get $\frac{\frac{1}{b^2}+c^2}{2f(bc)} = \frac{b^2+c^2}{2bc}$, i.e., $f(bc) = \frac{c+b^2c^3}{b(b^2+c^2)}$.

Since $f(bc) = bc$ or $f(bc) = \frac{1}{bc}$, if $f(bc) = bc$, then

$$bc = \frac{c+b^2c^3}{b(b^2+c^2)},$$

yielding $b^4c = c$ and $b = 1$, which is a contradiction. If $f(bc) = \frac{1}{bc}$, then

$$\frac{1}{bc} = \frac{c+b^2c^3}{b(b^2+c^2)},$$

yielding $b^2c^4 = b^2$ and $c = 1$, which is also a contradiction.

Thus, either $f(x) = x$ for all $\dot{x} \in (0,+\infty)$ or $f(x) = \frac{1}{x}$ for all $x \in (0,+\infty)$. Upon verification, both $f(x) = x$ and $f(x) = \frac{1}{x}$ satisfy the given conditions.

【Score Situation】 This particular problem saw the following distribution of scores among contestants: 227 contestants scored 7 points, 27 contestants scored 6 points, no contestant scored 5 points, 128 contestants scored 4 points, 4 contestants scored 3 points, no contestant scored 2 points, 80 contestants scored 1 point, and 69 contestants scored 0 point. The average score for this problem is 4.402, indicating that it was simple.

Among the top five teams in the team scores, the scores of this problem are as follows: the China team scored 42 points (with a total team score of 217 points), the Russia team scored 42 points (with a total team score of 199 points), the United States team scored 41 points (with a total team score of 190 points), the South Korea team scored 42 points (with a total team score of 188 points), and the Iran team scored 42 points (with a total team score of 181 points).

The gold medal cutoff for this IMO was set at 31 points (with 47 contestants earning gold medals), the silver medal cutoff was 22 points (with 100 contestants earning silver medals), and the bronze medal cutoff was 15 points (with 120 contestants earning bronze medals).

In this IMO, only three contestants achieved a perfect score of 42 points, namely Xiaosheng Mu and Dongyi Wei from China, and Alex Zhai from the United States.

Problem 2.22 (IMO 50-5, proposed by France). Determine all functions f from the set of positive integers to the set of positive integers such that, for all positive integers a and b, there exists a non-degenerate triangle with the sides of lengths

$$a, \quad f(b), \quad \text{and} \quad f(b+f(a)-1).$$

Here, a triangle is non-degenerate if its vertices are not collinear.

Solution 1. The only f that satisfies the conditions is $f(n) = n$ for $n \in$ $\mathbf{N}_+$. From the discrete nature of integers, for any $a, b \in \mathbf{N}_+$,

$$f(b) + f(b + f(a) - 1) - 1 \geq a, \tag{1}$$

$$f(b) + a - 1 \geq f(b + f(a) - 1), \tag{2}$$

$$f(b + f(a) - 1) + a - 1 \geq f(b). \tag{3}$$

Setting $a = 1$ in (2) and (3), we find that $f(b) = f(b + f(1) - 1)$ for any $b \in \mathbf{N}_+$. If $f(1) \neq 1$, then the above equation implies that f is periodic. Considering the domain of f, we are sure that f is bounded. Let M be a positive integer such that $M \geq f(n)$ for all $n \in \mathbf{N}_+$ (i.e., M is an upper bound of f). Setting $a = 2M$ in (1) leads to a contradiction. Therefore $f(1) = 1$.

Setting $b = 1$ in (1) and (2), we get $f(f(n)) = n$ for $n \in \mathbf{N}_+$. If there exists $t \in \mathbf{N}_+$ such that $f(t) < t$, then clearly $t \geq 2$ and $f(t) \leq t - 1$. Setting $a = f(t)$ in (2) yields

$$f(b + t - 1) = f(b + f(a) - 1)$$
$$\leq f(b) + a - 1$$
$$\leq f(b) + t - 2.$$

Let $M = (t - 1) \cdot \max_{1 \leq i \leq t - 1} f(i)$. For any integer $n > M$, let n_0 be the unique positive integer such that

$$1 \leq n_0 \leq t - 1, \quad n_0 \equiv n (\mathrm{mod} (t - 1)).$$

Then

$$f(n) \leq f(n_0) + \frac{t - 2}{t - 1}(n - n_0) \leq \frac{M}{t - 1} + \frac{(t - 2)n}{t - 1} < n.$$

Therefore, $f(n) < n$ for integers $n > M$. Choose $n_1 \in \mathbf{N}_+$ such that $n_1 > M$ and n_1 is not equal to any of $f(1), f(2), \ldots, f(M)$. From $f(f(n_1)) = n_1$, we know $f(n_1) > M$, and from the above conclusion, $n_1 > f(n_1) > f(f(n_1)) = n_1$, a contradiction.

This implies that $f(t) \geq t$ for any $t \in \mathbf{N}_+$ and further $t = f(f(t)) \geq f(t) \geq t$, implying the equality throughout, i.e., $f(n) = n$ for $n \in \mathbf{N}_+$.

Upon verification, $f(n) = n$ for $n \in \mathbf{N}_+$ satisfies the conditions. Therefore, the solution is $f(n) = n$ for $n \in \mathbf{N}_+$.

Solution 2. As in Solution 1, we know $f(1) = 1$ and $f(f(n)) = n$. It is easy to deduce that f is surjective. Hence f is a bijection from $\mathbf{N}_+$ to $\mathbf{N}_+$. Otherwise, if $f(n_1) = f(n_2)(n_1 \neq n_2)$, then $n_1 = f(f(n_1)) = f(f(n_2)) = n_2$, a contradiction.

It is evident that $f(2) > 1$, and 2, $f(b)$, $f(b+f(2)-1)$ can form a triangle. Suppose $t = f(2) - 1 > 0$, since f is a bijection, $0 < |f(b+t) - f(b)| < 2$. Thus for any $b \in \mathbf{N}_+$,

$$f(b+t) - f(b) = \pm 1.$$

By mathematical induction and since f is a bijection, $f(b+nt) = f(b) \pm n$ for any $n \in \mathbf{N}_+$. As $f(b+nt) > 0$, it must be that $f(b+nt) = f(b) + n$, i.e., $f(b+t) - f(b) = 1$ for any $b \in \mathbf{N}_+$.

Therefore, $f(1), f(1+t), f(1+2t), \ldots, f(1+nt), \ldots$ cover all positive integers. Since f is a bijection, $1, 1+t, 1+2t, \ldots$ must also cover all positive integers, so $t = 1$, implying $f(b+1) = f(b) + 1$ for any $b \in \mathbf{N}_+$.

Since $f(1) = 1$, we deduce that $f(n) = n$ for any $n \in \mathbf{N}_+$. In conclusion, $f(n) = n$ is the only solution.

Note. There are several similar problems:

- **(Chinese Team Selection Test 2016, Problem 18).** Find all functions $f : \mathbf{R}_+ \to \mathbf{R}_+$ satisfying the following condition: for any three distinct real numbers a, b, c, a triangle can be formed with side lengths a, b, c if and only if a triangle can be formed with side lengths $f(a), f(b), f(c)$.
- **(British Mathematical Olympiad 2014, 1st Round, Problem 6).** Determine all functions $f(n)$ from the positive integers to the positive integers which satisfy the following condition: whenever a, b, and c are positive integers such that $\frac{1}{a} + \frac{1}{b} = \frac{1}{c}$,

$$\frac{1}{f(a)} + \frac{1}{f(b)} = \frac{1}{f(c)}.$$

【Score Situation】 This particular problem saw the following distribution of scores among contestants: 153 contestants scored 7 points, 7 contestants scored 6 points, 4 contestants scored 5 points, 6 contestants scored 4 points, 33 contestants scored 3 points, 50 contestants scored 2 points, 42 contestants scored 1 point, and 270 contestants scored 0 point. The average score for this problem is 2.474, indicating that it had a certain level of difficulty.

Among the top five teams in the team scores, the scores of this problem are as follows: the China team scored 42 points (with a total team score of 221 points), the Japan team scored 42 points (with a total team score of 212 points), the Russia team scored 42 points (with a total team score of 203 points), the South Korea team scored 42 points (with a total team score of 188 points), and the North Korea team scored 36 points (with a total team score of 183 points).

The gold medal cutoff for this IMO was set at 32 points (with 49 contestants earning gold medals), the silver medal cutoff was 24 points (with 98 contestants earning silver medals), and the bronze medal cutoff was 14 points (with 135 contestants earning bronze medals).

In this IMO, only two contestants achieved a perfect score of 42 points, namely Dongyi Wei from China and Makoto Soejima from Japan.

Problem 2.23 (IMO 51-1, proposed by France). Determine all functions $f : \mathbf{R} \to \mathbf{R}$ such that

$$f(\lfloor x \rfloor y) = f(x)\lfloor f(y) \rfloor \tag{1}$$

for all $x, y \in \mathbf{R}$. Here, $\lfloor z \rfloor$ denotes the greatest integer less than or equal to z.

Solution. The solution is $f(x) = C$, where C is a constant and either $C = 0$ or $1 \leq C < 2$.

Setting $x = 0$ in (1), we get

$$f(0) = f(0)\lfloor f(y) \rfloor \tag{2}$$

for all $y \in \mathbf{R}$. Thus, there are two cases:

Case 1: $f(0) \neq 0$.

From (2), $\lfloor f(y) \rfloor = 1$ for all $y \in \mathbf{R}$. Therefore, (1) becomes $f(\lfloor x \rfloor y) = f(x)$. Setting $y = 0$ gives $f(x) = f(0) = C \neq 0$.

Furthermore, $\lfloor f(y) \rfloor = 1 = \lfloor C \rfloor$, and it follows that $1 \leq C < 2$.

Case 2: $f(0) = 0$.

If there exists $0 < \alpha < 1$ such that $f(\alpha) \neq 0$, then setting $x = \alpha$ in (1) gives

$$0 = f(0) = f(\alpha)\lfloor f(y) \rfloor$$

for all $y \in \mathbf{R}$, implying $\lfloor f(y) \rfloor = 0$ for all $y \in \mathbf{R}$. Setting $x = 1$ in (1), we find $f(y) = f(1)\lfloor f(y) \rfloor = 0$ for all $y \in \mathbf{R}$, which contradicts $f(\alpha) \neq 0$.

Therefore, $f(\alpha) = 0$ for $0 \leq \alpha < 1$. For any $z \in \mathbf{R}$, there exists an integer N such that $\alpha = \frac{z}{N} \in [0, 1)$. From (1),

$$f(z) = f(\lfloor N \rfloor \alpha) = f(N)\lfloor f(\alpha) \rfloor = 0$$

for all $z \in \mathbf{R}$.

Upon verification, $f(x) = C$, where C is a constant and either $C = 0$ or $1 \leq C < 2$, satisfies the given conditions.

【Score Situation】 This particular problem saw the following distribution of scores among contestants: 294 contestants scored 7 points, 54 contestants scored 6 points, 35 contestants scored 5 points, 34 contestants scored 4 points, 16 contestants scored 3 points, 27 contestants scored 2 points, 17 contestants scored 1 point, and 39 contestants scored 0 point. The average score for this problem is 5.450, indicating that it was simple.

Among the top five teams in the team scores, the scores of this problem are as follows: the China team scored 41 points (with a total team score of 197 points), the Russia team scored 41 points (with a total team score of 169 points), the United States team scored 40 points (with a total team score of 168 points), the South Korea team scored 42 points (with a total team score of 156 points), the Kazakhstan team scored 42 points (with a total team score of 148 points), and the Thailand team scored 42 points (with a total team score of 148 points).

The gold medal cutoff for this IMO was set at 27 points (with 47 contestants earning gold medals), the silver medal cutoff was 21 points (with 103 contestants earning silver medals), and the bronze medal cutoff was 15 points (with 115 contestants earning bronze medals).

In this IMO, only one contestant achieved a perfect score of 42 points, namely Zipei Nie from China.

Problem 2.24 (IMO 53-4, proposed by South Africa). Find all functions $f : \mathbf{Z} \to \mathbf{Z}$ such that, for all integers a, b, c that satisfy $a + b + c = 0$, the following equality holds:

$$f(a)^2 + f(b)^2 + f(c)^2 = 2f(a)f(b) + 2f(b)f(c) + 2f(c)f(a).$$

Here, $\mathbf{Z}$ denotes the set of integers.

Solution. Setting $a = b = c = 0$ gives $3f(0)^2 = 6f(0)^2$. Thus $f(0) = 0$.

Setting $b = -a$ and $c = 0$ yields $(f(a) - f(-a))^2 = 0$, implying f is an even function, i.e., $f(a) = f(-a)$ for all $a \in \mathbf{Z}$.

Setting $b = a$ and $c = -2a$, we obtain

$$2f(a)^2 + f(2a)^2 = 2f(a)^2 + 4f(a)f(2a).$$

Therefore, for all $a \in \mathbf{Z}$,

$$f(2a) = 0 \quad \text{or} \quad f(2a) = 4f(a). \tag{1}$$

If $f(r) = 0$ for some $r \geq 1$, then set $b = r$ and $c = -a - r$ to get

$$(f(a + r) - f(a))^2 = 0,$$

implying f is a periodic function with period r, i.e., $f(a + r) = f(a)$ for all $a \in \mathbf{Z}$. Particularly, if $f(1) = 0$, then f is a constant function, so $f(a) = 0$

for all $a \in \mathbf{Z}$. This function clearly satisfies the given conditions. Next, assume $f(1) = k \neq 0$.

From (1), we have $f(2) = 0$ or $f(2) = 4k$.

If $f(2) = 0$, then f is a periodic function with period 2. Hence for all $n \in \mathbf{Z}$,

$$f(2n) = 0 \quad \text{and} \quad f(2n+1) = k.$$

If $f(2) = 4k \neq 0$, then from (1), we know $f(4) = 0$ or $f(4) = 16k$. Suppose $f(4) = 0$, the function f is periodic with period 4, and

$$f(3) = f(-1) = f(1) = k.$$

Hence $f(4n) = 0$, $f(4n+1) = f(4n+3) = k$, and $f(4n+2) = 4k$ for all $n \in \mathbf{Z}$.

Suppose $f(4) = 16k \neq 0$, setting $a = 1$, $b = 2$, and $c = -3$ gives

$$(f(3))^2 - 10kf(3) + 9k^2 = 0,$$

implying $f(3) \in \{k, 9k\}$. Setting $a = 1$, $b = 3$, and $c = -4$ yields

$$(f(3))^2 - 34kf(3) + 225k^2 = 0,$$

implying $f(3) \in \{9k, 25k\}$. Thus $f(3) = 9k$.

Next, we use mathematical induction to prove $f(x) = kx^2$ for $x \in \mathbf{Z}$. For $x \in \{0, 1, 2, 3, 4\}$, the proposition already holds. Assume the proposition holds for $x \in \{0, 1, \ldots, n\}(n \geq 4)$.

Setting $a = n$, $b = 1$, and $c = -n - 1$, we get

$$f(n+1) \in \{k(n+1)^2, k(n-1)^2\}.$$

Setting $a = n - 1$, $b = 2$, and $c = -n - 1$, we obtain

$$f(n+1) \in \{k(n+1)^2, k(n-3)^2\}.$$

Since $k(n-1)^2 \neq k(n-3)^2$ for $n \neq 2$, it follows that $f(n+1) = k(n+1)^2$, proving $f(x) = kx^2$ for non-negative integers x. As f is even, $f(x) = kx^2$ for $x \in \mathbf{Z}$.

In conclusion, the possible functions are

$$f_1(x) = 0,$$

$$f_2(x) = kx^2,$$

$$f_3(x) = \begin{cases} 0, & x \equiv 0 \pmod 2, \\ k, & x \equiv 1 \pmod 2, \end{cases}$$

$$f_4(x) = \begin{cases} 0, & x \equiv 0 \pmod 4, \\ k, & x \equiv 1 \pmod 2, \\ 4k, & x \equiv 2 \pmod 4, \end{cases}$$

where k is any non-zero integer.

It is easy to verify that f_1 and f_2 satisfy the given conditions. For f_3, if a, b, c are all even integers, then

$$f(a) = f(b) = f(c) = 0,$$

satisfying the conditions. If a, b, c include one even and two odd integers, then the left side equals $2k^2$ as does the right side, meeting the conditions.

For f_4, by considering the symmetry and $a + b + c = 0$, it suffices to consider the cases where $(f(a), f(b), f(c))$ are $(0, k, k)$, $(4k, k, k)$, $(0, 0, 0)$, and $(0, 4k, 4k)$, all of which evidently satisfy the given conditions.

【Score Situation】 This particular problem saw the following distribution of scores among contestants: 143 contestants scored 7 points, 44 contestants scored 6 points, 26 contestants scored 5 points, 47 contestants scored 4 points, 74 contestants scored 3 points, 95 contestants scored 2 points, 65 contestants scored 1 point, and 53 contestants scored 0 point. The average score for this problem is 3.766, indicating that it was relatively straightforward.

Among the top five teams in the team scores, the scores of this problem are as follows: the South Korea team scored 39 points (with a total team score of 209 points), the China team scored 31 points (with a total team score of 195 points), the United States team scored 38 points (with a total team score of 194 points), the Russia team scored 41 points (with a total team score of 177 points), the Canada team scored 39 points (with a total team score of 159 points), and the Thailand team scored 39 points (with a total team score of 159 points).

The gold medal cutoff for this IMO was set at 28 points (with 51 contestants earning gold medals), the silver medal cutoff was 21 points (with 88 contestants earning silver medals), and the bronze medal cutoff was 14 points (with 137 contestants earning bronze medals).

In this IMO, only one contestant achieved a perfect score of 42 points, namely Jeck Lim from Singapore.

Problem 2.25 (IMO 56-5, proposed by Albania). Let $\mathbf{R}$ be the set of real numbers. Determine all functions $f : \mathbf{R} \to \mathbf{R}$ satisfying the equation

$$f(x + f(x + y)) + f(xy) = x + f(x + y) + yf(x)$$

for all real numbers x and y.

Solution. Denote the equation in the problem as $P(x, y)$. Assume f is a function that satisfies the condition. Considering $P(x, 1)$, we have

$$f(x + f(x + 1)) = x + f(x + 1). \tag{1}$$

Hence, $x + f(x + 1)$ is a fixed point of f for any $x \in \mathbf{R}$. Next, we study the following two cases.

Case 1: $f(0) \neq 0$.

Considering $P(0, y)$, we have

$$f(f(y)) + f(0) = f(y) + yf(0).$$

If y_0 is a fixed point of f, then setting $y = y_0$ in the above equation gives $y_0 = 1$. Thus, $x + f(x+1) = 1$, implying $f(x) = 2 - x$ for all $x \in \mathbf{R}$. It is easy to verify that $f(x) = 2 - x$ is a function that satisfies the conditions.

Case 2: $f(0) = 0$.

Considering $P(x+1, 0)$, we have

$$f(x + f(x+1) + 1) = x + f(x+1) + 1. \tag{2}$$

Considering $P(1, y)$, we obtain

$$f(1 + f(y+1)) + f(y) = 1 + f(y+1) + yf(1). \tag{3}$$

Setting $x = -1$ in (1) gives $f(-1) = -1$, and then setting $y = -1$ in (3) yields $f(1) = 1$. Thus, the equation (3) can be rewritten as

$$f(1 + f(y+1)) + f(y) = 1 + f(y+1) + y. \tag{4}$$

If both y_0 and $y_0 + 1$ are fixed points of f, then setting $y = y_0$ in (4) implies that $y_0 + 2$ is also a fixed point of f. Therefore, from (1) and (2), $x + f(x+1) + 2$ is a fixed point of f for any $x \in \mathbf{R}$, i.e.,

$$f(x + f(x+1) + 2) = x + f(x+1) + 2.$$

Replacing x with $x - 2$ in the above equality, we get

$$f(x + f(x-1)) = x + f(x-1).$$

Then considering $P(x, -1)$ gives

$$f(x + f(x-1)) = x + f(x-1) - f(x) - f(-x).$$

The above two equalities imply $f(-x) = -f(x)$, i.e., f is an odd function.

Considering $P(-1, -y)$ and using $f(-1) = -1$, we have

$$f(-1 + f(-y-1)) + f(y) = -1 + f(-y-1) + y.$$

Since f is an odd function, the above equality can be rewritten as

$$-f(1 + f(y + 1)) + f(y) = -1 - f(y + 1) + y,$$

and adding this equality to (4), we know $f(y) = y$ for any $y \in \mathbf{R}$. It is easy to verify that $f(x) = x$ is a function that satisfies the conditions.

In conclusion, there are two functions that satisfy the conditions: $f(x) = x$ and $f(x) = 2 - x$.

Note. There is a similar problem:

- **(Japan Team Selection Test 2022, Problem 9).** Find all functions $f : \mathbf{R} \to \mathbf{R}$ such that $f(-1) = -1$ and

$$f(x + f(y)) + f(x)f(y) = f(x + y) + yf(x) \quad \text{for all } x, y \in \mathbf{R}.$$

【Score Situation】 This particular problem saw the following distribution of scores among contestants: 30 contestants scored 7 points, 3 contestants scored 6 points, 4 contestants scored 5 points, 8 contestants scored 4 points, 90 contestants scored 3 points, 34 contestants scored 2 points, 255 contestants scored 1 point, and 153 contestants scored 0 point. The average score for this problem is 1.513, indicating that it was relatively challenging.

Among the top five teams in the team scores, the scores of this problem are as follows: the United States team scored 25 points (with a total team score of 185 points), the China team scored 22 points (with a total team score of 181 points), the South Korea team scored 25 points (with a total team score of 161 points), the North Korea team scored 22 points (with a total team score of 156 points), and the Vietnam team scored 28 points (with a total team score of 151 points).

The gold medal cutoff for this IMO was set at 26 points (with 39 contestants earning gold medals), the silver medal cutoff was 19 points (with 100 contestants earning silver medals), and the bronze medal cutoff was 14 points (with 143 contestants earning bronze medals).

In this IMO, only one contestant achieved a perfect score of 42 points, namely Zhuo Qun Alex Song from Canada.

Problem 2.26 (IMO 58-2, proposed by Albania). Let $\mathbf{R}$ be the set of real numbers. Determine all functions $f : \mathbf{R} \to \mathbf{R}$ such that for all real numbers x and y,

$$f(f(x)f(y)) + f(x + y) = f(xy).$$

Solution 1. If f is a solution, then $-f$ is also a solution. Therefore, we can assume $f(0) \leq 0$. Let the given equation be denoted as $P(x, y)$. From

$P(0,0)$,

$$f(f(0)^2) = 0. \tag{1}$$

For any real number $x \neq 1$, there exists a $y \in \mathbf{R}$ satisfying $x + y = xy$, i.e., $y = \frac{x}{x-1}$. From $P\left(x, \frac{x}{x-1}\right)$,

$$f\left(f(x)f\left(\frac{x}{x-1}\right)\right) = 0 \quad \text{for } x \neq 1. \tag{2}$$

Obviously, f has at least one zero, namely $(f(0))^2$.

Case 1: $f(0) = 0$.

From $P(x,0)$ it follows that $f(x) = 0$ for all $x \in \mathbf{R}$, so f is identically zero.

Case 2: $f(0) < 0$.

Conclusion 1: $f(a) = 0$ if and only if $a = 1$.

Indeed, from (1) we know there exists an $a \in \mathbf{R}$ such that $f(a) = 0$. If $a \neq 1$, then setting $x = a$ in (2) yields $f(0) = 0$, contradicting the assumption that $f(0) \neq 0$. From Conclusion 1 and (1), $(f(0))^2 = 1$ and $f(0) = -1$. From $P(x,1)$,

$$f(0) + f(x+1) = f(x),$$

i.e., $f(x+1) = f(x) + 1$. By mathematical induction, it is easy to show that for any integer n and any $x \in \mathbf{R}$,

$$f(x+n) = f(x) + n. \tag{3}$$

Conclusion 2: f is injective.

By contradiction, suppose there exist $a \neq b$ such that $f(a) = f(b)$. From (3), for any integer N, we have $f(a + N + 1) = f(b + N) + 1$. Choosing an integer $N > -b$, we see that there exist $x_0, y_0 \in \mathbf{R}$ satisfying $x_0 + y_0 = a + N + 1$ and $x_0 y_0 = b + N$. Since $a \neq b$, we have $(x_0 - 1)(y_0 - 1) \neq 0$. From $P(x_0, y_0)$,

$$f(f(x_0)f(y_0)) + f(a + N + 1) = f(b + N),$$

and thus $f(f(x_0)f(y_0)) + 1 = f(f(x_0)f(y_0) + 1) = 0$.

From Conclusion 1, $f(x_0)f(y_0) + 1 = 1$, i.e., $f(x_0)f(y_0) = 0$. However, $x_0 \neq 1$ and $y_0 \neq 1$, and from Conclusion 1, $f(x_0) \neq 0$ and $f(y_0) \neq 0$, a contradiction.

For any $t \in \mathbf{R}$, from $P(t, -t)$,

$$f(f(t)f(-t)) + f(0) = f(-t^2).$$

Thus, $f(f(t)f(-t)) = f(-t^2) + 1 = f(-t^2 + 1)$. As f is injective,

$$f(t)f(-t) = -t^2 + 1. \tag{4}$$

From $P(t, 1 - t)$,

$$f(f(t)f(1 - t)) + f(1) = f(t(1 - t)).$$

Thus, $f(f(t)f(1 - t)) = f(t(1 - t))$. As f is injective,

$$f(t)f(1 - t) = t(1 - t). \tag{5}$$

Since $f(1 - t) = 1 + f(-t)$, comparing (4) and (5) gives $f(t) = t - 1$. If $f(0) = 1$, then $f(t) = 1 - t$.

It is easy to verify that $f_1(x) = 0$, $f_2(x) = x - 1$, and $f_3(x) = 1 - x$ all satisfy the conditions.

Solution 2. After Conclusion 1 in Solution 1, the following process can replace Conclusion 2.

Define $g(x) = f(x + 1) = f(x) + 1$, and replace x and y with $x + 1$ and $y + 1$ respectively in the original functional equation. For any $x, y \in \mathbf{R}$,

$$f(f(x + 1)f(y + 1)) + f(x + y + 2) = f(xy + x + y + 1),$$

i.e.,

$$g(g(x)g(y)) + g(x + y) = g(xy + x + y). \tag{6}$$

By Conclusion 1, "0 is the only zero of g," we prove that $g(x) = x$.

Conclusion 3: Let $n \in \mathbf{Z}$ and $x \in \mathbf{R}$. Then

(i) $g(x + n) = g(x) + n$, and $g(x) = n$ if and only if $x = n$;
(ii) $g(nx) = ng(x)$.

For (i), note that $g(x + n) = g(x) + n$ is the same as (3) in Solution 1. Thus

$$g(x) = n \Leftrightarrow g(x - n) = 0 \Leftrightarrow x - n = 0.$$

For (ii), when $x = 0$, it is obvious. For $x \neq 0$, substitute $y = \frac{n}{x}$ into (6), yielding

$$g\left(g(x)g\left(\frac{n}{x}\right)\right) + g\left(x + \frac{n}{x}\right) = g\left(n + x + \frac{n}{x}\right)$$

$$\Leftrightarrow g\left(g(x)g\left(\frac{n}{x}\right)\right) = n \Leftrightarrow g(x)g\left(\frac{n}{x}\right) = n,$$

i.e., $g(x) = \frac{n}{g(\frac{n}{x})}$ for $x \neq 0$. Taking $n = 1$, we get $g\left(\frac{1}{x}\right) = \frac{1}{g(x)}$, and replacing x with nx in the above equation, we obtain $g(nx) = \frac{n}{g(\frac{1}{x})} = ng(x)$. Thus proving the conclusion.

Conclusion 4: g is an additive function, i.e., for any $a, b \in \mathbf{R}$,

$$g(a + b) = g(a) + g(b).$$

First, replace x with $-x$ and y with $-y$ in (6), and substitute $n = -1$ into Conclusion 3(ii), we know that g is an odd function. Thus

$$g(g(x)g(y)) - g(x + y) = -g(-xy + x + y).$$

Subtracting this from (6), we get

$$2g(x + y) = g(xy + x + y) + g(-xy + x + y).$$

Then, according to Conclusion 3(ii) (for $n = 2$), we rewrite the above equation as: for $\alpha = xy + x + y$ and $\beta = -xy + x + y$,

$$g(\alpha + \beta) = g(\alpha) + g(\beta).$$

Note that for any $\alpha, \beta \in \mathbf{R}$ satisfying

$$\left(\frac{\alpha + \beta}{2}\right)^2 - 4 \cdot \frac{\alpha - \beta}{2} \geq 0,$$

we can always find x and y such that

$$x + y = \frac{\alpha + \beta}{2}, \quad xy = \frac{\alpha - \beta}{2}.$$

Thus for any $a, b \in \mathbf{R}$ we can always find an integer n (which can be 1 or -1) such that

$$g(na) + g(nb) = g(na + nb) \Leftrightarrow ng(a) + ng(b) = ng(a + b).$$

Hence, the conclusion holds.

Therefore, taking $y = 1$ in (6) and using Conclusion 3, we obtain

$$g(g(x)g(1)) + g(x + 1) = g(2x + 1)$$

$$\Leftrightarrow g(g(x)) + g(x) + 1 = 2g(x) + 1$$

$$\Leftrightarrow g(g(x)) = g(x).$$

Since g is additive, $g(g(x) - x) = 0$, and with the assumption "0 is the only zero of g," it follows that $g(x) = x$.

【Score Situation】 This particular problem saw the following distribution of scores among contestants: 61 contestants scored 7 points, 8 contestants scored 6 points, 10 contestants scored 5 points, 79 contestants scored 4 points, 138 contestants scored 3 points, 26 contestants scored 2 points, 110 contestants scored 1 point, and 183 contestants scored 0 point. The average score for this problem is 2.304, indicating that it had a certain level of difficulty.

Among the top five teams in the team scores, the scores of this problem are as follows: the South Korea team scored 39 points (with a total team score of 170 points), the China team scored 25 points (with a total team score of 159 points), the Vietnam team scored 36 points (with a total team score of 155 points), the United States team scored 29 points (with a total team score of 148 points), and the Iran team scored 32 points (with a total team score of 142 points).

The gold medal cutoff for this IMO was set at 25 points (with 48 contestants earning gold medals), the silver medal cutoff was 19 points (with 90 contestants earning silver medals), and the bronze medal cutoff was 16 points (with 153 contestants earning bronze medals).

In this IMO, no contestant achieved a perfect score of 42 points.

Problem 2.27 (IMO 60-1, proposed by South Africa). Let $\mathbf{Z}$ be the set of integers. Determine all functions $f : \mathbf{Z} \to \mathbf{Z}$ such that, for all integers a and b,

$$f(2a) + 2f(b) = f(f(a+b)).$$

Solution. Let's denote the equation given in the problem as $P(a,b)$. From $P(0,x)$,

$$f(0) + 2f(x) = f(f(x)). \tag{1}$$

From $P(x,0)$,

$$f(2x) + 2f(0) = f(f(x)). \tag{2}$$

Comparing (1) and (2), we get

$$f(2x) = 2f(x) - f(0). \tag{3}$$

Substituting (1) and (3) into $P(a,b)$, we have

$$2f(a) - f(0) + 2f(b) = f(0) + 2f(a+b). \tag{4}$$

Let $g(x) = f(x) - f(0)$. Then $g(0) = 0$, and (4) can be rewritten as

$$g(a+b) = g(a) + g(b). \tag{5}$$

By Cauchy's functional equation, $g(n) = g(1) \cdot n$ for any integer n. Thus f can be expressed as $f(x) = kx + c$, where $k, c \in \mathbf{Z}$. Substituting this back into $P(a,b)$, we see that for all integers a and b,

$$2k(a+b) + 3c = k^2(a+b) + (k+1)c.$$

This holds if and only if $2k = k^2$, and $3c = (k+1)c$. Therefore, $k = 2$ or 0. If $k = 2$, then c can be any integer. If $k = 0$, then $c = 0$.

Thus, the functions that satisfy the conditions are $f(x) = 0$ or $f(x) = 2x + c$, where c is any integer.

Note. There are several similar problems:

- Find all continuous functions $f : \mathbf{R} \to \mathbf{R}$ such that

$$f(f(x+y)) = f(x) + f(y) \quad \text{for all } x, y \in \mathbf{R}.$$

- **(Japan Team Selection Test 2021, Problem 12).** Find all functions $f : \mathbf{Z} \to \mathbf{Z}$ satisfying

$$f^{(a^2+b^2)}(a+b) = af(a) + bf(b) \quad \text{for all } a, b \in \mathbf{Z}.$$

Here, $f^{(0)}(n) = n$ and $f^{(k)}(n) = f(f^{(k-1)}(n))$.

【Score Situation】 This particular problem saw the following distribution of scores among contestants: 382 contestants scored 7 points, 52 contestants scored 6 points, 5 contestants scored 5 points, 14 contestants scored 4 points, 24 contestants scored 3 points, 6 contestants scored 2 points, 65 contestants scored 1 point, and 73 contestants scored 0 point. The average score for this problem is 5.179, indicating that it was simple.

Among the top five teams in the team scores, the scores of this problem are as follows: the China team scored 40 points (with a total team score of 227 points), the United States team scored 42 points (with a total team score of 227 points), the South Korea team scored 42 points (with a total team score of 226 points), the North Korea team scored 41 points (with a total team score of 187 points), and the Thailand team scored 42 points (with a total team score of 185 points).

The gold medal cutoff for this IMO was set at 31 points (with 52 contestants earning gold medals), the silver medal cutoff was 24 points (with 94 contestants earning silver medals), and the bronze medal cutoff was 17 points (with 156 contestants earning bronze medals).

In this IMO, a total of six contestants achieved a perfect score of 42 points.

Problem 2.28 (IMO 63-2, proposed by the Netherlands). Let $\mathbf{R}_+$ denote the set of positive real numbers. Find all functions $f : \mathbf{R}_+ \to \mathbf{R}_+$ such that for each $x \in \mathbf{R}_+$, there is exactly one $y \in \mathbf{R}_+$ satisfying

$$xf(y) + yf(x) \leq 2.$$

Solution 1. The unique function that satisfies the conditions is $f(x) = \frac{1}{x}$ for all $x \in \mathbf{R}_+$.

First, we verify that the function $f(x) = \frac{1}{x}$ meets the conditions. For any $x, y \in \mathbf{R}_+$, using the AM-GM inequality, we know

$$xf(y) + yf(x) = \frac{x}{y} + \frac{y}{x} \geq 2,$$

with equality if and only if $x = y$. This indicates that the only $y \in \mathbf{R}_+$ satisfying $xf(y) + yf(x) \leq 2$ is $y = x$.

Next, we prove that if f satisfies the conditions, then $f(x) = \frac{1}{x}$.

If a pair of positive real numbers (x, y) satisfies $xf(y) + yf(x) \leq 2$, then we call it a "good pair." If (x, y) is a good pair, then (y, x) is also a good pair.

We assert that for any good pair (x, y), it must be that $x = y$. Otherwise, suppose $x \neq y$. Then the conditions imply that neither (x, x) nor (y, y) are good pairs, i.e., $xf(x) > 1$ and $yf(y) > 1$. Using the AM-GM inequality, we have

$$xf(y) + yf(x) \geq 2\sqrt{xf(y) \cdot yf(x)} = 2\sqrt{xf(x) \cdot yf(y)} > 2,$$

contradicting the assumption that (x, y) is a good pair.

From this assertion, it follows that for each $x \in \mathbf{R}_+$, we see that (x, x) is the only good pair that includes it. Thus $xf(x) \leq 1$, i.e., $f(x) \leq \frac{1}{x}$ for all $x \in \mathbf{R}_+$. Let $y = \frac{1}{f(x)}$. Then

$$xf(y) + yf(x) \leq x \cdot \frac{1}{y} + \frac{1}{f(x)} \cdot f(x) = xf(x) + 1 \leq 2,$$

indicating that $\left(x, \frac{1}{f(x)}\right)$ is also a good pair.

By the assertion, $x = \frac{1}{f(x)}$, and thus $f(x) = \frac{1}{x}$.

Solution 2. The verification for $f(x) = \frac{1}{x}$ is the same as in Solution 1. Next, we use another approach to proving that if f satisfies the conditions, then $f(x) = \frac{1}{x}$.

For each $x \in \mathbf{R}_+$, denote the unique $y \in \mathbf{R}_+$ satisfying $xf(y) + yf(x) \leq 2$ as $\sigma(x)$. First, we prove two conclusions:

(i) f is strictly decreasing, i.e., $f(y_1) > f(y_2)$ for any $y_1 < y_2$.

Otherwise, suppose $f(y_1) \leq f(y_2)$. Let $x = \sigma(y_2)$. Then

$$xf(y_1) + y_1 f(x) < xf(y_2) + y_2 f(x) \leq 2,$$

contradicting the uniqueness of $\sigma(x)$.

(ii) For any $x \in \mathbf{R}_+$, let $y = \sigma(x)$. Then $xf(y) + yf(x) = 2$.
Otherwise, suppose $xf(y) + yf(x) < 2$. Let $y' = y + \varepsilon$, where

$$\varepsilon = \frac{2 - xf(y) - yf(x)}{f(x)}$$

is a positive number. Using conclusion (i), we have

$$xf(y') + y'f(x) < xf(y) + yf(x) + \varepsilon f(x) = 2,$$

contradicting the uniqueness of $\sigma(x)$.

From conclusion (ii), for any $x, y \in \mathbf{R}_+$,

$$xf(y) + yf(x) \geq 2.$$

Taking $y = x$ in the above inequality gives $f(x) \geq \frac{1}{x}$, and therefore

$$2 = f(x)\sigma(x) + f(\sigma(x))x \geq \frac{\sigma(x)}{x} + \frac{x}{\sigma(x)},$$

which is only possible if $\sigma(x) = x$.

Substituting $\sigma(x) = x$ into conclusion (ii) yields $f(x) = \frac{1}{x}$.

Note. There is a similar problem:

- **(Korean Mathematical Olympiad 2023, Final Round, Problem 2).** Let $\mathbf{R}_+$ denote the set of positive real numbers. Suppose a function $f : \mathbf{R}_+ \to \mathbf{R}_+$ satisfies the following condition:
  For each $x \in \mathbf{R}_+$, there exists a $y \in \mathbf{R}_+$ such that

$$(x + f(y))(y + f(x)) \leq 4,$$

and the number of y is finite. Prove that $f(x) > f(y)$ for any positive real numbers $x < y$.

【Score Situation】 This particular problem saw the following distribution of scores among contestants: 303 contestants scored 7 points, 19 contestants scored 6 points, 3 contestants scored 5 points, 7 contestants scored 4 points, 41 contestants scored 3 points, 23 contestants scored 2 points, 89 contestants scored 1 point, and 104 contestants scored 0 point. The average score for this problem is 4.306, indicating that it was simple.

Among the top five teams in the team scores, the scores of this problem are as follows: the China team scored 42 points (with a total team score of 252 points), the South Korea team scored 42 points (with a total team score of 208 points), the United States team scored 42 points (with a total team score of 207 points), the Vietnam team scored 42 points (with a total team score of 196 points), and the Romania team scored 41 points (with a total team score of 194 points).

The gold medal cutoff for this IMO was set at 34 points (with 44 contestants earning gold medals), the silver medal cutoff was 29 points (with 101 contestants earning silver medals), and the bronze medal cutoff was 23 points (with 140 contestants earning bronze medals).

In this IMO, a total of 10 contestants achieved a perfect score of 42 points.

2.3 Summary

Functional equations constitute a rich and broadly applied branch of mathematics. In 2000, one of the seven Millennium Prize Problems released by the Clay Mathematics Institute in the United States concerned the issue of smoothness and existence of solutions for the three-dimensional Navier-Stokes equations.

In the first 64 IMOs, there were a total of 28 function problems. These problems can be broadly categorized into three types, as depicted in Figure 2.1. The score details for these problems are presented in Table 2.2. Due to the smaller number of participating teams and missing contestant score information in early IMOs, there are several blanks in Table 2.2.

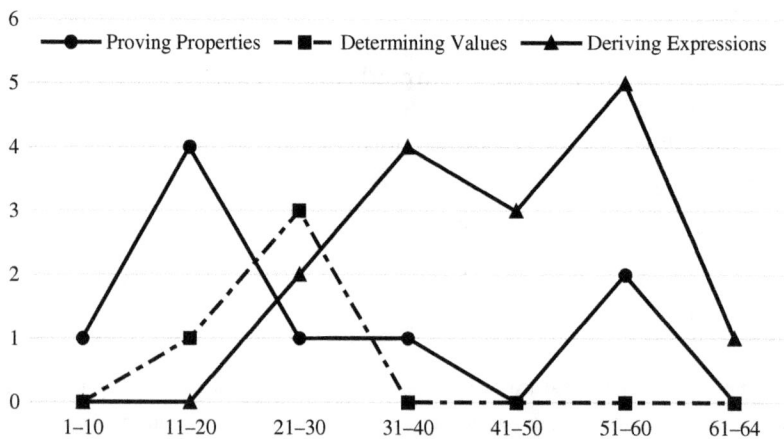

Figure 2.1   Numbers of Function Problems in the First 64 IMOs

Problems 2.1–2.9 focus on "proving properties of functions;" among these nine problems, the one with the lowest average score is Problem 2.8 (IMO 52-3), proposed by Belarus. Problems 2.10–2.13 deal with "determining numerical values of function variables or outputs;" among these four problems, the one with the lowest average score is Problem 2.13 (IMO 29-3), proposed by the United Kingdom. Problems 2.14–2.28 are about "deriving expressions for functions that meet specific conditions;" among these 15 problems, the one with the lowest average score is Problem 2.19 (IMO 40-6), proposed by Japan.

These 28 problems were proposed by 17 countries, with the United Kingdom contributing the most, totaling six problems. Bulgaria proposed three problems, while India, France, South Africa, and Albania each contributed two problems.

From Table 2.2, it can be observed that in the first 64 IMOs, there were four function problems with an average score of 1–2 points; seven problems with an average score of 2–3 points; eight problems with an average score of 3–4 points; nine problems with an average score above 4 points. Overall, the function problems were relatively simple, and the contestants scored relatively high.

In the 24th–64th IMOs, there were a total of 20 function problems. Among these, four had an average score of 1–2 points; six had an average score of 2–3 points; five had an average score of 3–4 points; five had an average score above 4 points. Further analysis of the problem numbers of these 20 function problems, as shown in Table 2.3, reveals that these problems frequently appeared as the 1st/4th or 2nd/5th problem. The majority of these problems, totaling 15, were of the type deriving expressions for functions that meet specific conditions. The other two types of function problems were less frequent, with only five problems appearing.

Excluding Problem 2.15 (IMO 24-1), the function problems from the 25th–64th IMOs are arranged in order of their average scores, from left to right, and a scatter plot of the score details is presented, as shown in Figure 2.2.

From Table 2.2 and Figure 2.2, it is observable that in the function problems, the average score of the top five teams generally exceeds the average score of the problem by 2.5 points, the average score of the 6th–15th teams typically surpasses the average score by 2 points, and the average score of the 16th–25th teams usually exceeds the average score by 1.5 points. This characteristic is more pronounced in difficult problems. However, in simpler problems, the disparity between the average scores of the top five teams, the 6th–15th teams, and the 16th–25th teams is not significant, as seen in problems such as Problem 2.28 (IMO 63-2), Problem 2.21 (IMO 49-4), Problem 2.27 (IMO 60-1), and Problem 2.23 (IMO 51-1).

From Figure 2.2, it can also be observed that there are several function problems where the average score is higher than the average score of the 16th–25th teams, such as Problem 2.13 (IMO 29-3), Problem 2.16

Table 2.2 Score Details of Function Problems in the First 64 IMOs

Problem	2.1	2.2	2.3	2.4	2.5	2.6	2.7	2.8
Full points	7.000	7.000	7.000	6.000	8.000	7.000	7.000	7.000
Average score	4.593	4.366	2.818	3.376	3.054	3.523	3.383	1.055
Top five mean		5.725	5.175	4.975		6.667	6.400	4.867
6th–15th mean				2.588		5.800	5.333	3.242
16th–25th mean						2.983	4.426	1.091
Problem number in IMO	10-5	11-2	14-5	15-5	19-6	28-4	34-5	52-3
Proposing country	The German Democratic Republic	Hungary	Bulgaria	Poland	Bulgaria	Vietnam	Germany	Belarus

Problem	2.9	2.10	2.11	2.12	2.13	2.14	2.15	2.16
Full points	7.000	8.000	7.000	7.000	7.000	7.000	7.000	7.000
Average score	2.452	3.313	6.510	4.613	1.731	3.575	4.152	2.955
Top five mean	6.633	5.025	6.900	6.700	4.167		6.400	6.067
6th–15th mean	5.983	2.921	6.096	5.525	2.864		5.450	3.833
16th–25th mean	3.700		4.492	4.175	1.276		3.655	2.926
Problem number in IMO	54-5	20-3	22-6	23-1	29-3	24-1	27-5	31-4
Proposing country	Bulgaria	The United Kingdom	Finland	The United Kingdom	The United Kingdom	The United Kingdom	The United Kingdom	Turkey

(Continued)

Table 2.2 (*Continued*)

Problem	2.17	2.18	2.19	2.20	2.21	2.22	2.23	2.24
Full points	7.000	7.000	7.000	7.000	7.000	7.000	7.000	7.000
Average score	2.963	3.221	1.151	2.267	4.402	2.474	5.450	3.766
Top five mean	5.667	6.467	3.200	6.467	6.967	6.800	6.889	6.306
6th–15th mean	4.350	5.100	2.083	4.550	6.383	6.212	6.648	5.333
16th–25th mean	3.817	4.067	1.136	3.500	6.417	5.467	6.667	5.450
Problem number in IMO	33-2	35-5	40-6	43-5	49-4	50-5	51-1	53-4
Proposing country	India	The United Kingdom	Japan	India	South Korea	France	France	South Africa

Problem	2.25	2.26	2.27	2.28
Full points	7.000	7.000	7.000	7.000
Average score	1.513	2.304	5.179	4.306
Top five mean	4.067	5.367	6.900	6.967
6th–15th mean	3.067	4.318	6.864	6.800
16th–25th mean	2.133	3.870	6.741	6.467
Problem number in IMO	56-5	58-2	60-1	63-2
Proposing country	Albania	Albania	South Africa	The Netherlands

Note. Top five mean = Total score of the top five teams ÷ Total number of contestants from the top five teams,
6th–15th mean = Total score of the 6th–15th teams ÷ Total number of contestants from the 6th–15th teams,
16th–25th mean = Total score of the 16th–25th teams ÷ Total number of contestants from the 16th–25th teams.

Table 2.3 Numbers of Function Problems in the 24th–64th IMOs

Function Problem	Problem Number			Number of Problems in the First 64 IMOs
	1, 4	2, 5	3, 6	
Proving properties	1	2	1	9
Determining values	0	0	1	4
Deriving expressions	6	8	1	15
Total	7	10	3	28

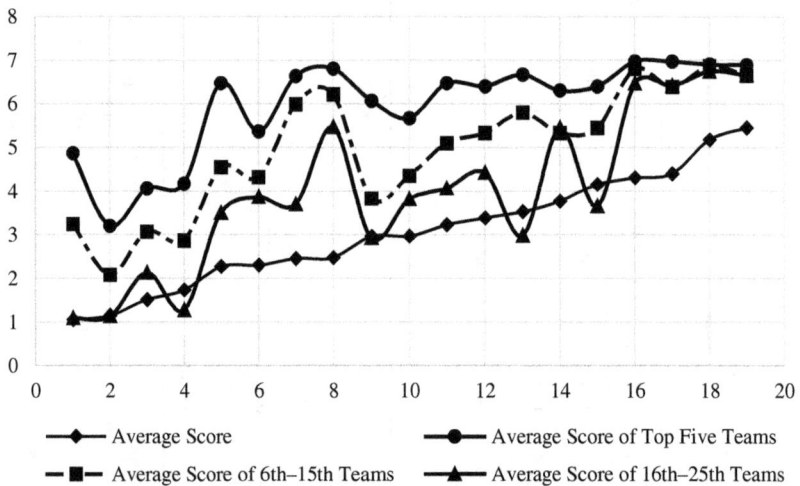

Figure 2.2 Score Details of Function Problems in the 25th–64th IMOs

(IMO 31-4), Problem 2.6 (IMO 28-4), and Problem 2.15 (IMO 27-5). This phenomenon is due to the smaller number of participating teams in early IMOs. It was not until the 30th IMO in 1989 that the number of participating teams exceeded 50. Therefore, it is common to see situations where the average score is close to or even higher than the average score of the 16th–25th teams during this period.

Chapter 3

Sequence Problems

When it comes to sequences, one might immediately think of the story of the renowned mathematician Gauss discovering the formula for the sum of an arithmetic sequence. However, the appearance of formulas for the sum of arithmetic and geometric sequences dates back much earlier.

The invention of numbers naturally led ancient scholars to study sequences. The *Nine Chapters on the Mathematical Art* recorded such a problem: To calculate the total distance traveled by a good horse over fifteen days, this distance can be determined by multiplying 15 with the sum of the first day's distance and seven times the daily incremental distance.

Here, the daily incremental distance is the constant additional distance d the horse travels each day beyond the previous day's distance, and let the first day's distance be a_1. Then, the total distance traveled by the good horse in n days is given by $S_n = \left(a_1 + \frac{(n-1)d}{2} \right) n$. This indicates that Chinese mathematicians of that time were already aware of the formula for the sum of an arithmetic sequence.

By the 19th century, people began to focus on more types of sequence summation problems and derived sum formulas. These efforts are mainly reflected in two lengthy entries, "Algebra" and "Series," written by the Scottish mathematician William Wallace for the Encyclopedia Britannica.

In "Algebra," Wallace used the method of adding in reverse order to derive the formula for the sum of an arithmetic sequence and used the method of displacement subtraction to derive the formula for the sum of a geometric sequence. In "Series," he also derived sum formulas for other sequences, such as sequences with terms

$$\frac{k \times (k+1) \times \cdots \times (k+p-1)}{1 \times 2 \times \cdots \times p}, \ k^p, \ \text{and} \ \frac{1}{k \times (k+1) \times \cdots \times (k+p-1)}.$$

After the emergence of the concept of functions, sequences can be regarded as a special type of functions whose domains are the set of natural numbers. Viewing sequences from this perspective facilitates the study of their monotonicity, boundedness, and relationships with more advanced mathematical concepts. For instance, a problem from the 1980 British Mathematical Olympiad demonstrates the application of the concept of "limits" in this context: Find the set of real numbers a_0 for which the infinite sequence $\{a_n\}$ of real numbers defined by $a_{n+1} = 2^n - 3a_n (n \geq 0)$ is strictly increasing.

Additionally, one can approach the divisibility properties of sequences from a number theory perspective. For instance, a problem from the 29th IMO Shortlist states: An integer sequence is defined by $a_0 = 0$, $a_1 = 1$, and $a_n = 2a_{n-1} + a_{n-2}(n > 1)$. Prove that $2^k | a_n$ if and only if $2^k | n$. Furthermore, sequences frequently appear in problems related to combinatorics, inequalities, and other topics.

In the first 64th IMOs, there had been a total of 14 sequence problems, approximately accounting for 13.8% of all algebra problems. These problems can be primarily categorized into three types: (1) determining the value of a specific term or the number of terms, totaling four problems; (2) addressing problems related to the existence of sequences, totaling five problems; (3) proving quantitative relationships satisfied by sequences, totaling five problems. The statistical distribution of these three types of problems in the previous IMOs is presented in Table 3.1.

It can be observed that there were not many sequence problems, with only 1–3 appearing in every 10 IMOs. To some extent, this is because sequence problems are typically intertwined with number theory and combinatorics, rarely focusing on the algebraic properties of sequences, and hence not being included in Table 3.1.

Although the number of sequence problems is not particularly large, this does not imply they are always straightforward. In fact, many are

Table 3.1 Numbers of Sequence Problems in the First 64 IMOs

Content	Session							
	1–10	11–20	21–30	31–40	41–50	51–60	61–64	Total
Determining values	1	0	0	1	0	2	0	4
Existence problems	0	2	1	1	0	0	1	5
Proving quantitative relationships	0	1	1	1	1	1	0	5
Algebra problems	20	20	14	13	15	13	6	101
The percentage of sequence problems among the algebra problems	5.0%	15.0%	14.3 %	23.1%	6.7%	23.1%	16.7%	13.8%

quite challenging. For example, the average score of Problem 3.14 (IMO 51-6) is merely 0.368 points, with as many as 470 contestants scoring zero.

Addressing sequence problems typically involves two primary methods. The first is the recursive method, such as the use of mathematical induction. The second is the algebraic method, which demands proficiency in techniques including substitution, identity transformations, and finding the general term of a sequence. In mathematics competitions, many sequence problems often start with algebraic transformations.

This chapter will be divided into three parts. The first part introduces some common types and properties of sequences. Although periodic sequences are usually associated with divisibility and congruence, their properties are also reviewed here. Subsequently, commonly used methods for finding the general term and sum of sequences are discussed, along with several theorems that are easily overlooked.

The second part revolves around three types of problems: "determining the value of a specific term or the number of terms," "addressing problems related to the existence of sequences," and "proving quantitative relationships satisfied by sequences." These problems are presented in chronological order, and some problems include various solutions, generalizations, and similar problems.

It is important to note that for each problem, the solutions are followed by information on the scores, including the number of contestants in each score range, the average score, and the scores of the top five teams. However, early IMOs often lacked information on contestant scores, so the number of contestants in each score range only represents the counted number of contestants, and some problems lack scores of the top five teams.

The third part provides a brief summary of this chapter.

3.1 Common Theorems, Formulas, and Methods

3.1.1 *Common sequences*

(1) *Arithmetic sequences*

An arithmetic sequence $\{a_n\}$ is defined such that, starting from the second term, each term's difference with its preceding term equals a constant. This constant is referred to as the common difference of the arithmetic sequence and is usually denoted by d. The general term is given by $a_n = a_1 + (n-1)d$.

(2) *Geometric sequences*

A geometric sequence $\{a_n\}$ is defined such that, starting from the second term, each term's ratio with its preceding term equals a constant. This constant is referred to as the common ratio of the geometric sequence and is usually denoted by q. The general term is given by $a_n = a_1 \cdot q^{n-1}$.

Let the sum of the first n terms of the sequence $\{a_n\}$ be denoted as S_n. The following propositions hold:

Proposition 3.1. *If* $S_n = An^2 + Bn + C$, *then* $\{a_n\}$ *is an arithmetic sequence if and only if* $C = 0$.

Proposition 3.2. $\{a_n\}$ *is an arithmetic sequence with a common difference* d *if and only if* $\{a^{a_n}\}$ *is a geometric sequence with a common ratio* a^d, *where* $a > 0$ *and* $a \neq 1$.

Proposition 3.3. $\{a_n\}$ *is an arithmetic sequence if and only if* $S_n = \frac{n(a_1+a_n)}{2}$.

Proposition 3.4. $\{a_n\}$ *is an arithmetic sequence with a common difference* d *if and only if* $\left\{\frac{S_n}{n}\right\}$ *is an arithmetic sequence with a common difference* $\frac{d}{2}$.

Proposition 3.5. *If* $\{a_n\}$ *is an arithmetic sequence, then* $S_n, S_{2n} - S_n, S_{3n} - S_{2n}, \ldots$ *form an arithmetic sequence.*

Proposition 3.6. *If $\{a_n\}$ is an arithmetic sequence, then $S_{m+n} = \frac{(S_n - S_m)(n+m)}{n-m}$.*

Furthermore, if $S_n = m$ and $S_m = n$, then $S_{m+n} = -(m+n)$. If $S_n = S_m (m \neq n)$, then $S_{m+n} = 0$.

Proposition 3.7. *If $m + n = p + q$, where m, n, p, q are positive integers, then $a_m + a_n = a_p + a_q$.*

Furthermore, if $m + n = 2p$, then $a_m + a_n = 2a_p$.

Proposition 3.8. *If $\{a_n\}$ and $\{b_n\}$ are both arithmetic sequences, then $\{ka_n + b\}$ and $\{ma_n + nb_n\}$ are also arithmetic sequences, where k, b, m, n are constants.*

Proposition 3.9. *Let A_n and B_n denote the sums of the first n terms of arithmetic sequences $\{a_n\}$ and $\{b_n\}$ respectively. Then $\frac{a_n}{b_n} = \frac{A_{2n-1}}{B_{2n-1}}$.*

Proposition 3.10. *If sequences $\{a_n\}$ and $\{b_n\}$ satisfy the condition that for $n \geq 1$,*

$$b_n = \frac{a_1 + 2a_2 + \cdots + na_n}{1 + 2 + \cdots + n},$$

then $\{b_n\}$ is an arithmetic sequence if and only if $\{a_n\}$ is an arithmetic sequence.

Proposition 3.11. *If real numbers a and d satisfy $d \neq 0$ and $ad \geq 0$, then the arithmetic sequence*

$$a + d, a + 2d, \ldots, a + nd, \ldots$$

has three terms in a geometric progression if and only if $\frac{a}{d}$ is a rational number.

Proposition 3.12. *If $S_n = Aq^n + B(AB \neq 0, q \neq 0, 1)$, then $\{a_n\}$ is a geometric sequence if and only if $A + B = 0$.*

Proposition 3.13. *$\{a_n\}$ is a geometric sequence with a positive common ratio q if and only if $\{\log_a a_n\}$ is an arithmetic sequence with a common difference $\log_a q$, where $a > 0$ and $a \neq 1$.*

Proposition 3.14. $\{a_n\}$ *is a geometric sequence with a common ratio q if and only if $S_{n+1} = a_1 + qS_n$.*

Proposition 3.15. *If $\{a_n\}$ is a geometric sequence, then $S_n, S_{2n} - S_n, S_{3n} - S_{2n}, \ldots$ also form a geometric sequence.*

Proposition 3.16. *If $m + n = p + q$, where m, n, p, q are positive integers, then $a_m \cdot a_n = a_p \cdot a_q$.*

Furthermore, if $m + n = 2p$, then $a_m \cdot a_n = a_p^2$.

Proposition 3.17. *If $\{a_n\}$ is a geometric sequence, then $a_m, a_{m+k}, a_{m+2k}, \ldots, a_{m+nk}, \ldots$ is also a geometric sequence, where m, k are positive integers.*

(3) High-order arithmetic sequences

For a given sequence $\{a_n\}$, taking the differences between the adjacent terms, a new sequence is obtained:

$$a_2 - a_1, a_3 - a_2, \ldots, a_{n+1} - a_n, \ldots;$$

this sequence is called the first order difference sequence of $\{a_n\}$.

If this sequence is denoted as $\{b_n\}$, where $b_n = a_{n+1} - a_n$, and the differences between the adjacent terms of $\{b_n\}$ are taken, then a sequence is obtained:

$$b_2 - b_1, b_3 - b_2, \ldots, b_{n+1} - b_n, \ldots;$$

this sequence is termed the second order difference sequence of the original sequence $\{a_n\}$.

Following this pattern, for any $p \in \mathbf{N}_+$, the pth order difference sequence of $\{a_n\}$ can be defined. If the pth order difference sequence of $\{a_n\}$ is a non-zero constant sequence, then $\{a_n\}$ is referred to as a pth order arithmetic sequence. Particularly, a first order arithmetic sequence is commonly known as an arithmetic sequence, while second and higher-order arithmetic sequences are collectively termed high-order arithmetic sequences. The following propositions hold:

Proposition 3.18. *If $\{a_n\}$ is a pth order arithmetic sequence, then its first order difference sequence is a $(p - 1)$th order arithmetic sequence.*

Proposition 3.19. *A sequence $\{a_n\}$ is pth order arithmetic if and only if its general term is a polynomial of degree p with respect to n.*

Proposition 3.20. *If $\{a_n\}$ is a pth order arithmetic sequence, then the sum of its first n terms is a degree $p+1$ polynomial with respect to n.*

The most common problems in high-order arithmetic sequences involve finding the general term formula and the sum of the first n terms.

(4) Recursive sequences

There are two prevalent methods for defining a sequence: one is by providing its general term formula, and the other is by specifying the initial terms along with the recurrence relation.

For any $n \in \mathbf{N}_+$, a sequence $\{a_n\}$ determined by the recurrence relation

$$a_{n+k} = f(a_{n+k-1}, a_{n+k-2}, \ldots, a_n)$$

is called a kth order recursive sequence. If f is linear, then the sequence is called a kth order linear recursive sequence; otherwise, it is called a kth order nonlinear recursive sequence.

Proposition 3.21. *A kth order arithmetic sequence $\{a_n\}$ is determined by the recurrence relation*

$$a_{n+k+1} = C_{k+1}^k a_{n+k} - C_{k+1}^{k-1} a_{n+k-1} + \cdots + (-1)^k a_n.$$

Proposition 3.22. *If a characteristic equation*

$$x^k = \lambda_1 x^{k-1} + \lambda_2 x^{k-2} + \cdots + \lambda_k (\lambda_k \neq 0)$$

has k distinct roots $x_1, x_2, \ldots, x_k$, then for the corresponding recurrence relation

$$a_{n+k} = \lambda_1 a_{n+k-1} + \lambda_2 a_{n+k-2} + \cdots + \lambda_k a_n,$$

the general term formula of the sequence $\{a_n\}$ is given by

$$a_n = c_1 x_1^n + c_2 x_2^n + \cdots + c_k x_k^n,$$

where $c_1, c_2, \ldots, c_k$ are the solutions of the following system of equations:

$$\begin{cases} c_1 x_1 + c_2 x_2 + \cdots + c_k x_k = a_1, \\ c_1 x_1^2 + c_2 x_2^2 + \cdots + c_k x_k^2 = a_2, \\ \cdots\cdots\cdots\cdots\cdots\cdots\cdots\cdots \\ c_1 x_1^k + c_2 x_2^k + \cdots + c_k x_k^k = a_k. \end{cases}$$

Proposition 3.23. *If a characteristic equation*

$$x^k = \lambda_1 x^{k-1} + \lambda_2 x^{k-2} + \cdots + \lambda_k (\lambda_k \neq 0)$$

has a k-fold root λ, then for the corresponding recurrence relation

$$a_{n+k} = \lambda_1 a_{n+k-1} + \lambda_2 a_{n+k-2} + \cdots + \lambda_k a_n,$$

the general term formula of the sequence $\{a_n\}$ is given by

$$a_n = (c_1 + c_2 n + \cdots + c_k n^{k-1}) \lambda^n,$$

where $c_1, c_2, \ldots, c_k$ are the solutions of the following system of equations:

$$\begin{cases} (c_1 + c_2 + \cdots + c_k) \lambda = a_1, \\ (c_1 + 2c_2 + \cdots + 2^{k-1} c_k) \lambda^2 = a_2, \\ \cdots\cdots\cdots\cdots\cdots\cdots\cdots\cdots \\ (c_1 + kc_2 + \cdots + k^{k-1} c_k) \lambda^k = a_k. \end{cases}$$

Proposition 3.24. *If a characteristic equation*

$$x^k = \lambda_1 x^{k-1} + \lambda_2 x^{k-2} + \cdots + \lambda_k (\lambda_k \neq 0)$$

has m distinct roots $x_1, x_2, \ldots, x_m$, where x_i is a k_i-fold root, and $\sum_{i=1}^m k_i = k$, then for the corresponding recurrence relation

$$a_{n+k} = \lambda_1 a_{n+k-1} + \lambda_2 a_{n+k-2} + \cdots + \lambda_k a_n,$$

the general term formula of the sequence $\{a_n\}$ is given by

$$a_n = p_1(n) x_1^n + p_2(n) x_2^n + \cdots + p_m(n) x_m^n,$$

where $n \in \mathbf{N}_+$, $p_i(n) = c_{i1} + c_{i2} n + \cdots + c_{ik_i} n^{k_i - 1} (1 \leq i \leq m)$, and $c_{ij} (1 \leq i \leq m, 1 \leq j \leq k_i)$ are determined by the following system of equations:

$$\begin{cases} p_1(1) x_1 + p_2(1) x_2 + \cdots + p_m(1) x_m = a_1, \\ p_1(2) x_1^2 + p_2(2) x_2^2 + \cdots + p_m(2) x_m^2 = a_2, \\ \cdots\cdots\cdots\cdots\cdots\cdots\cdots\cdots\cdots\cdots \\ p_1(k) x_1^k + p_2(k) x_2^k + \cdots + p_m(k) x_m^k = a_k. \end{cases}$$

Proposition 3.25. *If $\{a_n\}$ is a kth order linear recursive sequence with a characteristic equation*

$$f(x) = x^k - \lambda_1 x^{k-1} - \lambda_2 x^{k-2} - \cdots - \lambda_k = 0 (\lambda_k \neq 0),$$

and the sum of its first n terms is denoted as S_n, then the sequence $\{S_n\}$ is a $(k+1)$th order linear recursive sequence with the recurrence relation

$$S_{n+k+1} = (1+\lambda_1)S_{n+k} + (\lambda_2 - \lambda_1)S_{n+k-1} + \cdots + (\lambda_k - \lambda_{k-1})S_{n+1} - \lambda_k S_n,$$

and the characteristic equation is given by $(x-1)f(x) = 0$.

Example 3.1. Given a sequence $\{a_n\}$ such that $a_1 = 3$, $a_2 = 8$, and $a_n = 2a_{n-1} + 2a_{n-2}$ for $n = 3, 4, \ldots$, find the general term formula for $\{a_n\}$.

Solution. The characteristic equation corresponding to the recurrence relation is $x^2 = 2x + 2$, with two roots $x_1 = 1 + \sqrt{3}$ and $x_2 = 1 - \sqrt{3}$. Therefore,

$$a_n = c_1(1 + \sqrt{3})^n + c_2(1 - \sqrt{3})^n,$$

which leads to the system of equations

$$\begin{cases} c_1(1 + \sqrt{3}) + c_2(1 - \sqrt{3}) = 3, \\ c_1(1 + \sqrt{3})^2 + c_2(1 - \sqrt{3})^2 = 8. \end{cases}$$

Solving this system yields $c_1 = \frac{3+2\sqrt{3}}{6}$ and $c_2 = \frac{3-2\sqrt{3}}{6}$. Thus,

$$a_n = \frac{3 + 2\sqrt{3}}{6}(1 + \sqrt{3})^n + \frac{3 - 2\sqrt{3}}{6}(1 - \sqrt{3})^n.$$

In particular, for a recursive system given by

$$\begin{cases} x_n = ax_{n-1} + by_{n-1}, \\ y_n = cx_{n-1} + dy_{n-1}, \end{cases}$$

the characteristic equation is determined by

$$\lambda^2 - (a + d)\lambda + (ad - bc) = 0.$$

If the characteristic equation has two distinct roots λ_1 and λ_2, then the general term formulas for $\{x_n\}$ and $\{y_n\}$ are

$$\begin{cases} x_n = A\lambda_1^n + B\lambda_2^n, \\ y_n = C\lambda_1^n + D\lambda_2^n, \end{cases}$$

where the constants A, B, C, D are determined by the initial conditions:

$$\begin{cases} x_1 = A\lambda_1 + B\lambda_2, \\ y_1 = C\lambda_1 + D\lambda_2 \end{cases} \quad \text{and} \quad \begin{cases} x_2 = A\lambda_1^2 + B\lambda_2^2, \\ y_2 = C\lambda_1^2 + D\lambda_2^2. \end{cases}$$

If the characteristic equation has a double root λ, then the general term formulas for $\{x_n\}$ and $\{y_n\}$ become:

$$\begin{cases} x_n = (An + B)\lambda^n, \\ y_n = (Cn + D)\lambda^n, \end{cases}$$

where the constants A, B, C, D are determined by the initial conditions:

$$\begin{cases} x_1 = (A + B)\,\lambda, \\ y_1 = (C + D)\,\lambda \end{cases} \quad \text{and} \quad \begin{cases} x_2 = (2A + B)\,\lambda^2, \\ y_2 = (2C + D)\,\lambda^2. \end{cases}$$

Actually, by substituting $y_{n-1} = \frac{x_n - ax_{n-1}}{b}$ into $y_n = cx_{n-1} + dy_{n-1}$, we obtain

$$x_{n+1} = (a + d)x_n + (bc - ad)x_{n-1}.$$

(5) *Periodic sequences*

For a sequence $\{a_n\}$, if there exist specific positive integers T and n_0, such that $a_{n+T} = a_n$ for all $n \geq n_0$, then $\{a_n\}$ is called a periodic sequence with a period T starting from the n_0th term. If $n_0 = 1$, then the sequence $\{a_n\}$ is called purely periodic. If $n_0 \geq 2$, then the sequence $\{a_n\}$ is called eventually periodic or uttimately periodic.

Proposition 3.26. *The range of values for a periodic sequence is a finite set.*

Proposition 3.27. *If T is a period of a periodic sequence $\{a_n\}$, then for any positive integer k, the integer kT is also a period of $\{a_n\}$.*

Proposition 3.28. *Periodic sequences have a minimum period.*

The minimum period of a periodic sequence is called its least period or exact period.

Proposition 3.29. *If T is the least period of a periodic sequence $\{a_n\}$, and T_1 is another period of $\{a_n\}$, then $T|T_1$.*

Let $\{a_n\}$ be an integer sequence and m be a given positive integer. If b_n is the remainder when $a_n(n \geq 1)$ is divided by m, i.e., $b_n \equiv a_n(\mathrm{mod}\, m)$ and $b_n \in \{0, 1, 2, \ldots, m-1\}$, then the sequence $\{b_n\}$ is called the modular sequence of $\{a_n\}$ with respect to m, denoted as $\{a_n(\mathrm{mod}\, m)\}$.

For an integer sequence $\{a_n\}$, if there exist positive integers T and N such that $a_{n+T} \equiv a_n(\mathrm{mod}\, m)$ for all $n \geq N$, then $\{a_n(\mathrm{mod}\, m)\}$ is called a modular periodic sequence, and T is called a period of the modular periodic sequence.

The smallest value of T is referred to as the least period, denoted as $T(m)$. If $N = 1$, then the sequence $\{a_n(\mathrm{mod}\, m)\}$ is called purely modular periodic. If $N \geq 2$, then the sequence $\{a_n(\mathrm{mod}\, m)\}$ is called eventually or ultimately modular periodic.

Proposition 3.30. *If $T = T(m)$ is the least period of a modular periodic sequence, then T_1 is another period if and only if $T|T_1$.*

Proposition 3.31. *If $\{a_n(\mathrm{mod}\, m)\}$ and $\{b_n(\mathrm{mod}\, m)\}$ are both modular periodic sequences, then $\{a_n \pm b_n(\mathrm{mod}\, m)\}$ and $\{a_n b_n(\mathrm{mod}\, m)\}$ are also modular periodic sequences.*

Proposition 3.32. *Suppose $\{a_n(\mathrm{mod}\, m)\}$ is a modular periodic sequence. If $m'|m$, then $\{a_n(\mathrm{mod}\, m')\}$ is also a modular periodic sequence, and $T(m')|T(m)$.*

Proposition 3.33. *If $\{a_n(\mathrm{mod}\, m_1)\}$ and $\{a_n(\mathrm{mod}\, m_2)\}$ are both modular periodic sequences, then $\{a_n(\mathrm{mod}\, m)\}$ is also a modular periodic sequence, and $T(m) = [T(m_1), T(m_2)]$, where $m = [m_1, m_2]$. Here $[a, b]$ denotes the least common multiple of a and b.*

3.1.2 *Common methods for finding general term formulas*

(1) *Iteration method*

  (i) $a_{n+1} = a_n + f(n)$.
    For $n \geq 2$,

$$a_n = a_{n-1} + f(n-1) = a_{n-2} + f(n-2) + f(n-1) = \cdots = a_1 + \sum_{k=1}^{n-1} f(k).$$

  (ii) $a_{n+1} = a_n \cdot f(n)$.
    For $n \geq 2$,

$$a_n = a_{n-1} \cdot f(n-1) = a_{n-2} \cdot f(n-2) \cdot f(n-1) = \cdots = a_1 \prod_{k=1}^{n-1} f(k).$$

(iii) $a_{n+1} = pa_n + f(n)$, where $p \neq 0, 1$.

Let $b_n = \frac{a_n}{p^n}$ and $g(n) = \frac{f(n)}{p^{n+1}}$. Then $b_{n+1} = b_n + g(n)$, which can be transformed into (i).

(iv) $a_{n+1} = pa_n + q$, where p and q are constants, $pq \neq 0$, and $p \neq 1$.

Subtracting $a_n = pa_{n-1} + q$ from $a_{n+1} = pa_n + q$, we obtain $a_{n+1} - a_n = p(a_n - a_{n-1})$. Let $b_n = a_{n+1} - a_n$. Then it reduces to (ii), where $f(n) = p$ is a constant function.

Similarly, the following recurrence relations can also be reduced to the aforementioned types:

(v) $a_{n+1} = f(n)a_n + g(n)$.

(vi) $a_{n+1} = \frac{ma_n}{pa_n + q}$, where $p \neq 0$ and $m \neq 0$.

(vii) $a_{n+2} = pa_{n+1} + qa_n$, where $pq \neq 0$.

(viii) $a_{n+1} = pa_n^q$, where $p \neq 0$ and $q \neq 1$.

(2) Fixed-point method

(i) $a_{n+1} = pa_n + q$, where p and q are constants, $pq \neq 0$, and $p \neq 1$.

Consider the fixed point of the function $f(x) = px + q$, i.e., the solution of $px + q = x$ is $\lambda = \frac{q}{1-p}$. We have $a_{n+1} - \lambda = p(a_n - \lambda)$, and then

$$a_n - \lambda = p^{n-1}(a_1 - \lambda).$$

(ii) $a_{n+1} = \frac{a_n}{pa_n + q}$, where $pq \neq 0$ and $a_1 \neq 0$.

Taking the reciprocal of both sides of the recurrence relation yields $\frac{1}{a_{n+1}} = q \cdot \frac{1}{a_n} + p$. Let $b_n = \frac{1}{a_n}$. Then it reduces to (i).

(iii) $a_{n+1} = \frac{\alpha a_n + \beta}{a_n + \gamma}$, where α, β, γ are constants.

Consider the fixed points of the function $f(x) = \frac{\alpha x + \beta}{x + \gamma}$, i.e., the solutions of the equation $x = \frac{\alpha x + \beta}{x + \gamma}$ are x_1, x_2.

- If $x_1 \neq x_2$, then $\frac{a_n - x_1}{a_n - x_2} = \frac{a_1 - x_1}{a_1 - x_2} \cdot \left(\frac{\alpha - x_1}{\alpha - x_2} \right)^{n-1}$.
- If $x_1 = x_2$, then $\frac{1}{a_n - x_1} = \frac{1}{a_1 - x_1} + (n - 1) \cdot \frac{1}{\alpha - x_1}$.

Example 3.2. Consider a sequence $\{a_n\}$ defined as follows: $a_1 = 2$ and $a_{n+1} = \frac{2a_n + 6}{a_n + 1}$ for $n = 1, 2, \ldots$ Determine the general term formula for $\{a_n\}$.

Solution. Find the fixed points by setting $x = \frac{2x+6}{x+1}$, yielding $x_1 = 3$ and $x_2 = -2$. Therefore,

$$a_{n+1} - 3 = \frac{2a_n + 6}{a_n + 1} - 3 = \frac{-a_n + 3}{a_n + 1}, \quad a_{n+1} + 2 = \frac{2a_n + 6}{a_n + 1} + 2 = \frac{4a_n + 8}{a_n + 1}.$$

Dividing the two equations, we get

$$\frac{a_{n+1} - 3}{a_{n+1} + 2} = -\frac{a_n - 3}{4(a_n + 2)},$$

which implies that $\left\{ \frac{a_{n+1}-3}{a_{n+1}+2} \right\}$ is a geometric sequence with a common ratio of $-\frac{1}{4}$ and an initial term of $\frac{a_1-3}{a_1+2} = -\frac{1}{4}$. Consequently, $\frac{a_n-3}{a_n+2} = \left(-\frac{1}{4} \right)\left(-\frac{1}{4} \right)^{n-1} = \left(-\frac{1}{4} \right)^n$. Thus, $a_n = \frac{3(-4)^n + 2}{(-4)^n - 1}$.

(3) *Characteristic equation method*

For $a_{n+2} = pa_{n+1} + qa_n$, where $pq \neq 0$, let α, β be the roots of the characteristic equation $x^2 = px + q$.

- If $\alpha \neq \beta$, then $a_n = \frac{1}{\alpha - \beta} \left((a_2 - \beta a_1)\alpha^{n-1} - (a_2 - \alpha a_1)\beta^{n-1} \right)$.
- If $\alpha = \beta$, then $a_n = \left(\frac{a_1}{\alpha} + (n-1) \cdot \frac{a_2 - \alpha a_1}{\alpha^2} \right) \alpha^n$.

(4) *Vieta's formulas method*

For $a_{n+1} = pa_n + \sqrt{(p^2 - 1)a_n^2 + 1}$, where p is a positive integer, consider a_n as a constant. Then, a_{n-1} and a_{n+1} are the roots of the quadratic equation

$$x^2 - 2pa_n x + a_n^2 - 1 = 0.$$

Using Vieta's formulas, we have $a_{n+1} = 2pa_n - a_{n-1}$.

There is a related problem:

- **(British Mathematical Olympiad 2002, 2nd Round, Problem 3).** Prove that a sequence defined by

$$y_0 = 1 \quad \text{and} \quad y_{n+1} = \frac{1}{2} \left(3y_n + \sqrt{5y_n^2 - 4} \right) \quad \text{for} \quad n \geq 0$$

consists only of integers.

3.1.3 *Common methods for summations of sequences*

(1) *Formula method*

For some sequences, such as arithmetic and geometric sequences, there exist corresponding summation formulas for the sum of the first n terms, denoted as S_n.

(2) *Reverse summation method*

This method is generally applicable in cases where the sum of the kth term and the kth term from the end in a sequence is equal to the sum of the first and last terms. For instance, if $\alpha = a_1 + a_n = a_k + a_{n+1-k}$ with $1 \le k \le n$, then $S_n = \frac{n\alpha}{2}$.

(3) *Telescoping method*

This method is generally suitable when the general term a_n can be decomposed into the difference of two terms, such as $a_n = f(n) - f(n-k)$, where k is an integer. Common telescoping forms include:

(i) $a_n = \frac{1}{n(n+k)} = \frac{1}{k}\left(\frac{1}{n} - \frac{1}{n+k}\right)$.

(ii) $a_n = \frac{1}{\sqrt{n}+\sqrt{n+k}} = \frac{\sqrt{n+k}-\sqrt{n}}{k}$.

(iii) $a_n = \frac{1}{n(n+1)(n+2)} = \frac{1}{2}\left(\frac{1}{n(n+1)} - \frac{1}{(n+1)(n+2)}\right)$.

(iv) $a_n = \frac{(2n)^2}{(2n-1)(2n+1)} = 1 + \frac{1}{2}\left(\frac{1}{2n-1} - \frac{1}{2n+1}\right)$.

(v) $a_n = \frac{2^{n-1}}{(2^n+1)(2^{n+1}+1)} = \frac{1}{2}\left(\frac{1}{2^n+1} - \frac{1}{2^{n+1}+1}\right)$.

(vi) $a_n = \frac{n+2}{2^n n(n+1)} = \frac{1}{2^{n-1}n} - \frac{1}{2^n(n+1)}$.

Additionally, the telescoping method is closely related to the bounding method.

(vii) $\frac{1}{k^3} < \frac{1}{k^3-k} = \frac{1}{k(k^2-1)} = \frac{1}{k(k-1)(k+1)} = \frac{1}{2}\left(\frac{1}{k(k-1)} - \frac{1}{k(k+1)}\right)$.

(viii) $\frac{1}{k^2} < \frac{4}{4k^2-1} < \frac{1}{k^2-1} \le \frac{1}{k^2-k} < \frac{1}{k^2-2k}$.

(ix) $\frac{1}{k^2} > \frac{1}{k^2+k} > \frac{1}{k^2+2k} > \frac{1}{k^2+3k+2}$.

(x) $\frac{1}{\sqrt{k}} = \frac{2}{2\sqrt{k}} < \frac{2}{\sqrt{k}+\sqrt{k-1}} = 2(\sqrt{k} - \sqrt{k-1})$.

(xi) $\frac{1}{\sqrt{k}} = \frac{2}{2\sqrt{k}} > \frac{2}{\sqrt{k}+\sqrt{k+1}} = 2(\sqrt{k+1} - \sqrt{k}).$

(xii) $\frac{1}{k\sqrt{k}} < \frac{1}{k\sqrt{k-1}} = \frac{1}{\sqrt{k}\cdot\sqrt{k}\cdot\sqrt{k-1}} = \frac{1}{\sqrt{k}}\left(\frac{1}{\sqrt{k-1}} - \frac{1}{\sqrt{k}}\right)\frac{1}{\sqrt{k}-\sqrt{k-1}}$

$< 2\left(\frac{1}{\sqrt{k-1}} - \frac{1}{\sqrt{k}}\right).$

(xiii) $\frac{1}{k\sqrt{k}} > \frac{1}{k\sqrt{k+1}} = \frac{1}{\sqrt{k}\cdot\sqrt{k}\cdot\sqrt{k+1}} = \frac{1}{\sqrt{k}}\left(\frac{1}{\sqrt{k}} - \frac{1}{\sqrt{k+1}}\right)\frac{1}{\sqrt{k+1}-\sqrt{k}}$

$> 2\left(\frac{1}{\sqrt{k}} - \frac{1}{\sqrt{k+1}}\right).$

(4) *Shifting and subtracting method*

This method is typically employed when the general term is in the form of "arithmetic times geometric." If $a_n = (an + b)q^{n-1}$, then the sum of the first n terms is given by $S_n = (An + B)q^n - B$, where $A = \frac{a}{q-1}$ and $B = \frac{b-A}{q-1}.$

3.1.4 *Mathematical induction*

(1) *First principle of mathematical induction*

Let $P(n)$ be a proposition concerning all positive integers n. If

- $P(1)$ is true;
- assuming $P(k)$ is true, one can deduce that $P(k+1)$ is also true;

then the proposition $P(n)$ is true for any positive integer n.

(2) *Second principle of mathematical induction*

Let $P(n)$ be a proposition concerning all positive integers n. If

- $P(1)$ is true;
- assuming that $P(k)$ holds for all positive integers k less than n, one can deduce that $P(n)$ is also true;

then the proposition $P(n)$ is true for any positive integer n.

As the general term a_n of a sequence is related to the positive integer n, sequence problems often employ mathematical induction. This includes finding the general term a_n, proving that a_n is an integer, odd number, or a perfect square, and establishing recurrence relations or inequalities between adjacent terms of the sequence.

3.1.5 *Other important theorems*

Theorem 3.1 (Well-Ordering Principle). For any non-empty subset T of the set of positive integers $\mathbf{N}_+$, there exists a minimum element.

In other words, there exists a positive integer $t_0 \in T$ such that $t_0 \le t$ for any $t \in T$.

Furthermore, the proof by infinite descent, also known as Fermat's method of descent, is closely related to the Well-Ordering Principle. If the proposition $P(n)$ concerning positive integers n holds true for $n = n_0$, and it can be deduced that the proposition also holds for infinitely many integers $n = n_1, n_2, \ldots (n_0 > n_1 > n_2 \cdots)$, then the proposition is false.

Fermat used this method to prove that the equation $x^4 + y^4 = z^4$ has no positive integer solution. He assumed the existence of a set of positive integer solution (x, y, z) and then derived another smaller set of positive integer solution (x', y', z'), where $z' < z$ (or $x' + y' + z' < x + y + z$). By repeating this process, a new smaller solution can be obtained each time, and an infinitely decreasing sequence of positive integers

$$z > z' > z'' > \cdots$$

will be generated. However, a strictly decreasing sequence of positive integers can only have a finite number of terms, leading to a contradiction and demonstrating that the original equation has no positive integer solution.

Corollary 3.1. *Let M be a non-empty subset of the set of positive integers $\mathbf{N}_+$ and suppose that M has an upper bound, i.e., there exists an $a \in \mathbf{N}_+$ such that $x \le a$ for any $x \in M$, then M has a maximum element.*

Corollary 3.2. *Any finite set of real numbers necessarily contains both a minimum element and a maximum element.*

Theorem 3.2. A set M composed of n real numbers can be expressed as $M = \{x_1, x_2, \ldots, x_n\}$, where $x_1 < x_2 < \cdots < x_n$.

Theorem 3.3 (Monotone Convergence Theorem). A monotonically increasing (decreasing) sequence that is bounded above (below) has a limit, and this limit is the least upper (greatest lower) bound, also known as supremum (infimum).

Theorem 3.4 (Nested Interval Theorem). Given a series of closed intervals $\{[a_n, b_n]\}$ satisfying the conditions:

- $[a_{n+1}, b_{n+1}] \subset [a_n, b_n]$ for $n = 1, 2, \ldots$ and
- $\lim_{n \to +\infty} (b_n - a_n) = 0$,

there exists a real number $\xi = \lim_{n \to +\infty} a_n = \lim_{n \to +\infty} b_n$, and ξ is the unique common point of all the closed intervals.

Theorem 3.5 (Boundedness Theorem). If a sequence $\{a_n\}$ has a limit, then the sequence $\{a_n\}$ is bounded.

Theorem 3.6. A sequence $\{a_n\}$ has a limit if and only if every non-trivial subsequence of $\{a_n\}$ has the same limit.

3.2 Problems and Solutions

3.2.1 *Determining values*

Problem 3.1 (IMO 9-5, proposed by the Soviet Union). Consider a sequence $\{C_n\}$, where

$$C_1 = a_1 + a_2 + \cdots + a_8,$$
$$C_2 = a_1^2 + a_2^2 + \cdots + a_8^2,$$
$$\cdots\cdots\cdots\cdots\cdots\cdots\cdots$$
$$C_n = a_1^n + a_2^n + \cdots + a_8^n,$$
$$\cdots\cdots\cdots\cdots\cdots\cdots\cdots$$

in which $a_1, a_2, \ldots, a_8$ are real numbers not all equal to zero. Suppose that an infinite number of terms of the sequence $\{C_n\}$ are equal to zero. Find all positive integers n for which $C_n = 0$.

Solution. Assume that a_1 has the largest absolute value. Then there must exist some $a_i (i \neq 1)$ such that $a_i = -a_1$. Otherwise, for sufficiently large n, it holds that $\left(\frac{a_i}{a_1}\right)^n$ either are constantly 1 or tend to zero $(i \neq 1)$. Thus,

$$|C_n| = |a_1|^n \left(1 + \left(\frac{a_2}{a_1}\right)^n + \cdots + \left(\frac{a_8}{a_1}\right)^n\right) > 0,$$

which contradicts the given conditions.

Without loss of generality, let $a_2 = -a_1$. Since n that makes $C_n = 0$ must be odd, $a_1^n + a_2^n = 0$. Removing a_1^k and a_2^k from C_k, the remaining $\{C'_n\}$ still has infinitely many terms equal to zero. Following the above reasoning, we can assume $a_3 = -a_4$, $a_5 = -a_6$, and $a_7 = -a_8$. Therefore, when n is any positive odd numbers, $C_n = 0$.

Note. If the given n expressions are of the form $C_k = a_1^k + a_2^k + \cdots + a_t^k$, where t is any positive integer, then the solution remains the set of all positive odd numbers.

There are several similar problems:

- **(Germany Team Selection Test 2004, Problem 16).** Let n be a positive integer. Find all complex numbers $x_1, x_2, \ldots, x_n$ satisfying the following system of equations:

$$\begin{cases} x_1 + 2x_2 + \cdots + nx_n = 0, \\ x_1^2 + 2x_2^2 + \cdots + nx_n^2 = 0, \\ \cdots\cdots\cdots\cdots\cdots\cdots\cdots\cdots\cdots \\ x_1^n + 2x_2^n + \cdots + nx_n^n = 0. \end{cases}$$

- **(All-Russian Mathematical Olympiad 1996, Final Round, Grade 10, Problem 4).** Prove that if nonzero numbers $a_1, a_2, \ldots, a_m$ satisfy for each $k = 0, 1, \ldots, n (n < m - 1)$,

$$a_1 + 2^k a_2 + 3^k a_3 + \cdots + m^k a_m = 0,$$

then the sequence $a_1, a_2, \ldots, a_m$ contains at least $n+1$ pairs of consecutive terms having opposite signs.

- **(Japan Mathematical Olympiad 1995, Final Round, Problem 5).** Let k, n be integers such that $1 \leq k \leq n$, and let $a_1, a_2, \ldots, a_k$ be numbers satisfying the following equations:

$$\begin{cases} a_1 + a_2 + \cdots + a_k = n, \\ a_1^2 + a_2^2 + \cdots + a_k^2 = n, \\ \cdots\cdots\cdots\cdots\cdots\cdots\cdots\cdots \\ a_1^k + a_2^k + \cdots + a_k^k = n. \end{cases}$$

Prove that

$$(x + a_1)(x + a_2) \cdots (x + a_n) = x^k + C_n^1 x^{k-1} + C_n^2 x^{k-2} + \cdots + C_n^{k-1} x + C_n^k,$$

where C_n^i is a binomial coefficient which means

$$\frac{n \cdot (n - 1) \cdots\cdots (n - i + 2) \cdot (n - i + 1)}{i \cdot (i - 1) \cdots\cdots 2 \cdot 1}.$$

【Score Situation】 This particular problem saw the following distribution of scores among contestants: 13 contestants scored 7 points, no contestant scored 6 points, 2 contestants scored 5 points, 3 contestants scored 4 points, 3 contestants scored 3 points, 1 contestant scored 2 points, 8 contestants scored 1 point, and 7 contestants scored 0 point. The average score of this problem is 3.568, indicating that it was relatively straightforward.

Among the top five teams in the team scores, the Soviet Union team achieved a total score of 275 points, the German Democratic Republic team achieved a total score of 257 points, the Hungary team achieved a total score of 251 points, the United Kingdom team achieved a total score of 231 points, and the Romania team achieved a total score of 214 points.

The gold medal cutoff for this IMO was set at 38 points (with 11 contestants earning gold medals), the silver medal cutoff was 30 points (with 14 contestants earning silver medals), and the bronze medal cutoff was 22 points (with 26 contestants earning bronze medals).

In this IMO, a total of five contestants achieved a perfect score of 42 points.

Problem 3.2 (IMO 36-4, proposed by Poland). Find the maximum value of x_0 for which there exists a sequence $x_0, x_1, \ldots, x_{1995}$ of positive reals with $x_0 = x_{1995}$, such that for $i = 1, 2, \ldots, 1995$,

$$x_{i-1} + \frac{2}{x_{i-1}} = 2x_i + \frac{1}{x_i}.$$

Solution. Rewrite the recurrence relation as

$$(x_{i-1} - 2x_i) + \frac{1}{x_i x_{i-1}}(2x_i - x_{i-1}) = 0.$$

Therefore, $x_i = \frac{x_{i-1}}{2}$ or $x_i = \frac{1}{x_{i-1}}$.

Suppose, starting from x_0, there are $k - t$ transformations of the first kind and t transformations of the second kind to reach x_k. It is evident that $x_k = 2^s x_0^{(-1)^t}$, where $s \equiv k - t \pmod 2$.

Now consider the case $k = 1995$. If t is even, then $x_0 = x_{1995} = 2^s x_0$. Since $x_0 > 0$, it follows that $s = 0$, implying $0 \equiv 1995 - t \pmod 2$, which contradicts the assumption that t is even. Therefore, t must be odd, and $x_0 = x_{1995} = 2^s x_0^{-1}$, yielding $x_0 = 2^{\frac{s}{2}}$.

At this point, $s \equiv 1995 - t \equiv 0 \pmod 2$. Hence s is even. However, since the two kinds of transformations occur alternatively, there may be a cancelation effect, so $s \le |s| \le 1995$.

Thus, $s \le 1994$, and $x_0 \le 2^{997}$. When $x_0 = 2^{997}$, the recurrence can be realized, i.e.,

$$x_i = 2^{997-i}, \quad i = 0, 1, 2, \ldots, 1994, \quad x_{1995} = \frac{1}{x_{1994}} = 2^{997}.$$

In conclusion, the maximum value of x_0 is 2^{997}.

【Score Situation】 This particular problem saw the following distribution of scores among contestants: 168 contestants scored 7 points, 48 contestants scored 6 points, 37 contestants scored 5 points, 25 contestants scored 4 points, 18 contestants scored 3 points, 16 contestants scored 2 points, 57 contestants scored 1 point, and 43 contestants scored 0 point. The average score for this problem is 4.592, indicating that it was simple.

Among the top five teams in the team scores, the scores of this problem are as follows: the China team scored 36 points (with a total team score of 236 points), the Romania team scored 42 points (with a total team score of 230 points), the Russia team scored 42 points (with a total team score of 227 points), the Vietnam team scored 42 points (with a total team score of 220 points), and the Hungary team scored 42 points (with a total team score of 210 points).

The gold medal cutoff for this IMO was set at 37 points (with 30 contestants earning gold medals), the silver medal cutoff was 29 points (with 71 contestants earning silver medals), and the bronze medal cutoff was 19 points (with 100 contestants earning bronze medals).

In this IMO, a total of 14 contestants achieved a perfect score of 42 points.

Problem 3.3 (IMO 58-1, proposed by South Africa). For each integer $a_0 > 1$, define a sequence $a_0, a_1, a_2, \ldots$ by

$$a_{n+1} = \begin{cases} \sqrt{a_n}, & \text{if } \sqrt{a_n} \text{ is an integer,} \\ a_n + 3, & \text{otherwise,} \end{cases} \quad \text{for each } n \geq 0.$$

Determine all values of a_0 for which there is a number A such that $a_n = A$ for infinitely many values of n.

Solution. The values of a_0 satisfying the conditions are all multiples of 3.

Since a_{n+1} is determined solely by a_n, the sequence $\{a_n\}$ has infinitely many equal terms if and only if it is a periodic sequence, which is also equivalent to the sequence being bounded.

Notably, perfect squares leave a residue of 0 or 1 when divided by 3. If a term $a_k \equiv 2 \pmod 3$, then a_k is not a perfect square, and $a_{k+1} = a_k + 3$, with $a_{k+1} \equiv 2 \pmod 3$. Recursively, $a_{m+1} = a_m + 3$ for $m \geq k$, and the sequence strictly increases from the kth term onward, resulting in an unbounded sequence. In particular, if $a_0 \equiv 2 \pmod 3$, then such an a_0 does not satisfy the conditions.

If $3 | a_k$, then whether $a_{k+1} = \sqrt{a_k}$ or $a_{k+1} = a_k + 3$, it holds that $3 | a_{k+1}$. Recursively, a_m is always a multiple of 3 for $m \geq k$. In particular, if $3 | a_0$, then every term in the sequence is divisible by 3.

Assume $3 | a_0$, and choose a perfect square $N^2 > a_0$ such that $3 | N$. We claim that $a_n \leq N^2$ for every n. By contradiction, assume there exists a term greater than N^2, and take the smallest k such that $a_k > N^2$. Since $3|a_k$, we have $a_k \geq N^2 + 3$. However, $a_{k-1} \leq N^2$ and $a_k - a_{k-1} \leq 3$, so $a_{k-1} = N^2$. According to the definition, $a_k = N$, leading to a contradiction. Therefore, there is no $a_k > N^2$, and the sequence $\{a_n\}$ is bounded, satisfying the conditions for such a_0.

Lastly, consider the case $a_0 \equiv 1 \pmod 3$. It is easy to see that if a_k is not divisible by 3 and $a_k > 1$, then 3 does not divide a_{k+1} and $a_{k+1} > 1$. Thus, every term in the sequence $\{a_n\}$ is not divisible by 3 and greater than 1. Suppose $\{a_n\}$ is bounded, i.e., it eventually becomes periodic. Then each term in the sequence is congruent to 1 modulo 3. Taking the largest term a_k in the period, we see that a_k is a perfect square; otherwise, $a_{k+1} = a_k + 3 > a_k$.

Let $a_k = N^2$. Then $a_{k+1} = N$, where $N \equiv 1 \pmod 3$ and $N > 1$, i.e., $N \geq 4$. As there exists $j > k$ such that $a_j = a_k = N^2 > (N-2)^2$ and $a_{k+1} = N \leq (N-2)^2$, take the smallest $l > k+1$ such that $a_l > (N-2)^2$, and then $a_{l-1} \leq (N-2)^2 < a_l$, implying $a_l = a_{l-1} + 3$. Since $a_l \equiv a_{l-1} \equiv 1 \pmod 3$, it follows that $a_{l-1} = (N-2)^2 \equiv 1 \pmod 3$, and by definition, $a_l = N - 2 \equiv 2 \pmod 3$, resulting in a contradiction. Therefore, $a_0 \equiv 1 \pmod 3$ does not satisfy the conditions.

In conclusion, the values of a_0 satisfying the conditions are all multiples of 3.

Note. There are several similar problems:

- **(Indian National Mathematical Olympiad 2016, Problem 3).** Let $\mathbf{N}$ denote the set of all natural numbers. Define a function $T : \mathbf{N} \to \mathbf{N}$ by $T(2k) = k$ and $T(2k+1) = 2k + 2$. We write $T^2(n) = T(T(n))$ and in general $T^k(n) = T^{k-1}(T(n))$ for any $k > 1$.

  (a) Show that for each $n \in \mathbf{N}$, there exists k such that $T^k(n) = 1$.
  (b) For $k \in \mathbf{N}$, let c_k denote the number of elements in the set $\{n | T^k(n) = 1\}$. Prove that $c_{k+2} = c_{k+1} + c_k$, for $k \geq 1$.

- **(IMO Shortlist 2015, Number Theory, Problem 4).** Suppose that $a_0, a_1, \ldots$ and $b_0, b_1, \ldots$ are two sequences of positive integers satisfying $a_0, b_0 \geq 2$ and

$$a_{n+1} = \gcd(a_n, b_n) + 1, \quad b_{n+1} = \operatorname{lcm}(a_n, b_n) - 1$$

  for all $n \geq 0$. Prove that the sequence $\{a_n\}$ is eventually periodic; in other words, there exist integers $N \geq 0$ and $t > 0$ such that $a_{n+t} = a_n$ for all $n \geq N$.

Here, $\gcd(x, y)$ means the greatest common divisor of x and y and $\text{lcm}(x, y)$ means the least common multiple of x and y.

- **(Balkan Mathematical Olympiad 2006, Problem 4).** Let m be a positive integer. Find all positive integers a such that the sequence $\{a_n\}$ defined by $a_0 = a$ and

$$a_{n+1} = \begin{cases} \frac{a_n}{2}, & \text{if } a_n \text{ is even}, \\ a_n + m, & \text{if } a_n \text{ is odd}, \end{cases} \quad \text{for } n = 0, 1, 2, \ldots$$

is periodic (there exists an integer $d > 0$ such that $a_{n+d} = a_n$ for all n).

- **(All-Russian Mathematical Olympiad 2000, 4th Round, Grade 10, Problem 6).** Given a positive integer a_0, construct a sequence $\{a_n\}$ as follows:

$$a_{n+1} = \begin{cases} a_n^2 - 5, & \text{if } a_n \text{ is odd}, \\ \frac{a_n}{2}, & \text{if } a_n \text{ is even}. \end{cases}$$

Prove that for any odd integer $a_0 > 5$ and real number $M > 0$, there is a positive integer k such that $a_k > M$.

- **(Baltic Way Mathematical Contests 1997, Problem 5).** In a sequence $u_0, u_1, \ldots$ of positive integers, u_0 is arbitrary, and for any non-negative integer n,

$$u_{n+1} = \begin{cases} \frac{u_n}{2}, & \text{for even } u_n, \\ a + u_n, & \text{for odd } u_n, \end{cases}$$

where a is a fixed odd positive integer. Prove that the sequence is periodic from a certain step.

【Score Situation】 This particular problem saw the following distribution of scores among contestants: 446 contestants scored 7 points, 25 contestants scored 6 points, 54 contestants scored 5 points, 12 contestants scored 4 points, 5 contestants scored 3 points, 17 contestants scored 2 points, 16 contestants scored 1 point, and 40 contestants scored 0 point. The average score for this problem is 5.943, indicating that it was simple.

Among the top five teams in the team scores, the scores of this problem are as follows: the South Korea team scored 42 points (with a total team score of 170 points), the China team scored 42 points (with a total team score of 159 points), the Vietnam team scored 42 points (with a total team score of 155 points), the United States team scored 42 points (with a total team score of 148 points), and the Iran team scored 42 points (with a total team score of 142 points).

The gold medal cutoff for this IMO was set at 25 points (with 48 contestants earning gold medals), the silver medal cutoff was 19 points (with 90 contestants earning silver medals), and the bronze medal cutoff was 16 points (with 153 contestants earning bronze medals).

In this IMO, no contestant achieved a perfect score of 42 points.

Problem 3.4 (IMO 59-2, proposed by Slovakia). Find all integers $n \geq 3$ for which there exist real numbers $a_1, a_2, \ldots, a_{n+2}$, such that $a_{n+1} = a_1$ and $a_{n+2} = a_2$, and

$$a_i a_{i+1} + 1 = a_{i+2}$$

for $i = 1, 2, \ldots, n$.

Solution 1. The values of n satisfying the conditions are all multiples of 3.

On the one hand, if n is a multiple of 3, then let $n = 3k$. Choose $a_{3i-2} = a_{3i-1} = -1$ and $a_{3i} = 2$ for $i = 1, 2, \ldots, k$, along with $a_{n+1} = a_{n+2} = -1$. It can be easily verified that for $i = 1, 2, \ldots, n$, the relation $a_i a_{i+1} + 1 = a_{i+2}$ satisfies the requirements.

On the other hand, assume there exist $a_1, a_2, \ldots, a_{n+2}$ satisfying the conditions. Extend them into an infinite sequence with period n at both ends. Then, $a_i a_{i+1} + 1 = a_{i+2}$ for any integer i. We successively prove the following conclusions:

(i) There is no i such that $a_i > 0$ and $a_{i+1} > 0$.

  If there exists such an i with $a_i > 0$ and $a_{i+1} > 0$, then $a_{i+2} = a_i a_{i+1} + 1 > 1$. By induction, $a_m > 1$ for all $m \geq i + 2$, implying

$$a_{m+2} = a_m a_{m+1} + 1 > a_m a_{m+1} > a_{m+1}.$$

  Thus, $a_{m+1} < a_{m+2} < a_{m+3} < \cdots < a_{m+n+1}$. However, this contradicts $a_{m+n+1} = a_{m+1}$.

(ii) There is no i such that $a_i = 0$.

  If there exists such an i with $a_i = 0$, then $a_{i+1} = a_{i-1} a_i + 1 = 1$ and $a_{i+2} = a_i a_{i+1} + 1 = 1$, which contradicts (i).

(iii) There is no i such that $a_i < 0$, $a_{i+1} < 0$, and $a_{i+2} < 0$.

  If there exists such an i with $a_i < 0$, $a_{i+1} < 0$, and $a_{i+2} < 0$, then $a_{i+2} = a_i a_{i+1} + 1 > 0$, a contradiction.

(iv) There is no i such that $a_i > 0$, $a_{i+1} < 0$, and $a_{i+2} > 0$.

  If there exists such an i with $a_i > 0$, $a_{i+1} < 0$, and $a_{i+2} > 0$, then from (i), $a_{i-1} < 0$ and $a_{i+3} < 0$. Since $0 < a_{i+2} = 1 + a_i a_{i+1} < 1$ and $|a_{i+1} a_{i+2}| = |a_{i+3} - 1| > 1$, we have $|a_{i+1}| > 1$, implying $a_{i+1} < -1$.

Moreover, $|a_i a_{i+1}| = |a_{i+2} - 1| < 1$ implies $|a_i| < 1$, yielding $0 < a_i < 1$. Then, from $a_{i-2} a_{i-1} = a_i - 1 < 0$, we get $a_{i-2} > 0$. Since $a_{i+1} < 0$ and $0 < a_i < 1$,

$$a_{i-1} = \frac{a_{i+1} - 1}{a_i} < a_{i+1} - 1 < a_{i+1}.$$

Thus, from $a_i > 0$, $a_{i+1} < 0$, and $a_{i+2} > 0$, we deduce $a_{i-1} < 0$, $a_{i-2} > 0$, and $a_{i-1} < a_{i+1}$. By induction, it is easy to see that for $k \geq 0$, $a_{i-2k} > 0$, $a_{i-2k+1} < 0$, and

$$a_{i+1} > a_{i-1} > a_{i-3} > a_{i-5} > \cdots.$$

Particularly, $a_{i+1} > a_{i+1-2n}$, which contradicts the fact that the defined sequence $\{a_m\}\,(m \in \mathbf{Z})$ has n as its period.

Consequently, it is known from (i), (ii), (iii), and (iv) that in any consecutive three terms a_i, a_{i+1}, a_{i+2}, exactly one term is positive, and the other two are negative. Thus, the sign of this sequence follows a period of 3. Since there is at least one positive term, say $a_i > 0$, then $a_{i+n} > 0$, leading to $3|n$.

Solution 2. The example part follows Solution 1. Here, we present an alternative proof for $3|n$. Extend $a_1, a_2, \ldots, a_n$ into an infinite sequence with period n at both ends. Then, for any i,

$$a_i a_{i+1} a_{i+2} = (a_{i+2} - 1)\, a_{i+2} = a_{i+2}^2 - a_{i+2},$$

$$a_i a_{i+1} a_{i+2} = a_i\, (a_{i+3} - 1) = a_i a_{i+3} - a_i,$$

leading to $a_{i+2}^2 - a_{i+2} = a_i a_{i+3} - a_i$.

Summing up the above equalities for $i = 1, 2, \ldots, n$ (and noting the periodicity), we have

$$\sum_{i=1}^{n} a_i^2 = \sum_{i=1}^{n} a_i a_{i+3},$$

which is equivalent to $\sum_{i=1}^{n}(a_i - a_{i+3})^2 = 0$.

Therefore, $a_i = a_{i+3}$ for any i. Hence, the sequence $\{a_i\}$ also has a period of 3. If n is not a multiple of 3, then the sequence $\{a_i\}$ has a period of 1, making it a constant sequence. However, the equation $x^2 + 1 = x$ has no real solution, and there is no constant sequence satisfying the given conditions. Thus, $3|n$.

Note. There are several similar problems:

- **(MathPath Summer Program Qualifying Test 2024, Problem 6).** Given some positive real number x, let $\lfloor x \rfloor$ denote the integer part of x and $\langle x \rangle$ denote the decimal part. Next, starting with a positive real number a_0, define a sequence via

$$a_{k+1} = \lfloor a_k \rfloor \langle a_k \rangle + 1 \quad \text{for all } k \geq 0.$$

Show that regardless of the initial value a_0, the sequence eventually stabilizes, meaning that $a_n = a_{n+1} = a_{n+2} = \cdots$ for some n onwards.

- **(United States of America Mathematical Olympiad 2021, Problem 5).** Let $n \geq 4$ be an integer. Find all positive real solutions to the following system of $2n$ equations:

$$a_1 = \frac{1}{a_{2n}} + \frac{1}{a_2}, \quad a_2 = a_1 + a_3,$$

$$a_3 = \frac{1}{a_2} + \frac{1}{a_4}, \quad a_4 = a_3 + a_5,$$

$$a_5 = \frac{1}{a_4} + \frac{1}{a_6}, \quad a_6 = a_5 + a_7,$$

$$\cdots\cdots\cdots\cdots\cdots\cdots\cdots$$

$$a_{2n-1} = \frac{1}{a_{2n-2}} + \frac{1}{a_{2n}}, \quad a_{2n} = a_{2n-1} + a_1.$$

- **(British Mathematical Olympiad 2020, 2nd Round, Problem 4).** A sequence $b_1, b_2, b_3 \ldots$ of nonzero real numbers has the property that

$$b_{n+2} = \frac{b_{n+1}^2 - 1}{b_n}$$

for all positive integers n.

Suppose that $b_1 = 1$ and $b_2 = k$, where $1 < k < 2$. Show that there is some constant B, depending on k, such that $-B \leq b_n \leq B$ for all n. Also show that, for some $1 < k < 2$, there is a value of n such that $b_n > 2020$.

【Score Situation】This particular problem saw the following distribution of scores among contestants: 157 contestants scored 7 points, 7 contestants scored 6 points, 16 contestants scored 5 points, 18 contestants scored 4 points, 66 contestants scored 3 points, 87 contestants scored 2 points, 85 contestants scored 1 point, and 158 contestants scored 0 point. The average score for this problem is 2.946, indicating that it had a certain level of difficulty.

Among the top five teams in the team scores, the scores of this problem are as follows: the United States team scored 41 points (with a total team score of 212 points), the Russia team scored 42 points (with a total team score of 201 points), the China team scored 37 points (with a total team score of 199 points), the Ukraine team scored 34 points (with a total team score of 186 points), and the Thailand team scored 38 points (with a total team score of 183 points).

The gold medal cutoff for this IMO was set at 31 points (with 48 contestants earning gold medals), the silver medal cutoff was 25 points (with 98 contestants earning silver medals), and the bronze medal cutoff was 16 points (with 143 contestants earning bronze medals).

In this IMO, only two contestants achieved a perfect score of 42 points, namely Agnijo Banerjee from the United Kingdom and James Lin from the United States.

3.2.2 Existence problems

Problem 3.5 (IMO 12-3, proposed by Sweden). Real   numbers $a_0, a_1, \ldots, a_n, \ldots$ satisfy the condition:

$$1 = a_0 \le a_1 \le a_2 \le \cdots \le a_n \le \cdots .$$

Numbers $b_1, b_2, \ldots, b_n, \ldots$ are defined by

$$b_n = \sum_{k=1}^{n} \left(1 - \frac{a_{k-1}}{a_k}\right) \frac{1}{\sqrt{a_k}}.$$

(a) Prove that $0 \le b_n < 2$ for all n.
(b) Given c with $0 \le c < 2$, prove that there exist numbers $a_0, a_1, \ldots$ with the above properties such that $b_n > c$ for n large enough.

Proof. (a) It is evident that $\frac{a_{k-1}}{a_k} \le 1$, implying $b_n \ge 0$. Moreover,

$$\left(1 - \frac{a_{k-1}}{a_k}\right) \frac{1}{\sqrt{a_k}} = \left(1 + \frac{\sqrt{a_{k-1}}}{\sqrt{a_k}}\right)\left(1 - \frac{\sqrt{a_{k-1}}}{\sqrt{a_k}}\right) \frac{1}{\sqrt{a_k}}$$

$$= \frac{\sqrt{a_{k-1}}}{\sqrt{a_k}}\left(1 + \frac{\sqrt{a_{k-1}}}{\sqrt{a_k}}\right)\left(\frac{1}{\sqrt{a_{k-1}}} - \frac{1}{\sqrt{a_k}}\right)$$

$$\le 2\left(\frac{1}{\sqrt{a_{k-1}}} - \frac{1}{\sqrt{a_k}}\right).$$

Thus,

$$b_n = \sum_{k=1}^{n} \left(1 - \frac{a_{k-1}}{a_k}\right) \frac{1}{\sqrt{a_k}} \le 2 \sum_{k=1}^{n} \left(\frac{1}{\sqrt{a_{k-1}}} - \frac{1}{\sqrt{a_k}}\right)$$

$$= 2 \left(\frac{1}{\sqrt{a_0}} - \frac{1}{\sqrt{a_n}}\right) = 2 \left(1 - \frac{1}{\sqrt{a_n}}\right) < 2.$$

(b) Choose an appropriate positive number $d \in (0,1)$ such that $a_k = d^{-2k}$, yielding $\frac{1}{\sqrt{a_k}} = d^k$. It follows that.

$$b_n = \sum_{k=1}^{n} \left(1 - \frac{a_{k-1}}{a_k}\right) \frac{1}{\sqrt{a_k}} = \sum_{k=1}^{n} \left(1 - d^2\right) d^k$$

$$= d \left(1 - d^2\right) \frac{1 - d^n}{1 - d} = d \left(1 + d\right) \left(1 - d^n\right).$$

For any positive number $c < 2$, let $d = \sqrt{\frac{c}{2}}$, and it is evident that $d(1 + d) > c$. Since $0 < d < 1$, we have $\lim_{n \to +\infty} (1 - d^n) = 1$, i.e., for n sufficiently large,

$$1 - d^n > \frac{2\sqrt{c}}{\sqrt{2} + \sqrt{c}} = \frac{c}{d(1 + d)}.$$

Therefore, there exist infinitely many n such that $b_n > c$.

Note. There are several similar problems:

- **(Turkey Mathematical Olympiad 1999, 2nd Round, Problem 4).** Find all sequences $a_1, a_2, \ldots, a_{2000}$ of real numbers with $\sum_{n=1}^{2000} a_n = 1999$ such that

$$\frac{1}{2} < a_n < 1 \quad \text{and} \quad a_n = a_{n-1}(2 - a_{n-1}) \quad \text{for all } n > 1.$$

- **(Chinese Northern Mathematical Olympiad 2006, Problem 8).** Given a sequence $\{a_n\}$ such that $a_0 = \frac{1}{2}$ and $a_{n+1} = a_n + \frac{1}{2006}a_n^2$ for $n \in \mathbf{N}$, prove that $1 - \frac{1}{2008} < a_{2006} < 1$.

【Score Situation】This particular problem saw the following distribution of scores among contestants: 5 contestants scored 8 points, no contestant scored 7 points, 1 contestant scored 6 points, no contestant scored 5 points, 1 contestant scored 4 points, no contestant scored 3 points, 1 contestant scored 2 points, 1 contestant scored 1 point, and 26 contestants scored 0 point. The average score of this problem is 1.514, indicating that it was relatively challenging.

Among the top five teams in the team scores, the Hungary team achieved a total score of 233 points, the German Democratic Republic team achieved a total score of 221 points, the Soviet Union team achieved a total score of 221 points, the Yugoslavia team achieved a total score of 209 points, and the Romania team achieved a total score of 208 points.

The gold medal cutoff for this IMO was set at 37 points (with 7 contestants earning gold medals), the silver medal cutoff was 30 points (with 11 contestants earning silver medals), and the bronze medal cutoff was 19 points (with 40 contestants earning bronze medals).

In this IMO, only three contestants achieved a perfect score of 40 points, namely Wolfgang Burmeister from the German Democratic Republic, Imre Ruzsa from Hungary, and Andrei Hodulev from the Soviet Union.

Problem 3.6 (IMO 15-6, proposed by Sweden). Let $a_1, a_2, \ldots, a_n$ be n positive numbers and q be a given real number such that $0 < q < 1$. Find n numbers $b_1, b_2, \ldots, b_n$ satisfying:

(a) $a_k < b_k$ for $k = 1, 2, \ldots, n$;
(b) $q < \frac{b_{k+1}}{b_k} < \frac{1}{q}$ for $k = 1, 2, \ldots, n-1$;
(c) $b_1 + b_2 + \cdots + b_n < \frac{1+q}{1-q}(a_1 + a_2 + \cdots + a_n)$.

Solution. For $k \in \{1, 2, \ldots, n\}$, let

$$b_k = a_1 q^{k-1} + a_2 q^{k-2} + \cdots + a_{k-1}q + a_k + a_{k+1}q + \cdots + a_n q^{n-k}.$$

(a) Clearly, $b_k > a_k$ for $k \in \{1, 2, \ldots, n\}$.
(b) For $k \in \{1, 2, \ldots, n-1\}$,

$$qb_k - b_{k+1} = (q^2 - 1)(a_{k+1} + \cdots + a_n q^{n-k-1}) < 0,$$

and

$$qb_{k+1} - b_k = \left(q^2 - 1\right)\left(a_1 q^{k-1} + \cdots + a_k\right) < 0,$$

so $q < \frac{b_{k+1}}{b_k} < \frac{1}{q}$.
(c) Note that

$$b_1 + b_2 + \cdots + b_n = \sum_{k=1}^{n}(a_1 q^{k-1} + a_2 q^{k-2} + \cdots + a_k + a_{k+1}q$$
$$+ \cdots + a_n q^{n-k})$$

$$< (a_1 + a_2 + \cdots + a_n)(1 + 2q + 2q^2 + \cdots + 2q^{n-1})$$

$$= (a_1 + a_2 + \cdots + a_n)\left(1 + 2q \cdot \frac{1 - q^{n-1}}{1 - q}\right)$$

$$< (a_1 + a_2 + \cdots + a_n)\left(1 + \frac{2q}{1 - q}\right)$$

$$= \frac{1 + q}{1 - q}(a_1 + a_2 + \cdots + a_n).$$

Thus, $b_1, b_2, \ldots, b_n$ satisfy the given conditions.

Note. This problem has certain connections with advanced mathematics. For n-dimensional vectors $\boldsymbol{A} = (a_1, a_2, \ldots, a_n)$ and $\boldsymbol{A}' = (a_1', a_2', \ldots, a_n')$, solutions satisfying the given conditions are $\boldsymbol{B} = (b_1, b_2, \ldots, b_n)$ and $\boldsymbol{B}' = (b_1', b_2', \ldots, b_n')$ respectively. It is evident that for vectors $\boldsymbol{A} + \boldsymbol{A}'$ and $C\boldsymbol{A}$ (where C is a positive real number), the solutions satisfying the given conditions are $\boldsymbol{B} + \boldsymbol{B}'$ and $C\boldsymbol{B}$ respectively. Therefore, if for unit vectors

$$\boldsymbol{u}_1 = (1, 0, 0, \ldots, 0),$$
$$\boldsymbol{u}_2 = (0, 1, 0, \ldots, 0),$$
$$\ldots\ldots\ldots\ldots\ldots\ldots\ldots$$
$$\boldsymbol{u}_n = (0, 0, 0, \ldots, 1),$$

the corresponding solutions $\boldsymbol{v}_1, \boldsymbol{v}_2, \ldots, \boldsymbol{v}_n$ can be found, then for the vector $(a_1, a_2, \ldots, a_n)$, numbers $b_1, b_2, \ldots, b_n$ satisfying the given conditions would be components of the vector $a_1\boldsymbol{v}_1 + a_2\boldsymbol{v}_2 + \cdots + a_n\boldsymbol{v}_n$.

【Score Situation】This particular problem saw the following distribution of scores among contestants: 10 contestants scored 8 points, 2 contestants scored 7 points, 2 contestants scored 6 points, 3 contestants scored 5 points, 3 contestants scored 4 points, 1 contestant scored 3 points, 2 contestants scored 2 points, 1 contestant scored 1 point, and 101 contestants scored 0 point. The average score of this problem is 1.128, indicating that it was relatively challenging.

Among the top five teams in the team scores, the scores of this problem are as follows: the Soviet Union team scored 30 points (with a total team score of 254 points), the Hungary team scored 11 points (with a total team score of 215 points), the German Democratic Republic team scored 14 points (with a total team score of 188 points), the Poland team scored 8 points (with a total team score of 174 points), and the United Kingdom team scored 25 points (with a total team score of 164 points).

The gold medal cutoff for this IMO was set at 35 points (with 5 contestants earning gold medals), the silver medal cutoff was 27 points (with 15 contestants earning silver medals), and the bronze medal cutoff was 17 points (with 48 contestants earning bronze medals).

In this IMO, only one contestant achieved a perfect score of 40 points, namely Sergei Konyagin from the Soviet Union.

Problem 3.7 (IMO 23-3, proposed by the Soviet Union).

Consider an infinite sequence $\{x_n\}$ of positive real numbers with the following properties:

$$x_0 = 1 \quad \text{and} \quad x_{i+1} \le x_i \quad \text{for all } i \ge 0.$$

(a) Prove that for every such sequence, there is an $n \ge 1$ such that

$$\frac{x_0^2}{x_1} + \frac{x_1^2}{x_2} + \cdots + \frac{x_{n-1}^2}{x_n} \ge 3.999.$$

(b) Find such a sequence for which

$$\frac{x_0^2}{x_1} + \frac{x_1^2}{x_2} + \cdots + \frac{x_{n-1}^2}{x_n} < 4.$$

Solution 1. (a) We consider two cases.

(i) If the sequence $\{x_n\}$ does not converge to 0, then there must exist a positive number $\delta > 0$ such that $x_i \ge \delta$ for all i. Therefore, $\frac{x_i^2}{x_{i+1}} \ge \frac{\delta^2}{x_{i+1}} \ge \delta^2$.

Choosing n such that $n\delta^2 \ge 3.999$ is sufficient.

(ii) If $\{x_n\}$ converges to 0, then

$$(x_i - 2x_{i+1})^2 = x_i^2 - 4x_i x_{i+1} + 4x_{i+1}^2 \ge 0,$$

implying $\frac{x_i^2}{x_{i+1}} \ge 4(x_i - x_{i+1})$. Thus,

$$\frac{x_0^2}{x_1} + \frac{x_1^2}{x_2} + \cdots + \frac{x_{n-1}^2}{x_n} \ge 4(x_0 - x_1) + \cdots + 4(x_{n-1} - x_n) = 4 - 4x_n.$$

Since $\lim_{n\to+\infty} x_n = 0$, there exists a positive integer N_0 such that $4x_n \le 0.001$ for $n \ge N_0$. Hence $4 - 4x_n \ge 3.999$.

(b) Choose $x_i = \frac{1}{2^i}$, as

$$\frac{x_0^2}{x_1} + \frac{x_1^2}{x_2} + \cdots + \frac{x_{n-1}^2}{x_n} = 2 + 1 + \frac{1}{2} + \cdots + \frac{1}{2^{n-2}} = 4 - \frac{1}{2^{n-2}} < 4.$$

Solution 2. We construct a positive sequence $\{C_n\}$ such that for each sequence satisfying

$$x_0 \geq x_1 \geq \cdots \geq x_n \geq \cdots > 0,$$

the following inequality holds:

$$\frac{x_0^2}{x_1} + \frac{x_1^2}{x_2} + \cdots + \frac{x_{n-1}^2}{x_n} \geq C_n x_0.$$

Hence,

$$\frac{x_0^2}{x_1} + \left(\frac{x_1^2}{x_2} + \cdots + \frac{x_{n-1}^2}{x_n} + \frac{x_n^2}{x_{n+1}} \right) \geq \frac{x_0^2}{x_1} + C_n x_1 \geq 2\sqrt{C_n} x_0,$$

implying $C_{n+1} = 2\sqrt{C_n}$.

Given that $\frac{x_0^2}{x_1} \geq x_0$, we can choose $C_1 = 1$. Consequently, the general term of $\{C_n\}$ is given by $C_n = 4 \times 2^{-\frac{1}{2^{n-2}}}$ for $n = 1, 2, \ldots$.

To show that $C_n \geq 3.999$, we need to find an $n \geq 1$ such that $\left(\frac{4}{3.999} \right)^{2^{n-2}} \geq 2$. Since

$$\left(\frac{4}{3.999} \right)^{2^{n-2}} > \left(1 + \frac{1}{4000} \right)^{2^{n-2}} > 1 + \frac{2^{n-2}}{4000},$$

when $n = 14$, we have $1 + \frac{2^{12}}{4000} = 1 + \frac{4096}{4000} > 2$.

Therefore, $\frac{x_0^2}{x_1} + \frac{x_1^2}{x_2} + \cdots + \frac{x_{n-1}^2}{x_n} \geq C_n x_0 = C_n > 2$ for $n \geq 14$. The construction for (b) is the same as in Solution 1.

Solution 3. Let $S_n = \frac{x_0^2}{x_1} + \frac{x_1^2}{x_2} + \cdots + \frac{x_{n-1}^2}{x_n}$. Since the sequence $\{x_n\}$ is a non-increasing sequence of positive real numbers,

$$S_2 = \frac{x_0^2}{x_1} + \frac{x_1^2}{x_2} \geq 2x_0 \sqrt{\frac{x_1}{x_2}} \geq 2x_0 = 2,$$

$$S_3 = \frac{x_0^2}{x_1} + \frac{x_1^2}{x_2} + \frac{x_2^2}{x_3} \geq \frac{x_0^2}{x_1}$$

$$+ 2x_1 \sqrt{\frac{x_2}{x_3}} \geq \frac{x_0^2}{x_1} + 2x_1 \geq 2\sqrt{2} x_0 = 2^{1 + \frac{1}{2}},$$

$$S_4 = \frac{x_0^2}{x_1} + \frac{x_1^2}{x_2} + \frac{x_2^2}{x_3} + \frac{x_3^2}{x_4} \geq \frac{x_0^2}{x_1} + \frac{x_1^2}{x_2} + 2x_2$$

$$\geq \frac{x_0^2}{x_1} + 2\sqrt{2}x_1 \geq 2\sqrt{2\sqrt{2}} = 2^{1+\frac{1}{2}+\frac{1}{4}},$$

and so on, leading to the general inequality $S_n \geq 2^{1+\frac{1}{2}+\cdots+\frac{1}{2^{n-2}}}$.

As $\lim_{n\to\infty} \left(1 + \frac{1}{2} + \cdots + \frac{1}{2^{n-2}}\right) = 2$, there exists a positive integer N such that $S_n > 3.999$ for $n > N$.

The construction for (b) is the same as in Solution 1.

Note. (i) For any positive sequence $\{x_n\}$ with $x_0 = 1$ and $x_{i+1} \leq x_i$ for $i \geq 0$, $\sum_{i=0}^{\infty} \frac{x_i^2}{x_{i+1}} \geq 4$ is valid.

(ii) For a positive sequence $\{x_n\}$ with $x_0 = 1$ and $x_{i+1} \leq x_i$ for $i \geq 0$,

$$\sum_{i=0}^{\infty} \frac{x_i^{r+t}}{x_{i+1}^t} \geq \left(\frac{r+t}{t}\right)^{\frac{t}{r}} \cdot \frac{r+t}{r}$$

is true, where $r, t \in \mathbf{R}_+$. The equality is achieved when $x_i = \left(\frac{t}{r+t}\right)^{\frac{i}{r}}$ for $i \geq 0$.

【Score Situation】This particular problem saw the following distribution of scores among contestants: 27 contestants scored 7 points, 5 contestants scored 6 points, 2 contestants scored 5 points, 9 contestants scored 4 points, 8 contestants scored 3 points, 36 contestants scored 2 points, 1 contestant scored 1 point, and 31 contestants scored 0 point. The average score for this problem is 3.042, indicating that it was relatively straightforward.

Among the top five teams in the team scores, the scores of this problem are as follows: the Germany team scored 27 points (with a total team score of 145 points), the Soviet Union team scored 23 points (with a total team score of 137 points), the German Democratic Republic team scored 28 points (with a total team score of 136 points), the United States team scored 27 points (with a total team score of 136 points), and the Vietnam team scored 20 points (with a total team score of 133 points).

The gold medal cutoff for this IMO was set at 37 points (with 10 contestants earning gold medals), the silver medal cutoff was 30 points (with 20 contestants earning silver medals), and the bronze medal cutoff was 21 points (with 31 contestants earning bronze medals).

In this IMO, only three contestants achieved a perfect score of 42 points, namely Bruno Haible from Germany, Grigori Perelman from the Soviet Union, and Lê Tự Quốc Thắng from Vietnam.

Problem 3.8 (IMO 32-6, proposed by the Netherlands). An infinite sequence $x_0, x_1, x_2, \ldots$ of real numbers is said to be bounded if there is a constant C such that $|x_i| \leq C$ for every $i \geq 0$.

Given any real number $a > 1$, construct a bounded infinite sequence $x_0, x_1, x_2, \ldots$ such that

$$|x_i - x_j| \geq |i - j|^{-a}$$

for every pair of distinct non-negative integers i and j.

Solution 1. From the problem requirement, the constructed sequence $\{x_i\}$ should be related to the index i to ensure that the absolute value of the difference between two numbers $|x_i - x_j|(i \neq j)$ is not too small relative to the difference in their indices $|i - j|$.

Here, using binary representation, let i be a non-negative integer represented as

$$i = b_0 + b_1 \cdot 2 + b_2 \cdot 2^2 + \cdots + b_r \cdot 2^r,$$

where $b_0, b_1, \ldots, b_r$ are either 0 or 1. Define

$$y_i = b_0 + b_1 \cdot 2^{-a} + b_2 \cdot 2^{-2a} + \cdots + b_r \cdot 2^{-ra} \quad \text{for } i = 0, 1, 2, \ldots.$$

It is evident that $|y_i| < \frac{1}{1-2^{-a}}$, making $\{y_i\}$ a bounded sequence. Next, estimate $|y_i - y_j|$ for $i \neq j$.

Assume $j = c_0 + c_1 \cdot 2 + c_2 \cdot 2^2 + \cdots + c_s \cdot 2^s$. Since $i \neq j$, there exists a non-negative integer t_0 such that $b_{t_0} \neq c_{t_0}$ and $b_t = c_t$ for $0 \leq t < t_0$. Consequently, $|i - j| \geq 2^{t_0}$, and

$$|y_i - y_j| > 2^{-t_0 a} - \left(2^{-(t_0+1)a} + 2^{-(t_0+2)a} + \cdots \right)$$

$$= 2^{-t_0 a} \cdot \frac{2^a - 2}{2^a - 1}$$

$$\geq |i - j|^{-a} \cdot \frac{2^a - 2}{2^a - 1}.$$

From the above expression, it is evident that choosing the sequence $x_i = \frac{2^a - 1}{2^a - 2} y_i$ for $i \geq 0$ satisfies the specific conditions.

Solution 2. We first prove that if the following lemma holds, then the problem is true for $a = 1$.

Lemma. *For any given real quadratic irrational number α (meaning α is not a rational number and is a root of some quadratic polynomial with integer coefficients), there exists a positive constant c (possibly dependent on α) such that for any integers $p \neq 0$ and q,*

$$|p\alpha - q| \geq \frac{c}{|p|}, \tag{1}$$

or equivalently $\left| \alpha - \frac{q}{p} \right| \geq \frac{c}{p^2}$.

This implies that such irrational numbers α cannot be approximated too closely by rational numbers $\frac{q}{p}$. If the lemma holds, then take the sequence

$$x_i = c^{-1} \left(\alpha i - \lfloor \alpha i \rfloor \right) \quad \text{for } i = 0, 1, 2, \ldots,$$

where $\lfloor t \rfloor$ denotes the greatest integer not exceeding the real number t. Clearly, $0 \leq x_i < c^{-1}$, making $\{x_i\}$ a bounded sequence. From (1), for $i \neq j$,

$$|x_i - x_j| = c^{-1} |\alpha(i - j) - (\lfloor \alpha i \rfloor - \lfloor \alpha j \rfloor)| \geq |i - j|^{-1},$$

which implies that the bounded infinite sequence $\{x_i\}$ satisfies the requirements of the given problem when $a = 1$.

Clearly, $|x_i - x_j| \geq |i - j|^{-1} > |i - j|^{-a}$ for $a > 1$.

Next, we prove that when $\alpha = \sqrt{2}$ and $c = \left(2 + \sqrt{2} \right)^{-1}$, inequality (1) holds for any integers $p \neq 0$ and q.

(i) Assume $1 \leq \frac{q}{p} \leq 2$. From $2p^2 - q^2 \neq 0$,

$$1 \leq |2p^2 - q^2| = \left| \sqrt{2}p + q \right| \cdot \left| \sqrt{2}p - q \right|$$

$$= |p| \cdot \left| \sqrt{2} + \frac{q}{p} \right| \cdot \left| \sqrt{2}p - q \right|$$

$$\leq \left(2 + \sqrt{2} \right) |p| \cdot \left| \sqrt{2}p - q \right|.$$

(ii) Assume $\frac{q}{p} < 1$. It is evident that

$$\left| \sqrt{2} - \frac{q}{p} \right| > \sqrt{2} - 1 > \left(2 + \sqrt{2} \right)^{-1} \geq \frac{(2 + \sqrt{2})^{-1}}{p^2}.$$

(iii) Assume $\frac{q}{p} > 2$. We have

$$\left| \sqrt{2} - \frac{q}{p} \right| > \left| 2 - \sqrt{2} \right| > \left(2 + \sqrt{2} \right)^{-1} \geq \frac{\left(2 + \sqrt{2} \right)^{-1}}{p^2}.$$

Thus, (1) holds. Therefore, we can take

$$x_i = (2 + \sqrt{2})(\sqrt{2}i - \lfloor \sqrt{2}i \rfloor) \quad \text{for } i = 0, 1, 2, \ldots.$$

Solution 3. For a real number $a > 1$, it is evident that there exists a finite constant C_a such that

$$\frac{1}{1^a} + \frac{1}{2^a} + \cdots + \frac{1}{n^a} < C_a \quad \text{for } n = 1, 2, 3, \ldots. \tag{2}$$

The definition of the sequence $x_0, x_1, x_2, \ldots$ will be inductive. Let $x_0 = 0$, fix an integer $k \geq 1$, and assume we have already defined real numbers $x_0, x_1, \ldots, x_{k-1}$ satisfying the conditions

$$|x_i| \leq C_a, |x_i - x_j| \cdot |i - j|^a \geq 1. \tag{3}$$

for $i, j \in \{0, 1, \ldots, k-1\}$ with $i \neq j$. Next, we need to find an x_k satisfying the conditions (3); this leads to

$$|x_k| \leq C_a, \quad |x_k - x_i| \cdot |k - i|^a \geq 1.$$

Thus, $x_k \notin (x_i - \delta_i, x_i + \delta_i)$, where $\delta_i = \frac{1}{(k-i)^a}$ for $i = 0, 1, \ldots, k - 1$.

From (2), consider the open intervals $I_i = (x_i - \delta_i, x_i + \delta_i)$. Their joint length does not exceed $2\delta_0 + 2\delta_1 + \cdots + 2\delta_{k-1}$, which is less than $2C_a$. Indeed, we can find a number $x_k \in [-C_a, C_a]$ that does not belong to any of the intervals $I_0, I_1, \ldots, I_{k-1}$. This means that conditions (3) are satisfied for $i, j \in \{0, 1, \ldots, k\}$ with $i \neq j$, completing the induction step.

Thus, we can obtain an infinite sequence $x_0, x_1, x_2, \ldots$.

Note. Solution 1's construction necessarily requires $a > 1$, while Solution 2 refines the problem by setting $a = 1$, and Solution 2 was proposed by Wei Luo from China.

In addition, the series

$$\sum_{n=1}^{\infty} \frac{1}{n^p} = 1 + \frac{1}{2^p} + \frac{1}{3^p} + \cdots + \frac{1}{n^p} + \cdots$$

in Solution 3 is called a p-series, and we can use

$$\frac{1}{k^p} < \frac{1}{p-1}\left(\frac{1}{(k-1)^{p-1}} - \frac{1}{k^{p-1}}\right), \quad \text{where } k \geq 2 \text{ is an integer,}$$

to prove that its sum is finite for $p > 1$. For $p \leq 1$, this sum is infinity.

Furthermore, for $p > 1$, the sum of the p-series is also known as the Riemann zeta function $\zeta(p)$, which is a monotone decreasing function of p. For any positive even integer $2n$, $\zeta(2n) = \frac{(-1)^{n+1}B_{2n}(2\pi)^{2n}}{2(2n)!}$, where B_{2n} is the $2n$th Bernoulli number.

There are several similar problems:

- **(From the "Problems in Mathematics" in the Hungarian journal *KöMaL*, October 2014, A.624).** (a) Prove that for every infinite sequence $x_1, x_2, \ldots \in [0,1]$ there exists some $C > 0$ such that for every positive integer r, there are positive integers n, m satisfying $|n - m| \geq r$ and $|x_n - x_m| < \frac{C}{|n-m|}$.

  (b) Show that for every $C > 0$ there exists an infinite sequence $x_1, x_2, \ldots \in [0,1]$ and a positive integer r such that $|x_n - x_m| > \frac{C}{|n-m|}$ for every pair n, m of positive integers with $|n - m| \geq r$.

- **(Turkey Team Selection Test 2014, Problem 1).** Find the number of $(a_1, a_2, \ldots, a_{2014})$ permutations of the $(1, 2, \ldots, 2014)$ such that $i + a_i \leq j + a_j$ for all $1 \leq i < j \leq 2014$.

- **(Turkey Team Selection Test 2005, Problem 4).** Show that for any integer $n \geq 2$ and all integers $a_1, a_2, \ldots, a_n$, the product $\prod_{i<j}(a_j - a_i)$ is divisible by $\prod_{i<j}(j - i)$.

- **(IMO Shortlist 2002, Algebra, Problem 2).** Let $a_1, a_2, \ldots$ be an infinite sequence of real numbers, for which there exists a real number c with $0 \leq a_i \leq c$ for all i, such that

$$|a_i - a_j| \geq \frac{1}{i+j} \quad \text{for all} \quad i, j \quad \text{with } i \neq j.$$

  Prove that $c \geq 1$.

- **(All-Russian Mathematical Olympiad 1998, 4th Round, Grade 11, Problem 8).** Given a sequence $\{a_n\}$ $(n \in \mathbf{N})$ of natural numbers such that every natural number occurs and for any distinct natural numbers m and n,

$$\frac{1}{1998} < \frac{|a_n - a_m|}{|n - m|} < 1998,$$

  prove that $|a_n - n| < 2000000$ for all natural numbers n.

- **(Canadian Mathematical Olympiad 1985, Problem 5).** Let $1 < x_1 < 2$ and define $x_{n+1} = 1 + x_n - \frac{1}{2}x_n^2$ for $n = 1, 2, \ldots$. Prove that

$$\left| x_n - \sqrt{2} \right| < \frac{1}{2^n} \quad \text{for } n \geq 3.$$

- **(All Soviet Union Mathematical Olympiad 1978).** Prove that there exists an infinite sequence $\{x_i\}$ such that for all m and all k ($k \neq m$), the inequality $|x_m - x_k| > \frac{1}{|m-k|}$ is valid.

【Score Situation】 This particular problem saw the following distribution of scores among contestants: 50 contestants scored 7 points, 3 contestants scored 6 points, 5 contestants scored 5 points, 10 contestants scored 4 points, 16 contestants scored 3 points, 27 contestants scored 2 points, 29 contestants scored 1 point, and 172 contestants scored 0 point. The average score for this problem is 1.808, indicating that it was relatively challenging.

Among the top five teams in the team scores, the scores of this problem are as follows: the Soviet Union team scored 42 points (with a total team score of 241 points), the China team scored 32 points (with a total team score of 231 points), the Romania team scored 32 points (with a total team score of 225 points), the Germany team scored 36 points (with a total team score of 222 points), and the United States team scored 31 points (with a total team score of 212 points).

The gold medal cutoff for this IMO was set at 39 points (with 20 contestants earning gold medals), the silver medal cutoff was 31 points (with 51 contestants earning silver medals), and the bronze medal cutoff was 19 points (with 84 contestants earning bronze medals).

In this IMO, a total of nine contestants achieved a perfect score of 42 points.

Problem 3.9 (IMO 64-3, proposed by Malaysia). For each integer $k \geq 2$, determine all infinite sequences of positive integers $a_1, a_2, \ldots$ for which there exists a polynomial P of the form

$$P(x) = x^k + c_{k-1}x^{k-1} + \cdots + c_1 x + c_0,$$

where $c_0, c_1, \ldots, c_{k-1}$ are non-negative integers, such that $P(a_n) = a_{n+1}a_{n+2}\cdots a_{n+k}$ for every integer $n \geq 1$.

Solution. All arithmetic sequences with a non-negative common difference meet the conditions.

On the one hand, if $\{a_n\}$ is an arithmetic sequence with a common difference d, then letting $P(x) = (x + d)(x + 2d) \cdots (x + kd)$ meets the conditions.

On the other hand, for any positive integer sequence $a_1, a_2, \ldots, a_n$ and a polynomial $P(x) = x^k + c_{k-1}x^{k-1} + \cdots + c_1 x + c_0$ ($c_0, c_1, \ldots, c_{k-1}$ are non-negative integers) satisfying the equation $P(a_n) = a_{n+1}a_{n+2} \cdots a_{n+k}$, there exists the following claim:

Claim. The sequence $\{a_n\}(n \in \mathbf{Z}_+)$ is non-decreasing.

Proof of the Claim. By contradiction, assume there exists an $m \in \mathbf{Z}_+$ such that $a_m > a_{m+1}$.

From the conditions,

$$\frac{P(a_{m+1})}{P(a_m)} = \frac{a_{m+2}a_{m+3} \cdots a_{m+k+1}}{a_{m+1}a_{m+2} \cdots a_{m+k}} = \frac{a_{m+k+1}}{a_{m+1}}.$$

Since each term of $P(x)$ has non-negative coefficients, $P(a_m) > P(a_{m+1})$. Thus $a_{m+1} > a_{m+k+1}$. Let $m_1 \geq m + 1$ be the smallest positive integer such that $a_{m+1} > a_{m_1+1}$. Then $a_{m_1} \geq a_{m+1} > a_{m_1+1}$. Denote m_1 as $f(m)$, and $f(m)$ is defined only when $a_m > a_{m+1}$. Let $f^{(t)}(m)$ denote $f(f(\cdots f(m) \cdots))$ with f iterated t times. The above analysis implies that $f^{(t)}(m)$ is well-defined for any $t \in \mathbf{Z}_+$, and

$$a_{m+1} > a_{f(m)+1} > a_{f^{(2)}(m)+1} > \cdots .$$

Thus, we have found an infinite strictly decreasing sequence of positive integers $\{a_{f^{(t)}(m)+1}\}(t \in \mathbf{Z}_+)$, where there is a contradiction. Hence the claim holds.

If $P(x) = x^k$, from the claim, $\{a_n\}$ is a constant sequence.

Suppose $P(x) \neq x^k$. For any $n \in \mathbf{Z}_+$ and $1 \leq i \leq k$, we have $a_{n+1}a_{n+2} \cdots a_{n+k} \geq a_{n+i} \cdot a_n^{k-1}$, and

$$a_{n+1}a_{n+2} \cdots a_{n+k} = P(a_n) = a_n^k + a_n^{k-1} \cdot c_{k-1} + \cdots + c_0$$

$$\leq a_n^{k-1} \left(a_n + \sum_{j=0}^{k-1} c_j \right).$$

Therefore, $a_{n+i} \leq a_n + c_0 + c_1 + \cdots + c_{k-1}$. Suppose

$$S_n = (a_{n+1} - a_n, a_{n+2} - a_n, \ldots, a_{n+k} - a_n).$$

Then $S_n \in \{0, 1, \ldots, M\}^k$, where $M = \sum_{j=0}^{k-1} c_j$.

By the Pigeonhole Principle and the claim, there exist $0 \leq d_1 \leq d_2 \leq \cdots \leq d_k$ such that there are infinitely many $n \in \mathbf{Z}_+$ for which $a_{n+i} = a_n + d_i$ for all $1 \leq i \leq k$. For these n, we have $P(a_n) = (a_n + d_1)(a_n + d_2) \cdots (a_n + d_k)$.

Since $P(x) \neq x^k$, it follows that among $a_{n+1}, a_{n+2}, \ldots, a_{n+k}$, there must be one greater than a_n, implying $\{a_n\}$ is unbounded. Therefore, there are infinitely many distinct values of a_n such that $P(a_n) = (a_n + d_1)(a_n + d_2) \cdots (a_n + d_k)$. Hence,

$$P(x) = (x + d_1)(x + d_2) \cdots (x + d_k).$$

If $S_m = (d_1', d_2', \ldots, d_k') \neq (d_1, d_2, \ldots, d_k)$, then there are only finitely many $m \in \mathbf{Z}_+$ satisfying

$$P(a_m) = \prod_{i=1}^{k} a_{m+i} = \prod_{i=1}^{k} (a_m + d_i') = \prod_{i=1}^{k} (a_m + d_i).$$

Hence, there exists $N \in \mathbf{Z}_+$ such that $S_n = \{d_1, d_2, \ldots, d_k\}$ for all $n \geq N$, i.e., for all $n \geq N$ and $1 \leq i \leq k$, it holds that $a_{n+i} = a_n + d_i$. Therefore,

$$d_i = a_{n+i} - a_n = \sum_{j=1}^{i} (a_{n+j} - a_{n+j-1}) = i d_1.$$

Thus, $\{a_n\}(n \geq N)$ is an arithmetic sequence with common difference d_1, implying there exists an integer a such that $a_n = a + n d_1$ for all $n \geq N$. For $n = N - 1$,

$$P(a_{N-1}) = a_N a_{N+1} \cdots a_{N+k-1} = P(a_N - d_1).$$

Since $P(x)$ is monotonically increasing on $(0, +\infty)$, we see that $a_{N-1} = a_N - d_1 = a + (N - 1)d_1$.

By mathematical induction, $a_n = a + n d_1$ for all $n \in \mathbf{Z}_+$.

Note. There are several similar problems:

- **(Turkey Mathematical Olympiad 2020, 2nd Round, Problem 5).** Find all polynomials $P(x)$ with real coefficients such that one can find an integer valued sequence $a_0, a_1, a_2, \ldots$ satisfying $\lfloor P(x) \rfloor = a_{\lfloor x^2 \rfloor}$ for all real numbers x.

  Here $\lfloor t \rfloor$ denotes the greatest integer not exceeding the real number t.

- **(Czech And Slovak Mathematical Olympiad, 3rd Round, Problem 6).** Let $n \geq 1$ be an integer and $a_1, a_2, \ldots, a_n$ be positive integers. Let $f : \mathbf{Z} \to \mathbf{R}$ be a function such that $f(x) = 1$ for each integer $x < 0$ and

$$f(x) = 1 - f(x - a_1)f(x - a_2) \cdots f(x - a_n)$$

for each integer $x \geq 0$. Show that there exist positive integers s and t such that $f(x + t) = f(x)$ for any integer $x > s$.

【Score Situation】 This particular problem saw the following distribution of scores among contestants: 73 contestants scored 7 points, 3 contestants scored 6 points, 6 contestants scored 5 points, 8 contestants scored 4 points, 23 contestants scored 3 points, 7 contestants scored 2 points, 102 contestants scored 1 point, and 396 contestants scored 0 point. The average score of this problem is 1.256, indicating that it was relatively challenging.

Among the top five teams in the team scores, the scores of this problem are as follows: the China team scored 42 points (with a total team score of 240 points), the United States team scored 42 points (with a total team score of 222 points), the South Korea team scored 42 points (with a total team score of 215 points), the Romania team scored 42 points (with a total team score of 208 points), and the Canada team scored 25 points (with a total team score of 183 points).

The gold medal cutoff for this IMO was set at 32 points (with 54 contestants earning gold medals), the silver medal cutoff was 25 points (with 90 contestants earning silver medals), and the bronze medal cutoff was 18 points (with 170 contestants earning bronze medals).

In this IMO, a total of five contestants achieved a perfect score of 42 points.

3.2.3 *Proving quantitative relationships*

Problem 3.10 (IMO 18-6, proposed by the United Kingdom). A sequence $\{u_n\}$ is defined by

$$u_0 = 2, \quad u_1 = \frac{5}{2}, \quad u_{n+1} = u_n(u_{n-1}^2 - 2) - u_1 \quad \text{for } n = 1, 2, \ldots.$$

Prove that for positive integers n,

$$\lfloor u_n \rfloor = 2^{\frac{2^n - (-1)^n}{3}},$$

where $\lfloor x \rfloor$ denotes the greatest integer not exceeding x.

Proof. We prove the following relation using mathematical induction:

$$u_n = 2^{\frac{2^n-(-1)^n}{3}} + 2^{-\frac{2^n-(-1)^n}{3}}. \tag{1}$$

Let $a_n = \frac{2^n-(-1)^n}{3}$. For $n = 0$, we have $u_0 = 2^0 + 2^0$. For $n = 1$, there is $u_1 = 2 + \frac{1}{2}$. The formula (1) holds for $n = 0, 1$. Assume the formula (1) holds for $n \le k$. Then for $n = k+1$,

$$u_{k+1} = (2^{a_k} + 2^{-a_k})((2^{a_{k-1}} + 2^{-a_{k-1}})^2 - 2) - \frac{5}{2}$$

$$= 2^{a_k+2a_{k-1}} + 2^{-a_k+2a_{k-1}} + 2^{a_k-2a_{k-1}} + 2^{-a_k-2a_{k-1}} - \frac{5}{2},$$

and note that

$$a_k + 2a_{k-1} = \frac{1}{3}(2^k - (-1)^k + 2^k - 2(-1)^{k-1})$$

$$= \frac{1}{3}(2^{k+1} - (-1)^{k+1})$$

$$= a_{k+1},$$

$$2a_{k-1} - a_k = \frac{1}{3}(2^k - 2(-1)^{k-1} - 2^k + (-1)^k)$$

$$= (-1)^k,$$

$$a_k - 2a_{k-1} = -(2a_{k-1} - a_k)$$

$$= (-1)^{k+1}.$$

Since $2^{(-1)^k} + 2^{(-1)^{k+1}} - \frac{5}{2} = 0$, there is $u_{k+1} = 2^{a_{k+1}} + 2^{-a_{k+1}}$. As $2^n \equiv (-1)^n \pmod 3$, we see that 2^{a_n} is always an integer. Clearly, $2^{-a_n} < 1$. Therefore, $\lfloor u_n \rfloor = 2^{a_n} = 2^{\frac{2^n-(-1)^n}{3}}$ for $n = 1, 2, \ldots$.

Note. If the expression for $\lfloor u_n \rfloor$ is not given in the problem statement and is asked to find the expression for u_n instead, one can identify a pattern by computing the first few terms of u_n and assume $u_n = 2^{a_n} + 2^{-a_n}$. Substituting this into $u_{n+1} = u_n(u_{n-1}^2 - 2) - u_1$, one can expand and obtain $a_{n+1} = a_n + 2a_{n-1}$, with the associated characteristic equation $x^2 - x - 2 = 0$. This leads to $a_n = \frac{2^n - (-1)^n}{3}$.

【Score Situation】 This particular problem saw the following distribution of scores among contestants: 67 contestants scored 7 points, 18 contestants scored 6 points, 5 contestants scored 5 points, 3 contestants scored 4 points, 4 contestants scored 3 points, 1 contestant scored 2 points, 10 contestants scored 1 point, and 31 contestants scored 0 point. The average score for this problem is 4.590, indicating that it was simple.

Among the top five teams in the team scores, the scores of this problem are as follows: the Soviet Union team scored 45 points (with a total team score of 250 points), the United Kingdom team scored 50 points (with a total team score of 214 points), the United States team scored 43 points (with a total team score of 188 points), the Bulgaria team scored 55 points (with a total team score of 174 points), and the Austria team scored 55 points (with a total team score of 167 points).

The gold medal cutoff for this IMO was set at 34 points (with 9 contestants earning gold medals), the silver medal cutoff was 23 points (with 28 contestants earning silver medals), and the bronze medal cutoff was 15 points (with 45 contestants earning bronze medals).

In this IMO, only one contestant achieved a perfect score of 40 points, namely Laurent Pierre from France.

Problem 3.11 (IMO 26-6, proposed by Sweden). For every real number x_1, construct a sequence $x_1, x_2, \ldots$ by setting

$$x_{n+1} = x_n \left(x_n + \frac{1}{n} \right) \quad \text{for each } n \geq 1.$$

Prove that there exists exactly one value of x_1 for which

$$0 < x_n < x_{n+1} < 1$$

for every n.

Proof 1. To prove $0 < x_n < x_{n+1} < 1$, it suffices to demonstrate $1 - \frac{1}{n} < x_n < 1$. Rewrite the recurrence relation as

$$x_n = \sqrt{x_{n+1} + \frac{1}{4n^2}} - \frac{1}{2n}. \tag{1}$$

If $1 - \frac{1}{n+1} < x_{n+1} < 1$, then evidently

$$x_n = \sqrt{x_{n+1} + \frac{1}{4n^2}} - \frac{1}{2n} < \sqrt{1 + \frac{1}{n} + \frac{1}{4n^2}} - \frac{1}{2n} = 1,$$

$$x_n > \sqrt{1 - \frac{1}{n+1} + \frac{1}{4n^2}} - \frac{1}{2n} > \sqrt{1 - \frac{1}{n} + \frac{1}{4n^2}} - \frac{1}{2n} = 1 - \frac{1}{n}.$$

Therefore, for any positive integer n, if $x_n \in \left(1 - \frac{1}{n}, 1\right)$, then it necessarily follows that

$$x_{n-1} \in \left(1 - \frac{1}{n-1}, 1\right),$$

$$x_{n-2} \in \left(1 - \frac{1}{n-2}, 1\right),$$

$$\cdots\cdots\cdots\cdots\cdots$$

$$x_1 \in (0, 1).$$

It is known that when x_k is given, $x_n (n > k)$ is determined. Consider x_n as a non-negative polynomial function $X_{n,k}(x)$ with the leading coefficient equal to 1 in terms of x_k. Clearly, $X_{n,k}(x)$ is a monotonically increasing function, and $x_n \in \left(1 - \frac{1}{n}, 1\right)$, $X_{n,k}(0) = 0$, and $X_{n,k}(1) \geq 1$. It follows that

$$1 - \frac{1}{k} < a_n < x_k < b_n < 1 \quad \text{for } n = k+1, k+2, \ldots, \tag{2}$$

where a_n and b_n are values of x_k when $x_n = 1 - \frac{1}{n}$ and $x_n = 1$ respectively. From $X_{n,k}(b_n) = 1$,

$$X_{n+1,k}(b_n) = X_{n,k}(b_n) \left(X_{n,k}(b_n) + \frac{1}{n}\right) > 1.$$

As $X_{n+1,k}(x)$ is an increasing function, $X_{n+1,k}(x) = 1$ has a solution $b_{n+1} < b_n$, indicating that $b_{k+1}, b_{k+2}, \ldots$ form a decreasing sequence.

Similarly, $a_{k+1}, a_{k+2}, \ldots$ form an increasing sequence, but always $a_n < b_n$. Hence $\{a_n\}$ has a limit a and $\{b_n\}$ has a limit b with $a \leq b$. To ensure (2) holds, i.e., x_k exists and is unique, it suffices to prove $a = b$. Furthermore,

$$b - a = \lim_{n \to +\infty} b_n - \lim_{n \to +\infty} a_n < \cdots < b_k - a_k < 1 - \left(1 - \frac{1}{k}\right) = \frac{1}{k},$$

so $b - a = 0$ from $k \to +\infty$.

Thus, we obtain a series of open intervals whose lengths tend to zero. For x_k within all open intervals (a_n, b_n), it is unique, which implies that x_1 is also uniquely determined.

Note. This problem is essentially a variation of the Nested Interval Theorem in advanced mathematics.

It is important to note that if $[a_n, b_n]$ is replaced with (a_n, b_n), then the existence of $\xi = \lim_{n \to +\infty} a_n = \lim_{n \to +\infty} b_n$ still holds, but it may not belong to any open interval. For instance, consider $(a_n, b_n) = \left(0, \frac{1}{n}\right)$. Nevertheless, if $a_n < a_{n+1}$ and $b_{n+1} < b_n$ hold for any n, then there still exists a unique ξ within all these open intervals.

Proof 2. Construct a sequence of polynomials $\{P_n(x)\}$ such that

$$P_1(x) = x, \quad P_{n+1}(x) = P_n(x)\left(P_n(x) + \frac{1}{n}\right) \quad \text{for } n \geq 1.$$

It is evident that $P_n(x)$ is a positive coefficient polynomial of degree 2^{n-1}. From Proof 1, the problem can be reduced to proving the existence of a unique positive real number t such that

$$1 - \frac{1}{n} < P_n(t) < 1, \quad n = 1, 2, \ldots.$$

Since $P_n(x)$ is an increasing function for $x \geq 0$ and $P_n(0) = 0$, we can find a unique pair $a_n, b_n (a_n < b_n)$ such that $P_n(a_n) = 1 - \frac{1}{n}$ and $P_n(b_n) = 1$.

Similar to Proof 1, it is known that $a_{n+1} > a_n$ and $b_{n+1} < b_n$. Also, $P_n(x)$ is a positive coefficient polynomial of degree 2^{n-1}, and $P_n''(x)$ is either zero or a positive coefficient polynomial of degree $2^{n-1} - 2$, i.e., $P_n''(x) \geq 0$ for $x \geq 0$. Thus, $P_n(x)$ is a convex function, and the graph of $P_n(x)$ for $0 \leq x \leq b_n$ lies below the line $y = \frac{1}{b_n}x$. Consequently, $P_n(x) \leq \frac{x}{b_n}$ for $0 \leq x \leq b_n$ and

$$P_n(a_n) = 1 - \frac{1}{n} \leq \frac{a_n}{b_n}.$$

Since $0 < b_n \leq 1$, we have $b_n - \frac{b_n}{n} \leq a_n$, implying that $b_n - a_n \leq \frac{b_n}{n} \leq \frac{1}{n}$ for any n.

The subsequent steps are identical to those in Proof 1.

【Score Situation】 This particular problem saw the following distribution of scores among contestants: 23 contestants scored 7 points, 4 contestants scored 6 points, 10 contestants scored 5 points, 9 contestants scored 4 points, 13 contestants scored 3 points, 19 contestants scored 2 points, 74 contestants scored 1 point, and 57 contestants scored 0 point. The average score for this problem is 2.019, indicating that it had a certain level of difficulty.

Among the top five teams in the team scores, the scores of this problem are as follows: the Romania team scored 34 points (with a total team score of 201 points), the United States team scored 40 points (with a total team score of 180 points), the Hungary team scored 25 points (with a total team score of 168 points), the Bulgaria team scored 23 points (with a total team score of 165 points), and the Vietnam team scored 12 points (with a total team score of 144 points).

The gold medal cutoff for this IMO was set at 34 points (with 14 contestants earning gold medals), the silver medal cutoff was 22 points (with 35 contestants earning silver medals), and the bronze medal cutoff was 15 points (with 52 contestants earning bronze medals).

In this IMO, only two contestants achieved a perfect score of 42 points, namely Géza Kós from Hungary and Daniel Tătaru from Romania.

Problem 3.12 (IMO 37-6, proposed by France). Let p, q, and n be three positive integers with $p + q < n$. Let $(x_0, x_1, \ldots, x_n)$ be an $(n+1)$-tuple of integers satisfying the following conditions:

(a) $x_0 = x_n = 0$.

(b) For each i with $1 \le i \le n$, either $x_i - x_{i-1} = p$ or $x_i - x_{i-1} = -q$.

Show that there exist indices $i < j$ with $(i, j) \ne (0, n)$, such that $x_i = x_j$.

Proof. First, assume without loss of generality that $(p, q) = 1$. This is because if $(p, q) = d$ with $p = dp_1$ and $q = dq_1$, then $(p_1, q_1) = 1$. We only need to consider $x_i' = \frac{x_i}{d}$ for $i = 0, 1, \ldots, n$ with $x_i' - x_{i-1}' = p_1$ or $-q_1$. Clearly, $n > p_1 + q_1$ and all conditions are satisfied.

Since $x_i - x_{i-1} = p$ or $-q$ for $i = 1, 2, \ldots, n$, if among these n differences $x_i - x_{i-1}$, a of them are p and b of them are $-q$, then $x_n = ap - bq$, where $a + b = n$. Thus, $ap = bq$.

From $(p, q) = 1$, we can assume $a = kq$ and $b = kp$. Moreover, since $n = a + b = k(p+q)$ and $n > p + q$, it follows that $k \ge 2$. Let $y_i = x_{i+p+q} - x_i$ for $i = 0, 1, \ldots, (k-1)(p+q)$.

As $x_i - x_{i-1} = p$ or $-q \equiv p(\mathrm{mod}(p+q))$, setting $i = 1, 2, \ldots$ and summing up, we obtain

$$x_i \equiv ip(\mathrm{mod}(p+q)).$$

Thus, $x_j - x_i \equiv (j - i)\, p(\mathrm{mod}\,(p+q))$. Particularly, $y_i \equiv 0(\mathrm{mod}\,(p+q))$, meaning each y_i is a multiple of $p + q$. Also

$$y_{i+1} - y_i = (x_{i+1+p+q} - x_{i+p+q}) - (x_{i+1} - x_i),$$

where the numbers in brackets are either p or $-q$, so $y_{i+1} - y_i$ can only be $p + q$, 0, or $-(p+q)$.

From $y_0 + y_{p+q} + y_{2(p+q)} + \cdots + y_{(k-1)(p+q)} = x_n = 0$, consider the sign of y_0. If $y_0 > 0$, then there must be a $y_{i(p+q)} < 0$. If $y_0 < 0$, then there must be a $y_{i(p+q)} > 0$.

Thus, $\frac{y_0}{p+q}, \frac{y_1}{p+q}, \ldots, \frac{y_{n-p-q}}{p+q}$ are all integers, and the difference between the adjacent terms can only be 1, 0, or -1. Moreover, it cannot be all positve or all negative. By the discrete version of the Intermediate Value Theorem, there must be a $y_t = 0$. Taking $i = t$ and $j = t+p+q$ meets the requirement.

【Score Situation】 This particular problem saw the following distribution of scores among contestants: 99 contestants scored 7 points, 10 contestants scored 6 points, 6 contestants scored 5 points, 3 contestants scored 4 points, 15 contestants scored 3 points, 16 contestants scored 2 points, 79 contestants scored 1 point, and 196 contestants scored 0 point. The average score for this problem is 2.243, indicating that it had a certain level of difficulty.

Among the top five teams in the team scores, the scores of this problem are as follows: the Romania team scored 30 points (with a total team score of 187 points), the United States team scored 36 points (with a total team score of 185 points), the Hungary team scored 36 points (with a total team score of 167 points), the Russia team scored 24 points (with a total team score of 162 points), and the United Kingdom team scored 30 points (with a total team score of 161 points).

The gold medal cutoff for this IMO was set at 28 points (with 35 contestants earning gold medals), the silver medal cutoff was 20 points (with 66 contestants earning silver medals), and the bronze medal cutoff was 12 points (with 99 contestants earning bronze medals).

In this IMO, only one contestant achieved a perfect score of 42 points, namely Ciprian Manolescu from Romania.

Problem 3.13 (IMO 50-3, proposed by the United States).

Suppose that $s_1, s_2, s_3, \ldots$ is a strictly increasing sequence of positive integers such that the subsequences

$$s_{s_1}, s_{s_2}, s_{s_3}, \ldots \quad \text{and} \quad s_{s_1+1}, s_{s_2+1}, s_{s_3+1}, \ldots$$

are both arithmetic progressions. Prove that the sequence $s_1, s_2, s_3, \ldots$ is itself an arithmetic progression.

Proof. From the conditions, it is evident that both $s_{s_1}, s_{s_2}, s_{s_3}, \ldots$ and $s_{s_1+1}, s_{s_2+1}, s_{s_3+1}, \ldots$ are strictly increasing sequences of positive integers. Let $s_{s_k} = a + (k-1)d_1$ and $s_{s_k+1} = b + (k-1)d_2$ for $k = 1, 2, \ldots$, where a, b, d_1, d_2 are positive integers. From $s_k < s_k + 1 \le s_{k+1}$ and the monotonicity of $\{s_n\}$, for any positive integer k,

$$s_{s_k} < s_{s_k+1} \le s_{s_{k+1}},$$

i.e.,

$$a + (k-1)d_1 < b + (k-1)d_2 \le a + kd_1,$$

which can be further simplified to

$$a - b < (k-1)(d_2 - d_1) \le a + d_1 - b.$$

From the arbitrariness of k, it follows that $d_2 - d_1 = 0$. Let $d = d_2 = d_1$, and $b - a = c \in \mathbf{N}_+$.

If $d = 1$, then $s_{s_{k+1}} = s_{s_k} + 1 \le s_{s_k+1}$ by the monotonicity of $\{s_n\}$. Hence $s_{k+1} \le s_k + 1$. Since $s_{k+1} > s_k$, it follows that $s_{k+1} = s_k + 1$, i.e., $\{s_n\}$ is an arithmetic sequence, and the conclusion holds.

Now, assume $d > 1$, and we prove that $s_{k+1} - s_k = c$ for any positive integer k. If not, then there are two cases:

Case 1: There exists a positive integer k such that $s_{k+1} - s_k < c$.

Since $s_{k+1} - s_k$ can only take integer values, there must be a minimum among them; assume $s_{i+1} - s_i = c_0$ is the minimum. Then,

$$
\begin{aligned}
s_{a+id} - s_{a+(i-1)d+1} &= s_{s_{s_{i+1}}} - s_{s_{s_i}+1} \\
&= (a + (s_{i+1} - 1)d) - (b + (s_i - 1)d) \\
&= c_0 d - c. \quad\quad (1)
\end{aligned}
$$

On the other hand, from $(a + id) - (a + (i-1)d + 1) = d - 1$,

$$s_{a+id} - s_{a+(i-1)d+1} \ge c_0(d-1)$$

(using the minimality of c_0). Comparing with (1), we get $c_0 \ge c$, a contradiction.

Case 2: There exists a positive integer k such that $s_{k+1} - s_k > c$.

Since $s_{k+1} - s_k$ can only take integer values, and for a given k,

$$s_{k+1} - s_k \le s_{s_{k+1}} - s_{s_k} = d,$$

there must be a maximum among them; assume $s_{j+1} - s_j = c_1$ is the maximum. Then,

$$
\begin{aligned}
s_{a+jd} - s_{a+(j-1)d+1} &= s_{s_{s_{j+1}}} - s_{s_{s_j}+1} \\
&= (a + (s_{j+1} - 1)d) - (b + (s_j - 1)\,d) \\
&= c_1 d - c. \quad\quad (2)
\end{aligned}
$$

On the other hand, from $(a + jd) - (a + (j-1)d + 1) = d - 1$,

$$s_{a+jd} - s_{a+(j-1)d+1} \leq c_1(d-1)$$

(using the maximality of c_1). Comparing with (2), we get $c_1 \leq c$, a contradiction.

Hence, $s_{k+1} - s_k = c$ for any positive integer k, i.e., $\{s_n\}$ is an arithmetic sequence.

Note. This problem can also be approached from the perspective of functions. The problem is equivalent to: Given that $f : \mathbf{Z}_+ \to \mathbf{Z}_+$ is a strictly increasing function, $b_n = f(f(n))$ is an arithmetic sequence, and $c_n = f(f(n) + 1)$ is also an arithmetic sequence. Prove that $a_n = f(n)$ is an arithmetic sequence.

【Score Situation】This particular problem saw the following distribution of scores among contestants: 51 contestants scored 7 points, 2 contestants scored 6 points, 5 contestants scored 5 points, 2 contestants scored 4 points, 5 contestants scored 3 points, 16 contestants scored 2 points, 127 contestants scored 1 point, and 357 contestants scored 0 point. The average score for this problem is 1.019, indicating that it was relatively challenging.

Among the top five teams in the team scores, the scores of this problem are as follows: the China team scored 42 points (with a total team score of 221 points), the Japan team scored 34 points (with a total team score of 212 points), the Russia team scored 31 points (with a total team score of 203 points), the South Korea team scored 22 points (with a total team score of 188 points), and the North Korea team scored 24 points (with a total team score of 183 points).

The gold medal cutoff for this IMO was set at 32 points (with 49 contestants earning gold medals), the silver medal cutoff was 24 points (with 98 contestants earning silver medals), and the bronze medal cutoff was 14 points (with 135 contestants earning bronze medals).

In this IMO, only two contestants achieved a perfect score of 42 points, namely Dongyi Wei from China and Makoto Soejima from Japan.

Problem 3.14 (IMO 51-6, proposed by Iran). Let $a_1, a_2, a_3, \ldots$ be a sequence of positive real numbers. Suppose that for some positive integer s,

$$a_n = \max\{a_k + a_{n-k} | 1 \leq k \leq n - 1\} \tag{1}$$

for all $n > s$. Prove that there exist positive integers l and N with $l \leq s$, such that $a_n = a_l + a_{n-l}$ for all $n \geq N$.

Proof. By the given conditions, a_n can be expressed as $a_n = a_{j_1} + a_{j_2}$ for each $n > s$, where $j_1, j_2 < n$ and $j_1 + j_2 = n$. If $j_1 > s$, then a_{j_1} can be further expressed as the sum of two terms in the sequence, and so on. Eventually, a_n can be represented as

$$a_n = a_{i_1} + \cdots + a_{i_k}, \tag{2}$$

$$1 \leq i_j \leq s, \quad i_1 + \cdots + i_k = n. \tag{3}$$

Let a_{i_1}, a_{i_2} be the two terms at the last step in expressing a_n in the form (2). Then $i_1 + i_2 > s$, and (3) becomes

$$1 \leq i_j \leq s, \quad i_1 + i_2 + \cdots + i_k = n, \quad i_1 + i_2 > s. \tag{4}$$

On the other hand, when indices $i_1, i_2, \ldots, i_k$ satisfy (4), let $s_j = i_1 + i_2 + \cdots + i_j$. From (1),

$$a_n = a_{s_k} \geq a_{s_{k-1}} + a_{i_k} \geq a_{s_{k-2}} + a_{i_{k-1}} + a_{i_k} \geq \cdots \geq a_{i_1} + a_{i_2} + \cdots + a_{i_k}.$$

Therefore, for each $n > s$,

$$a_n = \max \left\{ a_{i_1} + a_{i_2} + \cdots + a_{i_k} \mid (i_1, \ldots, i_k) \text{ satisfies (4)} \right\}.$$

Define $m = \max \left\{ \frac{a_i}{i} \mid 1 \leq i \leq s \right\}$, and let $l \leq s$ be a fixed positive integer such that $m = \frac{a_l}{l}$. Construct the sequence $\{b_n\}$:

$$b_n = a_n - mn \quad \text{for } n = 1, 2, \ldots,$$

so that $b_l = 0$. When $n \leq s$, by the definition of m, we have $b_n \leq 0$. When $n > s$,

$$
\begin{aligned}
b_n &= a_n - mn \\
&= \max \left\{ a_k + a_{n-k} \mid 1 \leq k \leq n - 1 \right\} - mn \\
&= \max \left\{ b_k + b_{n-k} + mn \mid 1 \leq k \leq n - 1 \right\} - mn \\
&= \max \left\{ b_k + b_{n-k} \mid 1 \leq k \leq n - 1 \right\} \leq 0,
\end{aligned}
$$

which implies $b_n \leq 0$ for $n = 1, 2, \ldots$, and for $n > s$,

$$b_n = \max\{b_k + b_{n-k} \mid 1 \leq k \leq n - 1\}.$$

If $b_k = 0$ for $k = 1, 2, \ldots, s$, then $b_n = 0$ for any positive integer n. Hence $a_n = mn$ for $n = 1, 2, \ldots$, and so the proposition holds.

If not, let $M = \max\limits_{1 \leq i \leq s} |b_i|$ and $\varepsilon = \min\{|b_i| : 1 \leq i \leq s, \ b_i < 0\}$. Thus, for $n > s$,

$$b_n = \max\{b_k + b_{n-k} | 1 \leq k \leq n - 1\} \geq b_l + b_{n-l} = b_{n-l},$$

leading to $0 \geq b_n \geq b_{n-l} \geq \cdots \geq -M$.

For the sequence $\{b_n\}$, by (2) and (3), each b_n belongs to the set

$$T = \{b_{i_1} + b_{i_2} + \cdots + b_{i_k} : 1 \leq i_1, i_2, \ldots, i_k \leq s\} \bigcap [-M, 0],$$

and T is a finite set. In fact, for any $x \in T$, let $x = b_{i_1} + \cdots + b_{i_k} (1 \leq i_1, \ldots, i_k \leq s)$. Then there are at most $\frac{M}{\varepsilon}$ non-zero terms among b_{i_j}. Otherwise, $x < \frac{M}{\varepsilon}(-\varepsilon) = -M$, a contradiction. Hence x has a finite number of representations of such sum.

Therefore, for each $t \in \{1, 2, \ldots, l\}$, the sequence

$$b_{s+t}, b_{s+t+l}, b_{s+t+2l}, \ldots$$

is increasing and takes a finite number of values, so it becomes constant from some index onwards. Hence, the sequence $\{b_n\}$ becomes periodic with period l from some index N_0 onwards, i.e.,

$$b_n = b_{n-l} = b_l + b_{n-l} \quad \text{for } n > N_0 + l,$$

so $a_n = b_n + mn = (b_l + ml) + (b_{n-l} + m(n-l)) = a_l + a_{n-l}$ for $n > N_0 + l$.

【Score Situation】 This particular problem saw the following distribution of scores among contestants: 15 contestants scored 7 points, 4 contestants scored 6 points, 6 contestants scored 5 points, no contestant scored 4 points, 3 contestants scored 3 points, 4 contestants scored 2 points, 14 contestants scored 1 point, and 470 contestants scored 0 point. The average score for this problem is 0.368, indicating that it was extremely difficult.

Among the top five teams in the team scores, the scores of this problem are as follows: the China team scored 25 points (with a total team score of 197 points), the Russia team scored 20 points (with a total team score of 169 points), the United States team scored 12 points (with a total team score of 168 points), the South Korea team scored 6 points (with a total team score of 156 points), the Kazakhstan team scored 12 points (with a total team score of 148 points), and the Thailand team scored 1 point (with a total team score of 148 points).

The gold medal cutoff for this IMO was set at 27 points (with 47 contestants earning gold medals), the silver medal cutoff was 21 points (with 103 contestants earning silver medals), and the bronze medal cutoff was 15 points (with 115 contestants earning bronze medals).

In this IMO, only one contestant achieved a perfect score of 42 points, namely Zipei Nie from China.

3.3 Summary

Through various definitions, we can derive many interesting sequences that are not only simple in form but also widely applicable. To a certain extent, sequences also bridge secondary school mathematics with university-level mathematics. For instance, the concepts of series and limits in caculus are closely related to sequences.

In the first 64 IMOs, there were a total of 14 sequence problems. These problems can be broadly categorized into three types, as depicted in Figure 3.1. The score details for these problems are presented in Table 3.2. Due to the smaller number of participating teams and missing contestant score information in early IMOs, there are several blanks in Table 3.2.

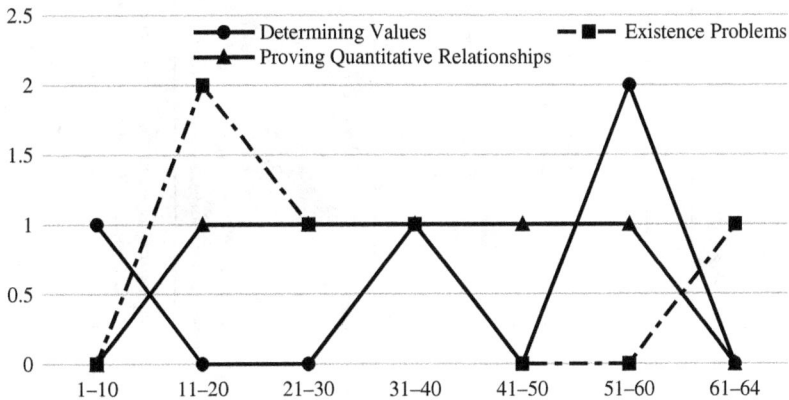

Figure 3.1 Numbers of Sequence Problems in the First 64 IMOs

Problems 3.1–3.4 focus on "determining the value of a specific term or the number of terms;" among these four problems, the one with the lowest average score is Problem 3.4 (IMO 59-2), proposed by Slovakia. Problems 3.5–3.9 deal with "addressing problems related to the existence of sequences;" among these five problems, the one with the lowest average score is Problem 3.6 (IMO 15-6), proposed by Sweden. Problems 3.10–3.14 are about "proving quantitative relationships satisfied by sequences;" among these five problems, the one with the lowest average score is Problem 3.14 (IMO 51-6), proposed by Iran.

Table 3.2 Score Details of Sequence Problems in the First 64 IMOs

Problem	3.1	3.2	3.3	3.4	3.5	3.6	3.7	3.8
Full points	7.000	7.000	7.000	7.000	8.000	8.000	7.000	7.000
Average score	3.568	4.592	5.943	2.946	1.514	1.128	3.042	1.808
Top five mean		6.800	7.000	6.400		2.200	6.250	5.767
6th–15th mean		6.433	6.909	6.150		0.663	3.925	3.233
16th–25th mean		5.967	6.778	5.133			1.750	1.833
Problem number in IMO	9-5	36-4	58-1	59-2	12-3	15-6	23-3	32-6
Proposing country	The Soviet Union	Poland	South Africa	Slovakia	Sweden	Sweden	The Soviet Union	The Netherlands

Problem	3.9	3.10	3.11	3.12	3.13	3.14
Full points	7.000	7.000	7.000	7.000	7.000	7.000
Average score	1.256	4.590	2.019	2.243	1.019	0.368
Top five mean	6.433	6.200	4.467	5.200	5.100	2.111
6th–15th mean	3.883	4.413	2.800	3.967	2.470	0.907
16th–25th mean	2.267		1.233	3.697	2.100	0.567
Problem number in IMO	64-3	18-6	26-6	37-6	50-3	51-6
Proposing country	Malaysia	The United Kingdom	Sweden	France	The United States	Iran

Note. Top five mean = Total score of the top five teams ÷ Total number of contestants from the top five teams,
6th–15th mean = Total score of the 6th–15th teams ÷ Total number of contestants from the 6th–15th teams,
16th–25th mean = Total score of the 16th–25th teams ÷ Total number of contestants from the 16th–25th teams.

These 14 problems were proposed by 11 countries. Sweden proposed three problems, while the Soviet Union contributed two problems.

From Table 3.2, it can be observed that in the first 64 IMOs, there was one sequence problem with an average score of 0–1 point; five problems with an average score of 1–2 points; three problems with an average score of 2–3 points; two problems with an average score of 3–4 points; three problems with an average score above 4 points. Overall, the sequence problems tended to be relatively difficult.

In the 24th–64th IMOs, there were a total of nine sequence problems. Among these, one had an average score of 0–1 point; three had an average score of 1–2 points; three had an average score of 2–3 points; no problem had an average score of 3–4 points; two had an average score above 4 points. Further analysis of the problem numbers of these nine sequence problems, as shown in Table 3.3, reveals that these problems frequently appeared as the 3rd/6th problem, with a relatively even distribution across different types of problems.

Table 3.3 Numbers of Sequence Problems in the 24th–64th IMOs

Sequence Problem	Problem Number			Number of Problems in the First 64 IMOs
	1, 4	2, 5	3, 6	
Determining values	2	1	0	4
Existence problems	0	0	2	5
Proving quantitative relationships	0	0	4	5
Total	2	1	6	14

From Table 3.2, it can be observed that in the sequence problems, the average score of the top five teams is generally about 3 points higher than the average score of the problem. However, in simpler problems, the differences in average scores among the top five teams, 6th–15th teams, and 16th–25th teams are not significant, as seen in problems such as Problem 3.2 (IMO 36-4), Problem 3.3 (IMO 58-1), and Problem 3.4 (IMO 59-2).

From Table 3.2, it can also be observed that in the sequence problems of the first 32 IMOs, the average score of the 16th–25th teams is close to or below the average score of the problem, such as Problem 3.7 (IMO 23-3), Problem 3.8 (IMO 32-6), and Problem 3.11 (IMO 26-6). However, after the 32nd IMO, the average score of the 16th–25th teams is about 1.5 points higher than the average score of the problem. This phenomenon is due to the smaller number of participating teams in early IMOs. It was not until the 30th IMO in 1989 that the number of participating teams exceeded 50. Therefore, during this period, it was common to see situations where the average score was close to or even higher than the average score of the 16th–25th teams.

Chapter 4

Inequality Problems

American mathematician Loren C. Larson once stated, "Inequalities are useful in virtually all areas of mathematics, and inequality problems are among the most beautiful." Indeed, mathematics, as a discipline studying the quantitative relationships and spatial forms of the real world, often manifests these relationships through inequalities, and they hold true in most mathematical domains.

The study of inequalities has a long history, primarily originating from geometry, with examples such as the triangle inequality, AM-GM inequality for $n = 2$, isoperimetric inequality, and others. Notably, the isoperimetric inequality was known to Archimedes and even earlier Greek mathematicians before the common era, but its first rigorous proof did not appear until the 19th century, over two millennia later. In particular, estimating the value of π was one of the significant contributions of inequalities to ancient mathematics.

However, before Newton and Cauchy, the development of inequalities progressed slowly, and only a few were known. Newton discovered $p_{r-1}p_{r+1} \leq p_r^2$, where $p_r = \sqrt[r]{\frac{\sum_{1 \leq a_1 < a_2 < \cdots < a_r \leq n} a_1 a_2 \cdots a_r}{C_n^r}}$, and Cauchy used induction to prove the AM-GM inequality. Subsequently, many mathematicians have studied inequalities, leading to numerous famous inequalities named after them.

Two significant events were noteworthy in the development of inequalities. One was the introduction of the Chebyshev inequality by mathematician P. L. Chebyshev in his 1882 paper. It states

$$\int_a^b f(x)g(x)p(x)\mathrm{d}x \int_a^b p(x)\mathrm{d}x \geq \int_a^b f(x)p(x)\mathrm{d}x \int_a^b g(x)p(x)\mathrm{d}x,$$

where $f(x), g(x), p(x)$ are integrable functions on $[a, b]$, with $p(x) \geq 0$, and $f(x), g(x)$ are both increasing or decreasing on $[a, b]$. Chebyshev also applied this inequality to probability theory.

The other pivotal moment was during the 1928 resignation speech of renowned mathematician Godfrey Harold Hardy as the president of the London Mathematical Society. He emphasized the elegance of inequalities and his fondness for them, leading to a high proportion of inequality-related papers in the society's journal. He proposed that mastering elementary inequalities is one of the necessary skills for the study of function theory. Along with John Edensor Littlewood and George Pólya, Hardy co-authored "Inequalities," which cataloged some of the famous historical inequalities and their origins, significantly impacting the field's development.

Subsequently, the study of inequalities gradually formed into a systematic scientific theory, significantly influencing other mathematical domains, and led to numerous publications on inequalities. Notably, the *Journal of Inequalities and Applications* was founded in 1997, and *Mathematical Inequalities and Applications* in 1998. These academic journals marked a new phase in the development of inequalities.

Due to the significant value and impact of inequalities, they frequently appear in various mathematics competitions. For example, inequality problems in the IMO are often special cases of some famous inequalities, with their difficulty significantly reduced compared to the original famous inequalities. Moreover, these problems typically have special, simpler solutions.

In the first 64 IMOs, there had been a total of 33 inequality problems, approximately accounting for 32.7% of all algebra problems. These problems can be primarily categorized into three types: (1) solving inequalities, totaling five problems; (2) proving inequalities, totaling 25 problems; (3) determining value ranges, totaling three problems. The statistical distribution of these three types of problems in the previous IMOs is presented in Table 4.1.

In the 11th–20th, 31st–40th, and 41st–50th IMOs, the proportion of inequality problems among all algebra problems exceeded 30%, especially in the 41st–50th IMOs, where the proportion of inequality problems surpassed 50%, making it a focal topic of the IMO. It can also be observed from the table that in early IMOs, there were problems involving solving inequalities, but this type of problems gradually disappeared over the time. However, the type of proving inequalities continued to thrive.

Remarkably, in recent IMOs, inequality problems with parameters have emerged, and this trend is evident in some National Mathematical

Table 4.1 Numbers of Inequality Problems in the First 64 IMOs.

Content	Session							Total
	1–10	11–20	21–30	31–40	41–50	51–60	61–64	
Solving inequalities	3	1	1	0	0	0	0	5
Proving inequalities	2	5	3	3	7	2	3	25
Determining value ranges	0	1	0	1	1	0	0	3
Algebra problems	20	20	14	13	15	13	6	101
The percentage of inequality problems among the algebra problems	25.0%	35.0%	28.6%	30.8%	53.3%	15.4%	50.0%	32.7%

Olympiads. Also, the difficulty and flexibility of proving inequalities have increased.

Among the inequality-proving problems, nine were related to three-variable inequalities, with four appearing in the 41st–50th IMOs. After the 51st IMO, such problems no longer appeared. The primary reason for this is the relatively mature methods available for proving three-variable inequalities, while n-variable inequalities pose more challenges.

Additionally, cyclic symmetric inequalities frequently appear in the IMO, where the assumption of symmetry plays a crucial role in these problems. When encountering problems with uncertain equality conditions, it is advisable to first seek simplification methods, such as adjusting to make the smallest number zero, or using partial derivatives to deduce certain properties.

This chapter will be divided into three parts. The first part introduces some famous inequalities, including the AM-GM inequality, Cauchy–Schwarz inequality, Hölder's inequality, rearrangement inequality, Chebyshev's inequality, etc. Simultaneously, there is a certain explanation of the applications of these inequalities. Following that, common methods for proving inequalities are presented, such as the comparison method, adjustment method, SOS method, local inequality method, etc.

The second part revolves around three types of problems: "solving inequalities," "proving inequalities," and "determining value ranges." These problems are presented in chronological order, and some problems include various solutions, generalizations, and similar problems.

It is important to note that for each problem, the solutions are followed by information on the scores, including the number of contestants in each score range, the average score, and the scores of the top five teams. However, early IMOs often lacked information on contestant scores, so the number of contestants in each score range only represents the counted number of contestants, and some problems lack scores of the top five teams.

The third part provides a brief summary of this chapter.

Inequalities come in a wide variety, and while this chapter introduces some methods for proving inequalities, ranging from straightforward basic approaches to techniques that require manipulation of the inequalities, this book is not a compendium of problem-solving methods, nor can it enumerate all possible strategies. Often, the most practical and effective methods are those discovered and understood by readers through their own exploration. Therefore, when appreciating problems from past IMOs, readers are encouraged to ponder the problems on their own before consulting the solutions. It's possible that the readers' own solutions may even be more elegant.

4.1 Common Theorems, Formulas, and Methods

4.1.1 *Solving inequalities*

Solving inequalities is similar to solving equations, where there might be extraneous solutions or potential loss of solutions. Therefore, while solving inequalities, it is essential to apply equivalent transformations. When dealing with inequalities involving radicals, exponentials, and logarithms, it is crucial to leverage the properties of functions.

For solving polynomial inequalities of one variable of higher degree, the "root plotting method" can be employed. Generally, consider a polynomial

$$f(x) = (x - a_1)(x - a_2) \cdots (x - a_n),$$

with its n real roots ordered as $a_1 < a_2 < \cdots < a_n$. The number line is divided into $n + 1$ intervals:

$$(-\infty, a_1), (a_1, a_2), \ldots, (a_{n-1}, a_n), (a_n, +\infty).$$

Counting these intervals from right to left, we obtain the following:

(i) $f(x) > 0$ in odd-numbered intervals;
(ii) $f(x) < 0$ in even-numbered intervals.

4.1.2 *Mean inequalities and the Cauchy–Schwarz inequality*

(1) *Mean inequalities*

Let $a_1, a_2, \ldots, a_n$ be n positive real numbers. Define

$$H_n = \frac{n}{\frac{1}{a_1} + \frac{1}{a_2} + \cdots + \frac{1}{a_n}}, \quad G_n = \sqrt[n]{a_1 a_2 \cdots a_n},$$

$$A_n = \frac{a_1 + a_2 + \cdots + a_n}{n}, \quad Q_n = \sqrt{\frac{a_1^2 + a_2^2 + \cdots + a_n^2}{n}},$$

where H_n, G_n, A_n, Q_n respectively represent the harmonic mean, geometric mean, arithmetic mean, and quadratic mean of these n numbers. These four means are related as follows:

$$H_n \leq G_n \leq A_n \leq Q_n,$$

with equality holding if and only if $a_1 = a_2 = \cdots = a_n$.

The inequality $a_1 + a_2 + \cdots + a_n \geq n \sqrt[n]{a_1 a_2 \cdots a_n}$, also known as the AM-GM inequality, is sometimes utilized in proving inequalities through a skillful manipulation of "terms" and "coefficients." Generally, the structure of the inequality determines the structure of the terms, while the condition for the equality determines the coefficients of the terms.

Example 4.1. Given $x, y, z > 0$ with $x + y + z = 3$, prove that

$$\frac{x^2}{(y+z)^3} + \frac{y^2}{(z+x)^3} + \frac{z^2}{(x+y)^3} \geq \frac{3}{8}.$$

Proof. By the AM-GM inequality,

$$\frac{1}{16} x(y+z) + \frac{1}{32} (y+z)^2 + \frac{x^2}{(y+z)^3} \geq \frac{3}{8} x,$$

and similarly for the other two inequalities. Therefore,

$$\frac{x^2}{(y+z)^3} + \frac{y^2}{(z+x)^3} + \frac{z^2}{(x+y)^3}$$

$$\geq \frac{3}{8}(x+y+z) - \frac{2(x^2 + y^2 + z^2) + 6(xy + yz + zx)}{32}.$$

Since $3(xy + yz + zx) \leq (x + y + z)^2$,

$$2(x^2 + y^2 + z^2) + 6(xy + yz + zx)$$

$$= 2(x+y+z)^2 + 2(xy + yz + zx) \leq 2(x+y+z)^2 + \frac{2}{3}(x+y+z)^2.$$

Substituting $x + y + z = 3$ into the above, the proposition holds.

When dealing with inequality problems, elimination and degree reduction are commonly used techniques. In this case, two terms $x(y+z)$ and $(y+z)^2$ are added to the original inequality, allowing the application of the AM-GM inequality. This approach eliminates the denominator while also utilizing the condition $x+y+z=3$. However, if only one term $(y+z)^3$ is added, leading to

$$\frac{1}{64}(y+z)^3 + \frac{x^2}{(y+z)^3} \geq \frac{1}{4}x,$$

then it becomes challenging to handle after rearrangement.

The coefficients for the two terms $x(y+z)$ and $(y+z)^2$ are determined by the equality condition $x=y=z=1$.

Furthermore, there are several common corollaries to the AM-GM inequality:

Corollary 4.1. *Let a, b, c, d be non-negative real numbers. Then*

$$ab + bc + cd + da \leq \frac{(a+b+c+d)^2}{4}.$$

Corollary 4.2. *Let a, b, c, d be non-negative real numbers. Then*

$$abc + bcd + cda + dab \leq \frac{(a+b+c+d)^3}{16}.$$

Corollary 4.3. *Given an integer $n \geq 4$, let $x_1, x_2, \ldots, x_n$ be non-negative real numbers. Then*

$$x_1 x_2 + x_2 x_3 + \cdots + x_n x_1 \leq \frac{(x_1 + x_2 + \cdots + x_n)^2}{4}.$$

Proof.   Without loss of generality, assume $x_1 = \max\{x_1, x_2, \ldots x_n\}$. Then

$$x_1 x_2 + x_2 x_3 + \cdots + x_n x_1 \leq x_1 x_2 + x_2 x_3 + x_3 x_4 + x_1(x_4 + x_5 + \cdots + x_n)$$
$$= x_1(x_2 + x_4 + x_5 + \cdots + x_n) + x_3(x_2 + x_4)$$
$$\leq (x_1 + x_3)(x_2 + x_4 + x_5 + \cdots + x_n)$$
$$\leq \left(\frac{x_1 + x_2 + \cdots + x_n}{2}\right)^2.$$

Example 4.2. Given an integer $n \geq 3$, let $a_1, a_2, \ldots, a_{2n}$ and $b_1, b_2, \ldots, b_{2n}$ be $4n$ non-negative real numbers satisfying

$$a_1 + a_2 + \cdots + a_{2n} = b_1 + b_2 + \cdots + b_{2n} > 0,$$

and $a_i a_{i+2} \geq b_i + b_{i+1}$ for any $i = 1, 2, \ldots, 2n$, where $a_{2n+1} = a_1$, $a_{2n+2} = a_2$, and $b_{2n+1} = b_1$.

Find the minimum value of $a_1 + a_2 + \cdots + a_{2n}$.

Solution. For $n = 3$,

$$(a_1 + a_3 + a_5)^2 \geq 3(a_1 a_3 + a_3 a_5 + a_5 a_1) \geq 3\sum_{i=1}^{6} b_i = 3\sum_{i=1}^{6} a_i,$$

$$(a_2 + a_4 + a_6)^2 \geq 3(a_2 a_4 + a_4 a_6 + a_6 a_2) \geq 3\sum_{i=1}^{6} b_i = 3\sum_{i=1}^{6} a_i.$$

Multiplying these two inequalities and taking the square root, we obtain

$$3\sum_{i=1}^{6} a_i \leq (a_1 + a_3 + a_5)(a_2 + a_4 + a_6) \leq \left(\frac{a_1 + a_2 + \cdots + a_6}{2}\right)^2,$$

and hence $\sum_{i=1}^{6} a_i \geq 12$. The equality holds when $a_i = b_i = 2$ for $i = 1, 2, 3, 4, 5, 6$.

For $n \geq 4$, from Corollary 4.1.3,

$$\frac{(a_1 + a_3 + \cdots + a_{2n-1})^2}{4} \geq a_1 a_3 + a_3 a_5 + \cdots + a_{2n-1} a_1 \geq \sum_{i=1}^{2n} b_i = \sum_{i=1}^{2n} a_i,$$

$$\frac{(a_2 + a_4 + \cdots + a_{2n})^2}{4} \geq a_2 a_4 + a_4 a_6 + \cdots + a_{2n} a_2 \geq \sum_{i=1}^{2n} b_i = \sum_{i=1}^{2n} a_i.$$

Multiplying these two inequalities and taking the square root, we get

$$\sum_{i=1}^{2n} a_i \leq \frac{(a_1 + a_3 + \cdots + a_{2n-1})(a_2 + a_4 + \cdots + a_{2n})}{4}$$

$$\leq \frac{(a_1 + a_2 + \cdots + a_{2n})^2}{16}.$$

Therefore $\sum_{i=1}^{2n} a_i \geq 16$. The equality holds when

$$a_1 = a_2 = a_3 = a_4 = 4, \quad a_i = 0(i = 5, 6, \ldots, 2n), \quad b_2 = 16,$$

$$b_i = 0(i \neq 2).$$

This was the second problem in the 2020 Chinese High School Mathematics League, illustrating a flexible application of the AM-GM inequality and related corollaries can greatly simplify problems. However, in some cases, it is necessary to transform the inequality first before using the AM-GM inequality.

Example 4.3. Given $a, b, c > 0$ with $a + b + c = 3$, prove that

$$\frac{a}{1+b^2} + \frac{b}{1+c^2} + \frac{c}{1+a^2} \geq \frac{3}{2}.$$

In fact, directly applying the AM-GM inequality to the denominators is not feasible, because it changes the direction of the inequality, i.e.,

$$\frac{a}{1+b^2} + \frac{b}{1+c^2} + \frac{c}{1+a^2} \leq \frac{a}{2b} + \frac{b}{2c} + \frac{c}{2a} \geq \frac{3}{2}.$$

If we transform the inequality first, then

$$\frac{a}{1+b^2} = a - \frac{ab^2}{1+b^2} \geq a - \frac{ab^2}{2b} = a - \frac{ab}{2}.$$

Since $3(ab + bc + ca) \leq (a + b + c)^2 = 9$,

$$\frac{a}{1+b^2} + \frac{b}{1+c^2} + \frac{c}{1+a^2} \geq a + b + c - \frac{ab + bc + ca}{2} \geq \frac{3}{2}.$$

This method of initially altering the sign of the expression and then applying the AM-GM inequality to the denominator or similar structures without changing the direction of the inequality is also known as the Cauchy Reverse Technique. Similarly, this technique can be used to solve the following problem:

Given $a, b, c, d > 0$ with $a + b + c + d = 4$, prove that

$$\frac{a}{1+b^2c} + \frac{b}{1+c^2d} + \frac{c}{1+d^2a} + \frac{d}{1+a^2b} \geq 2.$$

(2) *Weighted AM-GM inequality*

Let $a_1, a_2, \ldots, a_n$ and $\lambda_1, \lambda_2, \ldots, \lambda_n$ be positive real numbers, where $\sum_{i=1}^{n} \lambda_i = \lambda$. Then

$$\sqrt[\lambda]{a_1^{\lambda_1} a_2^{\lambda_2} \cdots a_n^{\lambda_n}} \leq \frac{\lambda_1 a_1 + \lambda_2 a_2 + \cdots + \lambda_n a_n}{\lambda},$$

with equality holding if and only if $a_1 = a_2 = \cdots = a_n$.

(3) Power mean inequality

Let $a_1, a_2, \ldots, a_n$ be n positive real numbers, and $\alpha > \beta > 0$. Then

$$\left(\frac{a_1^\beta + a_2^\beta + \cdots + a_n^\beta}{n} \right)^{\frac{1}{\beta}} \leq \left(\frac{a_1^\alpha + a_2^\alpha + \cdots + a_n^\alpha}{n} \right)^{\frac{1}{\alpha}},$$

with equality holding if and only if $a_1 = a_2 = \cdots = a_n$.

(4) Cauchy–Schwarz inequality

Let $a_1, a_2, \ldots, a_n, b_1, b_2, \ldots, b_n$ be real numbers. Then

$$(a_1 b_1 + a_2 b_2 + \cdots + a_n b_n)^2 \leq (a_1^2 + a_2^2 + \cdots + a_n^2)(b_1^2 + b_2^2 + \cdots + b_n^2),$$

and the equality holds if and only if $b_i = \lambda a_i$ for $i = 1, 2, \ldots, n$.

The Cauchy–Schwarz inequality is a well-known inequality, which can be extended to complex numbers. In fact, the Cauchy–Schwarz inequality has a clear geometric interpretation, deriving from the cosine of the angle between two vectors in an n-dimensional space or the area of a triangle in an n-dimensional space formed by two vectors with a common initial point.

There are various methods to prove the Cauchy–Schwarz inequality, and one approach is using a quadratic function.

Let $f(x) = (a_1 x - b_1)^2 + (a_2 x - b_2)^2 + \cdots + (a_n x - b_n)^2$, which can be rearranged as

$$f(x) = (a_1^2 + a_2^2 + \cdots + a_n^2)x^2 - 2(a_1 b_1 + a_2 b_2 + \cdots + a_n b_n)x$$
$$+ (b_1^2 + b_2^2 + \cdots + b_n^2).$$

It is clear that $f(x) \geq 0$. Hence its discriminant $\Delta \leq 0$, i.e.,

$$(a_1 b_1 + a_2 b_2 + \cdots + a_n b_n)^2 \leq (a_1^2 + a_2^2 + \cdots + a_n^2)(b_1^2 + b_2^2 + \cdots + b_n^2),$$

and the equality holds if and only if there exists x_0 such that

$$a_1 x_0 - b_1 = a_2 x_0 - b_2 = \cdots = a_n x_0 - b_n = 0.$$

Moreover, the Cauchy–Schwarz inequality has several straightforward corollaries:

Corollary 4.4. *For any real numbers $a_1, a_2, \ldots, a_n$ and positive real numbers $b_1, b_2, \ldots, b_n$,*

$$\frac{a_1^2}{b_1} + \frac{a_2^2}{b_2} + \cdots + \frac{a_n^2}{b_n} \geq \frac{(a_1 + a_2 + \cdots + a_n)^2}{b_1 + b_2 + \cdots + b_n}.$$

Corollary 4.5. *For any real numbers* $a_1, a_2, \ldots, a_n$ *and* $b_1, b_2, \ldots, b_n$,

$$\sqrt{a_1^2 + b_1^2} + \sqrt{a_2^2 + b_2^2} + \cdots + \sqrt{a_n^2 + b_n^2}$$
$$\geq \sqrt{(a_1 + a_2 + \cdots + a_n)^2 + (b_1 + b_2 + \cdots + b_n)^2}.$$

Corollary 4.6. *For any real numbers* $a_1, a_2, \ldots, a_n$,

$$(a_1 + a_2 + \cdots + a_n)^2 \leq n(a_1^2 + a_2^2 + \cdots + a_n^2).$$

In general, problems involving applications of the Cauchy–Schwarz inequality to find the minimum value are quite common, which sometimes leads to difficulty in associating the Cauchy–Schwarz inequality with problems seeking the maximum value.

Example 4.4. Given $a, b, c \geq 0$, prove that

$$\frac{a^2 - bc}{2a^2 + b^2 + c^2} + \frac{b^2 - ca}{2b^2 + c^2 + a^2} + \frac{c^2 - ab}{2c^2 + a^2 + b^2} \geq 0.$$

Proof. It is known that $\frac{a^2 - bc}{2a^2 + b^2 + c^2} = \frac{1}{2}\left(1 - \frac{(b+c)^2}{2a^2 + b^2 + c^2}\right)$, and thus the inequality is equivalent to proving

$$\frac{(b + c)^2}{2a^2 + b^2 + c^2} + \frac{(c + a)^2}{2b^2 + c^2 + a^2} + \frac{(a + b)^2}{2c^2 + a^2 + b^2} \leq 3.$$

Using the Cauchy–Schwarz inequality, we see that

$$\frac{(b + c)^2}{2a^2 + b^2 + c^2} \leq \frac{b^2}{a^2 + b^2} + \frac{c^2}{c^2 + a^2},$$

and similarly for the other two inequalities. Summing up these three inequalities completes the proof.

However, the last step $\frac{(b+c)^2}{2a^2+b^2+c^2} \leq \frac{b^2}{a^2+b^2} + \frac{c^2}{c^2+a^2}$ is easily overlooked.

(5) *Reverse of the Cauchy–Schwarz inequality*

Given real numbers $a_1, a_2, \ldots, a_n$, $b_1, b_2, \ldots, b_n$, and a, A, b, B satisfying

$$0 < a \leq a_i \leq A, \quad 0 < b \leq b_i \leq B, \quad i \in \{1, 2, \ldots, n\},$$

$$\left(\sum_{i=1}^n a_i^2\right)\left(\sum_{i=1}^n b_i^2\right) \leq \frac{1}{4}\left(\sqrt{\frac{AB}{ab}} + \sqrt{\frac{ab}{AB}}\right)^2 \left(\sum_{i=1}^n a_i b_i\right)^2,$$

and the equality holds if and only if $a = a_i = A$ and $b = b_i = B$.

Indeed, the Cauchy–Schwarz inequality is a special case of Hölder's inequality.

(6) *Hölder's inequality*

(i) Given $a_{i,j} > 0$ and $w_i > 0$ for $i \in \{1, 2, \ldots, n\}$, $j \in \{1, 2, \ldots, m\}$, and $\sum_{i=1}^{n} w_i = 1$,

$$\prod_{i=1}^{n} \left(\sum_{j=1}^{m} a_{i,j} \right)^{w_i} \geq \sum_{j=1}^{m} \left(\prod_{i=1}^{n} a_{i,j}^{w_i} \right),$$

and the equality holds if and only if the following n sequences are proportional:

$$(a_{11}, a_{12}, \ldots, a_{1m}), (a_{21}, a_{22}, \ldots, a_{2m}), \ldots, (a_{n1}, a_{n2}, \ldots, a_{nm}).$$

(ii) Given $a_i, b_i > 0$ for $i = 1, 2, \ldots, n$, and positive p, q satisfying $\frac{1}{p} + \frac{1}{q} = 1$,

$$a_1 b_1 + a_2 b_2 + \cdots + a_n b_n \leq (a_1^p + a_2^p + \cdots + a_n^p)^{\frac{1}{p}} (b_1^q + b_2^q + \cdots + b_n^q)^{\frac{1}{q}},$$

where the equality holds if and only if $a_i^p = \lambda b_i^q$ for $i = 1, 2, \ldots, n$.

Furthermore, there are several common corollaries to Hölder's inequality:

Corollary 4.7. *For positive real numbers* $a_1, a_2, \ldots, a_n, b_1, b_2, \ldots, b_n$ *and a positive integer* m,

$$\frac{a_1^{m+1}}{b_1^m} + \frac{a_2^{m+1}}{b_2^m} + \cdots + \frac{a_n^{m+1}}{b_n^m} \geq \frac{(a_1 + a_2 + \cdots + a_n)^{m+1}}{(b_1 + b_2 + \cdots + b_n)^m},$$

and the equality holds if and only if $a_i = \lambda b_i$ *for* $i = 1, 2, \ldots, n$.

Corollary 4.8. *For positive real numbers* $a_1, a_2, a_3, b_1, b_2, b_3, c_1, c_2, c_3$,

$$(a_1^3 + a_2^3 + a_3^3)(b_1^3 + b_2^3 + b_3^3)(c_1^3 + c_2^3 + c_3^3) \geq (a_1 b_1 c_1 + a_2 b_2 c_2 + a_3 b_3 c_3)^3.$$

Corollary 4.9. *For positive real numbers* $a_1, a_2, \ldots, a_n$,

$$(1 + a_1)(1 + a_2) \cdots (1 + a_n) \geq (1 + \sqrt[n]{a_1 a_2 \cdots a_n})^n.$$

4.1.3 *Other famous inequalities*

(1) *Triangle inequality*

For real numbers a and b,

$$||a| - |b|| \leq |a + b| \leq |a| + |b|,$$

and the first equality holds if and only if $ab \leq 0$ and the second equality holds if and only if $ab \geq 0$.

Theorem 4.1. *For real numbers* $a_1, a_2, \ldots, a_n$,

$$|a_1 + a_2 + \cdots + a_n| \leq |a_1| + |a_2| + \cdots + |a_n|.$$

This inequality also holds for complex numbers.

(2) Rearrangement inequality

For real numbers $a_1, a_2, \ldots, a_n, b_1, b_2, \ldots, b_n$ satisfying

$$a_1 \leq a_2 \leq \cdots \leq a_n \quad \text{and} \quad b_1 \leq b_2 \leq \cdots \leq b_n,$$

it holds that

$$a_1 b_n + a_2 b_{n-1} + \cdots + a_n b_1 \leq a_1 b_{t_1} + a_2 b_{t_2} + \cdots + a_n b_{t_n}$$

$$\leq a_1 b_1 + a_2 b_2 + \cdots + a_n b_n,$$

where $\{t_1, t_2, \ldots, t_n\} = \{1, 2, \ldots, n\}$.

That is, "reverse order sum" $\leq$ "mixed order sum" $\leq$ "same order sum."

(3) Chebyshev's inequality

For real numbers $a_1, a_2, \ldots, a_n, b_1, b_2, \ldots, b_n$ satisfying $a_1 \leq a_2 \leq \cdots \leq a_n$ and $b_1 \leq b_2 \leq \cdots \leq b_n$,

$$\sum_{i=1}^{n} a_i b_{n+1-i} \leq \frac{1}{n} \left(\sum_{i=1}^{n} a_i \right) \left(\sum_{i=1}^{n} b_i \right) \leq \sum_{i=1}^{n} a_i b_i.$$

Suppose it is required to prove an inequality expressed as a sum of fractions:

$$\frac{x_1}{y_1} + \frac{x_2}{y_2} + \cdots + \frac{x_n}{y_n} \geq 0,$$

where $x_1, x_2, \ldots, x_n$ are real numbers, and $y_1, y_2, \ldots, y_n$ are positive real numbers. Similar to the method of undetermined coefficients, if there exist positive real numbers $\alpha_1, \alpha_2, \ldots, \alpha_n$ such that $\alpha_1 x_1 \leq \alpha_2 x_2 \leq \cdots \leq \alpha_n x_n$ and $\alpha_1 y_1 \geq \alpha_2 y_2 \geq \cdots \geq \alpha_n y_n$, then we can employ Chebyshev's inequality,

$$\frac{x_1}{y_1} + \frac{x_2}{y_2} + \cdots + \frac{x_n}{y_n}$$

$$\geq \frac{1}{n} (\alpha_1 x_1 + \alpha_2 x_2 + \cdots + \alpha_n x_n) \left(\frac{1}{\alpha_1 y_1} + \frac{1}{\alpha_2 y_2} + \cdots + \frac{1}{\alpha_n y_n} \right).$$

Thus, it is only necessary to prove that $\alpha_1 x_1 + \alpha_2 x_2 + \cdots + \alpha_n x_n \geq 0$. The advantage of this method is that it eliminates the fractions in the inequality, and the values of $\alpha_1, \alpha_2, \ldots, \alpha_n$ can even be chosen appropriately such that $\alpha_1 x_1 + \alpha_2 x_2 + \cdots + \alpha_n x_n = 0$.

Example 4.5. Let $a, b, c, d > 0$ and satisfy $a + b + c + d = \frac{1}{a} + \frac{1}{b} + \frac{1}{c} + \frac{1}{d}$. Prove that

$$2(a + b + c + d) \geq \sqrt{a^2 + 3} + \sqrt{b^2 + 3} + \sqrt{c^2 + 3} + \sqrt{d^2 + 3}.$$

Proof. Assume without loss of generality that $a \geq b \geq c \geq d > 0$. From the given condition,

$$\frac{a^2 - 1}{a} + \frac{b^2 - 1}{b} + \frac{c^2 - 1}{c} + \frac{d^2 - 1}{d} = 0$$

$$\text{and} \quad 2a - \sqrt{a^2 + 3} = \frac{3(a^2 - 1)}{2a + \sqrt{a^2 + 3}}.$$

It is evident that $y = x - \frac{1}{x}$ and $y = \frac{x}{2x + \sqrt{x^2 + 3}}$ are monotonically increasing functions for $x > 0$. Therefore,

$$\sum_{\text{cyc}} \frac{a^2 - 1}{2a + \sqrt{a^2 + 3}} = \sum_{\text{cyc}} \left(\frac{a^2 - 1}{a} \cdot \frac{a}{2a + \sqrt{a^2 + 3}} \right)$$

$$\geq \frac{1}{4} \left(\sum_{\text{cyc}} \frac{a^2 - 1}{a} \right) \left(\sum_{\text{cyc}} \frac{a}{2a + \sqrt{a^2 + 3}} \right) = 0,$$

where $\sum_{\text{cyc}}$ denotes the cyclic sum, as in this case $\sum_{\text{cyc}} a = a + b + c + d$.

This method of using Chebyshev's inequality to prove other inequalities is also known as the Chebyshev Associate Technique.

(4) *Abel transformation*

For positive integers m, n with $m < n$,

$$\sum_{k=m}^{n} (A_k - A_{k-1}) b_k = A_n b_n - A_{m-1} b_m + \sum_{k=m}^{n-1} A_k (b_k - b_{k+1}),$$

which is known as Abel's summation by parts formula. In this formula, setting $A_0 = 0$ and $A_k = \sum_{i=1}^{k} a_i$ for $1 \leq k \leq n$, we obtain

$$\sum_{k=1}^{n} a_k b_k = b_n \sum_{k=1}^{n} a_k + \sum_{k=1}^{n-1} \left(\sum_{i=1}^{k} a_i \right) (b_k - b_{k+1}).$$

(5) *Abel's inequality*

Given $b_1 \geq b_2 \geq \cdots \geq b_n > 0$ and $m \leq \sum_{i=1}^{t} a_i \leq M$ for $t \in \{1, 2, \ldots, n\}$,

$$b_1 m \leq \sum_{k=1}^{n} a_k b_k \leq b_1 M.$$

Example 4.6. Let $a_1, a_2, \ldots, a_n$ be n non-negative real numbers and denote $S_k = \sum_{i=1}^{k} a_i$ for $1 \leq k \leq n$. Prove that

$$\sum_{i=1}^{n} \left(a_i S_i \sum_{j=i}^{n} a_j^2 \right) \leq \sum_{i=1}^{n} (a_i S_i)^2.$$

Proof.   Let $b_i = a_i S_i$ and $c_i = \sum_{j=i}^{n} a_j^2$ for $i = 1, 2, \ldots, n$. The original inequality is equivalent to

$$\sum_{i=1}^{n} b_i c_i \leq \sum_{i=1}^{n} b_i^2. \qquad (*)$$

Note that for $1 \leq i \leq n$,

$$B_i = b_1 + b_2 + \cdots + b_i = a_1 S_1 + a_2 S_2 + \cdots + a_i S_i$$

$$\leq (a_1 + a_2 + \cdots + a_i) S_i = S_i^2.$$

Thus, by Abel's summation by parts formula,

$$\sum_{i=1}^{n} b_i c_i = \sum_{i=1}^{n-1} B_i (c_i - c_{i+1}) + B_n c_n$$

$$\leq \sum_{i=1}^{n-1} a_i^2 S_i^2 + B_n c_n$$

$$\leq \sum_{i=1}^{n} a_i^2 S_i^2 = \sum_{i=1}^{n} b_i^2.$$

Hence, the inequality $(*)$ holds, and therefore, the original inequality holds.

(6) *Lagrange's identity*

For real numbers $a_1, a_2, \ldots, a_n, b_1, b_2, \ldots, b_n$,

$$\left(\sum_{i=1}^{n} a_i^2 \right) \left(\sum_{i=1}^{n} b_i^2 \right) - \left(\sum_{i=1}^{n} a_i b_i \right)^2 = \sum_{1 \leq i < j \leq n} (a_i b_j - a_j b_i)^2.$$

(7) *Minkowski's inequality*

For positive real numbers $a_1, a_2, \ldots, a_n, b_1, b_2, \ldots, b_n$ and $r > 1$,

$$\left(\sum_{i=1}^{n} (a_i + b_i)^r \right)^{\frac{1}{r}} \leq \left(\sum_{i=1}^{n} a_i^r \right)^{\frac{1}{r}} + \left(\sum_{i=1}^{n} b_i^r \right)^{\frac{1}{r}},$$

and the equality holds if and only if $a_i = \lambda b_i$ for $i = 1, 2, \ldots, n$.

This inequality reverses when $r < 1$.

(8) *Bernoulli's inequality*

Let $n \geq 2$ be an integer. Then for real numbers $a_1, a_2, \ldots, a_n$, all greater than -1 and of the same sign,

$$(1 + a_1)(1 + a_2) \cdots (1 + a_n) > 1 + a_1 + a_2 + \cdots + a_n.$$

In particular, if $x > -1$ and $n \geq 2$, then $(1 + x)^n > 1 + nx$.

Let $f(x)$ be a continuous function defined on the interval $[a, b]$. If for any $x_1, x_2 \in [a, b]$,

$$f\left(\frac{x_1 + x_2}{2} \right) \leq \text{ (or } \geq) \ \frac{f(x_1) + f(x_2)}{2},$$

then $f(x)$ is termed as a convex function (or concave function) on $[a, b]$.

If the second derivative $f''(x)$ exists and $f''(x) \geq 0$(or ≤ 0) for $x \in [a, b]$, then $f(x)$ is a convex function (or concave function) on $[a, b]$.

(9) *Jensen's inequality*

Let $f(x)$ be a convex function (or concave function) on $[a, b]$. Then for any $x_1, x_2, \ldots, x_n \in [a, b]$, it holds that

$$f\left(\frac{x_1 + x_2 + \cdots + x_n}{n} \right) \leq \text{ (or } \geq) \frac{f(x_1) + f(x_2) + \cdots + f(x_n)}{n},$$

and the equality holds if and only if $x_1 = x_2 = \cdots = x_n$.

More generally, if $f(x)$ is a convex function (or concave function) on $[a, b]$, then for any $x_1, x_2, \ldots, x_n \in [a, b]$, and any non-negative real numbers

$\lambda_1, \lambda_2, \ldots, \lambda_n$ satisfying $\lambda_1 + \lambda_2 + \cdots + \lambda_n = 1$,

$$f(\lambda_1 x_1 + \lambda_2 x_2 + \cdots + \lambda_n x_n)$$
$$\leq (\text{or} \geq) \lambda_1 f(x_1) + \lambda_2 f(x_2) + \cdots + \lambda_n f(x_n).$$

Example 4.7. Let a, b, x, y, z be positive real numbers such that $x + y + z = 1$. Prove that

$$\left(a + \frac{b}{x}\right)\left(a + \frac{b}{y}\right)\left(a + \frac{b}{z}\right) \geq (a + 3b)^3.$$

Proof. Let $f(x) = \ln(a + \frac{b}{x})$ for $x \in (0, +\infty)$. Then $f''(x) = \frac{2abx + b^2}{(ax^2 + bx)^2} > 0$, and hence $f(x)$ is a convex function. By Jensen's Inequality,

$$f(x) + f(y) + f(z) \geq 3f\left(\frac{x + y + z}{3}\right),$$

i.e., $\ln(a + \frac{b}{x}) + \ln(a + \frac{b}{y}) + \ln(a + \frac{b}{z}) \geq 3\ln(a + \frac{3b}{x+y+z})$, which implies

$$\left(a + \frac{b}{x}\right)\left(a + \frac{b}{y}\right)\left(a + \frac{b}{z}\right) \geq (a + 3b)^3.$$

(10) *Schur's inequality*

For non-negative real numbers x, y, z and $r > 0$,

$$\sum_{\text{cyc}} x^r (x - y)(x - z) \geq 0.$$

When $r = 1$, Schur's inequality becomes

$$\sum_{\text{cyc}} x(x - y)(x - z) \geq 0,$$

which is equivalent to

(i) $\sum_{\text{cyc}} x^3 - \sum_{\text{cyc}} x^2(y + z) + 3xyz \geq 0$;

(ii) $(\sum_{\text{cyc}} x)^3 - 4 \sum_{\text{cyc}} x \sum_{\text{cyc}} yz + 9xyz \geq 0$.

Example 4.8. Let x, y, z be positive real numbers such that $x + y + z = xyz$. Prove that

$$x^2 + y^2 + z^2 - 2(xy + yz + zx) + 9 \geq 0.$$

Proof. Since $x + y + z = xyz$, the original inequality is equivalent to

$$(x^2 + y^2 + z^2 - 2(xy + yz + zx))(x + y + z) + 9xyz \geq 0$$

$$\Leftrightarrow x^3 + y^3 + z^3 - (x^2y + y^2z + z^2x + xy^2 + yz^2 + zx^2) + 3xyz \geq 0,$$

which further simplifies to $\sum_{\text{cyc}} x^3 - \sum_{\text{cyc}} x^2(y + z) + 3xyz \geq 0$, that is (i). Hence, the original inequality holds.

4.1.4 *Common methods for proving inequalities*

(1) *Comparison method*

To prove $A \geq B$, it suffices to prove $A - B \geq 0$. If $B > 0$, then it also suffices to prove $\frac{A}{B} \geq 1$.

Example 4.9. Let a, b, c be positive real numbers. Prove that

$$\frac{a^2 + bc}{b + c} + \frac{b^2 + ca}{c + a} + \frac{c^2 + ab}{a + b} \geq a + b + c.$$

Proof. The inequality can be simplified to

$$\frac{a^2 + bc}{b + c} - a + \frac{b^2 + ca}{c + a} - b + \frac{c^2 + ab}{a + b} - c$$

$$= \frac{(a^2 - b^2)^2 + (b^2 - c^2)^2 + (c^2 - a^2)^2}{2(a + b)(b + c)(c + a)} \geq 0.$$

Example 4.10. Let a, b, c be positive real numbers. Prove that

$$a^{2a}b^{2b}c^{2c} \geq a^{b+c}b^{c+a}c^{a+b}.$$

Proof. Since the inequality is symmetric in a, b, c, assume without loss of generality that $a \geq b \geq c$. Thus,

$$\frac{a^{2a}b^{2b}c^{2c}}{a^{b+c}b^{c+a}c^{a+b}} = \left(\frac{a}{b}\right)^{a-b} \left(\frac{b}{c}\right)^{b-c} \left(\frac{a}{c}\right)^{a-c} \geq 1.$$

(2) *Analytical method*

Assume the inequality to be proven is valid. Then derive a series of equivalent inequalities from it until reaching a more easily provable one or an obviously true one.

Example 4.11. Let n be a positive integer. Prove that

$$\frac{1}{n + 1}\left(1 + \frac{1}{3} + \cdots + \frac{1}{2n - 1}\right) \geq \frac{1}{n}\left(\frac{1}{2} + \frac{1}{4} + \cdots + \frac{1}{2n}\right).$$

Proof.   To prove the inequality, it suffices to prove

$$n\left(1+\frac{1}{3}+\cdots+\frac{1}{2n-1}\right) \geq (n+1)\left(\frac{1}{2}+\frac{1}{4}+\cdots+\frac{1}{2n}\right),$$

which is reduced to proving

$$\frac{n}{2} \geq \frac{1}{2}+\frac{1}{4}+\cdots+\frac{1}{2n}$$

$$\text{and}\quad n\left(\frac{1}{3}+\cdots+\frac{1}{2n-1}\right) \geq n\left(\frac{1}{4}+\cdots+\frac{1}{2n}\right).$$

(3) *Proof by contradiction*

Assume the inequality to be proven is false. Then derive a contradiction, thereby confirming that the inequality is indeed true. The proof by contradiction is often suitable when a direct proof is difficult or when there are few conditions. By assuming the opposite, it effectively adds an additional condition, making it easier to approach the problem.

Example 4.12. Let x, y, z be positive real numbers. Prove that

$$\sqrt{x+\sqrt[3]{y+\sqrt[4]{z}}} \geq \sqrt[32]{xyz}.$$

Proof.   Using proof by contradiction, assume there exist x_0, y_0, z_0 such that

$$\sqrt{x_0+\sqrt[3]{y_0+\sqrt[4]{z_0}}} < \sqrt[32]{x_0 y_0 z_0}.$$

Then $\sqrt{x_0} < \sqrt[32]{x_0 y_0 z_0}$, $\sqrt[6]{y_0} < \sqrt[32]{x_0 y_0 z_0}$, and $\sqrt[24]{z_0} < \sqrt[32]{x_0 y_0 z_0}$, implying

$$\begin{cases} x_0^{16} < x_0 y_0 z_0, \\ y_0^{16} < (x_0 y_0 z_0)^3, \\ z_0^{16} < (x_0 y_0 z_0)^{12}. \end{cases}$$

Multiplying the above three inequalities yields $x_0^{16} \cdot y_0^{16} \cdot z_0^{16} < (x_0 y_0 z_0)^{16}$, leading to a contradiction. Therefore, the original inequality holds.

(4) *Mathematical induction*

When proving propositions related to positive integers n, mathematical induction is often used.

Example 4.13. Given a sequence of positive numbers $\{x_n\}$ satisfying $S_n \geq 2S_{n-1}$ for $n \in \{2, 3, \ldots\}$, where $S_n = x_1 + x_2 + \cdots + x_n$, prove that there exists a constant $C > 0$ such that $x_n \geq C \cdot 2^n$ for $n \in \{1, 2, 3, \ldots\}$.

Proof. We use mathematical induction with $C = \frac{1}{4}x_1$.

The conclusion is evident when $n \in \{1, 2\}$, i.e., $x_2 \geq x_1 = C \cdot 2^2$.

When $n \geq 3$, assume $x_k \geq C \cdot 2^k$ for $k \in \{1, 2, 3, \ldots, n-1\}$. From $S_n \geq 2S_{n-1}$,

$$
\begin{aligned}
x_n &\geq x_1 + (x_2 + \cdots + x_{n-1}) \\
&\geq x_1 + (C \cdot 2^2 + \cdots + C \cdot 2^{n-1}) \\
&\geq C(2^2 + 2^2 + 2^3 + \cdots + 2^{n-1}) \\
&= C \cdot 2^n.
\end{aligned}
$$

Thus, by mathematical induction, the conclusion holds.

(5) *Bounding method*

To prove an inequality $A \leq B$, one can introduce one or more intermediate quantities C. If it can be shown that $A \leq C$ and $C \leq B$, then $A \leq B$.

Example 4.14. Let x_1, x_2, x_3 be non-negative real numbers satisfying $x_1 + x_2 + x_3 = 1$. Prove that

$$
(x_1 + 3x_2 + 5x_3)\left(x_1 + \frac{x_2}{3} + \frac{x_3}{5}\right) \leq \frac{9}{5}.
$$

Proof. By the AM-GM inequality,

$$
\begin{aligned}
&(x_1 + 3x_2 + 5x_3)\left(x_1 + \frac{x_2}{3} + \frac{x_3}{5}\right) \\
&= \frac{1}{5}(x_1 + 3x_2 + 5x_3)\left(5x_1 + \frac{5x_2}{3} + x_3\right) \\
&\leq \frac{1}{20}\left((x_1 + 3x_2 + 5x_3) + \left(5x_1 + \frac{5x_2}{3} + x_3\right)\right)^2 \\
&= \frac{1}{20}\left(6x_1 + \frac{14}{3}x_2 + 6x_3\right)^2 \\
&\leq \frac{1}{20}(6x_1 + 6x_2 + 6x_3)^2 = \frac{9}{5}.
\end{aligned}
$$

(6) *Variable substitution method*

Consider a complex expression as a whole and replace it with a single letter to simplify the problem. When conditions or conclusions involve forms such as $x^2 + y^2 = r^2$, $\sqrt{r^2 - x^2}$, $|x| \leq 1$, etc., consider substitutions like

$x = r \sin \alpha$ and $x = r \cos \alpha$. When conditions or conclusions involve forms like $\sqrt{r^2 + x^2}$ and $\sqrt{x^2 - r^2}$, consider substitutions like $x = r \tan \alpha$ and $x = r \sec \alpha$. When making substitutions, it is important to note that the range of values for α is determined by the range of values for the original variable x.

Example 4.15. Given $0° \le \alpha \le 90°$, prove that $2 \le \sqrt{5 - 4 \sin \alpha} + \sin \alpha \le \frac{9}{4}$.

Proof. Let $x = \sqrt{5 - 4 \sin \alpha}$. Then $\sin \alpha = \frac{5 - x^2}{4}$. Since $0 \le \sin \alpha \le 1$, it follows that $1 \le x \le \sqrt{5}$. Define

$$y = \sqrt{5 - 4 \sin \alpha} + \sin \alpha$$

$$= x + \frac{5 - x^2}{4}$$

$$= -\frac{1}{4}(x - 2)^2 + \frac{9}{4}.$$

Given $1 \le x \le \sqrt{5}$, it follows that $2 \le y \le \frac{9}{4}$.

(7) *Method of undetermined coefficients*

This method involves introducing appropriate parameters, determining their values based on the characteristics of the expression in the problem, and using them to prove the inequality.

Example 4.16. Let x, y, z be three real numbers, not all zero. Find the maximum value of $\frac{xy + 2yz}{x^2 + y^2 + z^2}$.

Solution. It is required to prove the existence of a constant c such that

$$x^2 + y^2 + z^2 \ge \frac{1}{c}(xy + 2yz),$$

with equality holding for some values of x, y, z. Since the terms on the right side are xy and $2yz$, the term y^2 on the left side should be split into two terms αy^2 and $(1 - \alpha)y^2$. By using

$$x^2 + \alpha y^2 \ge 2\sqrt{\alpha} xy, \quad (1 - \alpha)y^2 + z^2 \ge 2\sqrt{1 - \alpha} yz,$$

and from $\frac{2\sqrt{1 - \alpha}}{2\sqrt{\alpha}} = 2$, we obtain $\alpha = \frac{1}{5}$.

Thus, $x^2 + y^2 + z^2 \ge \frac{2}{\sqrt{5}}(xy + 2yz)$, implying $c = \frac{\sqrt{5}}{2}$.

When dealing with non-symmetric inequalities, the coefficients of similar terms are often not the same. Introducing additional parameters to group terms can facilitate the use of common inequalities such as the AM-GM inequality, Cauchy–Schwarz inequality, etc.

(8) *Function method*

Prove inequalities by introducing functions and utilizing their properties such as monotonicity, convexity/concavity, boundedness, etc.

Example 4.17. Let a, b be non-negative real numbers. Prove that

$$3a^3 + 7b^3 \geq 9ab^2.$$

Proof. If $a = 0$ or $b = 0$, then the conclusion holds. If $a, b > 0$, then let $x = \frac{a}{b}$ and $f(x) = 3x^3 + 7 - 9x$ for $x > 0$. Clearly, $f'(x) = 9(x^2 - 1)$.

When $x \in (0, 1)$, since $f'(x) < 0$, $f(x)$ is strictly decreasing on $(0, 1)$. When $x \in (1, +\infty)$, from $f'(x) > 0$, we see that $f(x)$ is strictly increasing on $(1, +\infty)$. Therefore, $f(x)$ attains its minimum value at $x = 1$, i.e., $f(x) \geq f(1) = 1 > 0$.

(9) *Construction method*

We can prove inequalities by constructing identities, figures, functions, etc.

(i) Construct Identities

Example 4.18. Given $a^2 + b^2 + c^2 + d^2 = 1$, prove that

$$(a + b)^4 + (a + c)^4 + (a + d)^4 + (b + c)^4 + (b + d)^4 + (c + d)^4 \leq 6.$$

Proof. Consider the sum

$$(a - b)^4 + (a - c)^4 + (a - d)^4 + (b - c)^4 + (b - d)^4 + (c - d)^4,$$

and construct the identity

$$6(a^2 + b^2 + c^2 + d^2)^2$$
$$= (a + b)^4 + (a + c)^4 + (a + d)^4 + (b + c)^4 + (b + d)^4 + (c + d)^4$$
$$+ (a - b)^4 + (a - c)^4 + (a - d)^4 + (b - c)^4 + (b - d)^4 + (c - d)^4,$$

from which the inequality follows.

(ii) Construct Figures

Example 4.19. Let a, b, c be positive real numbers. Prove that

$$\sqrt{a^2 + ab + b^2} \leq \sqrt{a^2 - ac + c^2} + \sqrt{b^2 - bc + c^2}.$$

Proof. Construct $\triangle XYZ$ such that $XY = a$, $XZ = b$, and $\angle YXZ = 120°$.

Inside $\angle YXZ$, take a point W such that $\angle YXW$ and $\angle WXZ$ are both $60°$, and $XW = c$. By the law of cosines, it is evident that

$$YZ = \sqrt{a^2 + ab + b^2}, \quad YW = \sqrt{a^2 - ac + c^2}, \quad WZ = \sqrt{b^2 - bc + c^2}.$$

Clearly, $YZ \leq YW + WZ$, so the original inequality holds.

(iii) Construct Functions

Example 4.20. Given real numbers $x_1, x_2, \ldots, x_n, y_1, y_2, \ldots, y_n$ with

$$x_1^2 + x_2^2 + \cdots + x_n^2 \leq 1,$$

prove that

$$(x_1 y_1 + x_2 y_2 + \cdots + x_n y_n - 1)^2$$
$$\geq (x_1^2 + x_2^2 + \cdots + x_n^2 - 1)(y_1^2 + y_2^2 + \cdots + y_n^2 - 1).$$

Proof. If $x_1^2 + x_2^2 + \cdots + x_n^2 = 1$, then the original inequality holds. If $x_1^2 + x_2^2 + \cdots + x_n^2 < 1$, construct a quadratic function

$$f(x) = (x_1^2 + x_2^2 + \cdots + x_n^2 - 1)x^2 - 2(x_1 y_1 + x_2 y_2 + \cdots + x_n y_n - 1)x$$
$$+ (y_1^2 + y_2^2 + \cdots + y_n^2 - 1)$$
$$= (x_1 x - y_1)^2 + (x_2 x - y_2)^2 + \cdots + (x_n x - y_n)^2 - (x - 1)^2,$$

whose graph is a downward-opening parabola.

Since $f(1) = (x_1 - y_1)^2 + (x_2 - y_2)^2 + \cdots + (x_n - y_n)^2 \geq 0$, its discriminant must be non-negative, i.e.,

$$4(x_1 y_1 + x_2 y_2 + \cdots + x_n y_n - 1)^2$$
$$- 4(x_1^2 + x_2^2 + \cdots + x_n^2 - 1)(y_1^2 + y_2^2 + \cdots + y_n^2 - 1) \geq 0.$$

Thus, $(x_1 y_1 + x_2 y_2 + \cdots + x_n y_n - 1)^2 \geq (x_1^2 + x_2^2 + \cdots + x_n^2 - 1)(y_1^2 + y_2^2 + \cdots + y_n^2 - 1)$.

The tangent line method is also closely related to constructing functions. Suppose $a + b + c$ is a constant value, and we need to prove $f(a) + f(b) + f(c) \geq 0$. However, if the expression of the function $f(x)$ is complex, and there exists a tangent line $y = l(x)$ to $f(x)$ at some point x_0 such that the graph of the function $f(x)$ lies above the tangent line, and $l(a) + l(b) + l(c) \geq 0$, then

$$f(a) + f(b) + f(c) \geq l(a) + l(b) + l(c) \geq 0.$$

Example 4.21. Given real numbers $a, b, c > 0$, prove that

$$\frac{\sqrt{a+b+c} + \sqrt{a}}{b+c} + \frac{\sqrt{a+b+c} + \sqrt{b}}{c+a} + \frac{\sqrt{a+b+c} + \sqrt{c}}{a+b} \geq \frac{9 + 3\sqrt{3}}{2\sqrt{a+b+c}}.$$

Proof. Without loss of generality, assume $a + b + c = 1$. The inequality is equivalent to $\frac{1}{1-\sqrt{a}} + \frac{1}{1-\sqrt{b}} + \frac{1}{1-\sqrt{c}} \geq \frac{9+3\sqrt{3}}{2}$.

Define the function $f(x) = \frac{1}{1-\sqrt{x}}$, and its tangent line at $x = \frac{1}{3}$ is $y = \frac{9+6\sqrt{3}}{4}x + \frac{3}{4}$. When $0 < x < 1$,

$$\frac{1}{1 - \sqrt{x}} \geq \frac{9 + 6\sqrt{3}}{4}x + \frac{3}{4}$$

is equivalent to $(\sqrt{3x} - 1)^2((3 + 2\sqrt{3})\sqrt{x} + 1) \geq 0$. Therefore,

$$\frac{1}{1 - \sqrt{a}} + \frac{1}{1 - \sqrt{b}} + \frac{1}{1 - \sqrt{c}} \geq \frac{9 + 6\sqrt{3}}{4}(a + b + c) + \frac{9}{4} = \frac{9 + 3\sqrt{3}}{2}.$$

The advantage of this method is that it transforms the calculation of complex function values into the calculation of linear function values. This method can also be used to solve the following problems:

- (**Chinese Western Mathematic Invitational 2007, Problem 3**). Let a, b, c be real numbers with $a + b + c = 3$. Prove that

$$\frac{1}{5a^2 - 4a + 11} + \frac{1}{5b^2 - 4b + 11} + \frac{1}{5c^2 - 4c + 11} \leq \frac{1}{4}.$$

- (**Korean Mathematical Olympiad 2011, Final Round, Problem 4**). Let a, b, c be non-negative with $a+b+c = 1$. Find the maximum value of

$$\frac{1}{a^2 - 4a + 9} + \frac{1}{b^2 - 4b + 9} + \frac{1}{c^2 - 4c + 9}.$$

(10) Adjustment method

When dealing with multivariable inequalities, adjust some of the variables to "positions" that are easier to handle. Since lower-order symmetric or cyclic inequalities often have equalities when two numbers are equal or when one of the numbers is zero, these are ideal "positions" for using the adjustment method.

Example 4.22. Let a, b, c be positive real numbers. Prove that

$$\frac{63}{2} + \frac{(a+b+c)(a^2+b^2+c^2)}{abc} \geq \frac{27(a+b+c)}{2\sqrt[3]{abc}}.$$

Proof.   Without loss of generality, assume $abc = 1$, and let

$$f(a,b,c) = (a+b+c)(a^2+b^2+c^2) + \frac{63}{2} - \frac{27}{2}(a+b+c).$$

Then,

$$f(a,b,c) - f(a, \sqrt{bc}, \sqrt{bc})$$

$$= (\sqrt{b} - \sqrt{c})^2 \left((b + \sqrt{bc} + c)^2 + bc + a(\sqrt{b} + \sqrt{c})^2 + a^2 - \frac{27}{2} \right)$$

$$\geq (\sqrt{b} - \sqrt{c})^2 \left(10bc + 4a\sqrt{bc} + a^2 - \frac{27}{2} \right)$$

$$= (\sqrt{b} - \sqrt{c})^2 \left(\frac{10}{3a} + \frac{10}{3a} + \frac{10}{3a} + 2\sqrt{a} + 2\sqrt{a} + a^2 - \frac{27}{2} \right)$$

$$\geq (\sqrt{b} - \sqrt{c})^2 \left(6\sqrt[6]{4\left(\frac{10}{3}\right)^3} - \frac{27}{2} \right) \geq 0.$$

Thus, it suffices to prove $f(\frac{1}{t^2}, t, t) \geq 0$, which is equivalent to

$$(t-1)^2(8t^7 + 16t^6 - 30t^5 - 9t^4 + 12t^3 + 6t^2 + 4t + 2) \geq 0,$$

and this inequality can be easily proven using the AM-GM inequality.

   If the equality conditions of an inequality are known, using the adjustment method can be convenient. However, sometimes the equality conditions of an inequality are not readily apparent. Here, we introduce a method to find the equality conditions: the Lagrange multiplier method.

Example 4.23. Let a, b, c be positive real numbers, and $ab + bc + ca = 11$. Find the minimum value of $3a + 4b + 5c$.

Solution. Define $f(a, b, c) = 3a + 4b + 5c - k(ab + bc + ca - 11)$, where k is an undetermined coefficient. Set the derivatives with respect to each variable to zero:

$$f_a(a, b, c) = 3 - k(b + c) = 0,$$

$$f_b(a, b, c) = 4 - k(a + c) = 0,$$

$$f_c(a, b, c) = 5 - k(a + b) = 0,$$

yielding $k = \frac{3}{b+c} = \frac{4}{c+a} = \frac{5}{a+b}$. From $ab + bc + ca = 11$, we get $a = 3$, $b = 2$, and $c = 1$, which should be the equality conditions for $3a + 4b + 5c$ to attains its minimum value.

We return to the original problem. From $c = \frac{11-ab}{a+b}$,

$$3a + 4b + 5c = 3a + 4b + \frac{55 - 5ab}{a+b}$$

$$= \frac{55 + 2ab + 3a^2 + 4b^2 - 22(a+b)}{a+b} + 22$$

$$= \frac{(a+b-5)^2 + 2(a-3)^2 + 3(b-2)^2}{a+b} + 22$$

$$\geq 22.$$

This method involves knowledge of advanced mathematics and is provided for reference only.

(11) *Sum of squares method*

If an expression concerning a, b, c can be transformed into

$$S = f(a, b, c) = S_c(a - b)^2 + S_b(c - a)^2 + S_a(b - c)^2,$$

where S_a, S_b, S_c are expressions in terms of a, b, c, then the inequality $S \geq 0$ holds if at least one of the following conditions is satisfied:

(i) $S_a \geq 0$, $S_b \geq 0$, and $S_c \geq 0$.
(ii) $a \geq b \geq c$, $S_b \geq 0$, $S_b + S_a \geq 0$, and $S_b + S_c \geq 0$.
(iii) $a \geq b \geq c$, $S_a \geq 0$, $S_c \geq 0$, $S_a + 2S_b \geq 0$, and $S_c + 2S_b \geq 0$.
(iv) $a \geq b \geq c$, $S_b \geq 0$, $S_c \geq 0$, and $a^2 S_b + b^2 S_a \geq 0$.
(v) $a \geq b \geq c$ are the side lengths of a triangle, and $S_a \geq 0$, $S_b \geq 0$, and $b^2 S_b + c^2 S_c \geq 0$.
(vi) $S_a + S_b + S_c \geq 0$ and $S_a S_b + S_b S_c + S_c S_a \geq 0$.

The proofs for these conditions are not difficult, and conditions (ii) and (iv) are illustrated here. For condition (ii), it is easy to see that $(c - a)^2 \geq (a - b)^2 + (b - c)^2$, and thus,

$$S_c(a - b)^2 + S_b(c - a)^2 + S_a(b - c)^2$$

$$\geq (S_c + S_b)(a - b)^2 + (S_a + S_b)(b - c)^2 \geq 0.$$

For condition (iv), it is evident that $\frac{a-c}{b-c} \geq \frac{a}{b}$, from which,

$$S_c(a - b)^2 + S_b(c - a)^2 + S_a(b - c)^2$$

$$\geq S_a(b - c)^2 + S_b(c - a)^2$$

$$\geq S_a(b - c)^2 + \frac{a^2(b - c)^2}{b^2} S_b$$

$$= \frac{b^2 S_a + a^2 S_b}{b^2}(b - c)^2$$

$$\geq 0.$$

Example 4.24 (Iran Team Selection Test, 1996). Given non-negative real numbers x, y, z, prove that

$$\frac{1}{(x + y)^2} + \frac{1}{(y + z)^2} + \frac{1}{(z + x)^2} \geq \frac{9}{4(xy + yz + zx)}.$$

Proof. Let $a = y + z$, $b = z + x$, and $c = x + y$. Then the inequality transforms into

$$(2ab + 2bc + 2ca - a^2 - b^2 - c^2)\left(\frac{1}{a^2} + \frac{1}{b^2} + \frac{1}{c^2}\right) \geq 9,$$

which is equivalent to $\sum_{cyc}\left(\frac{2a}{b} + \frac{2b}{a} - 4\right) - \sum_{cyc}\frac{a^2 + b^2 - 2ab}{c^2} \geq 0$, i.e.,

$$\left(\frac{2}{ab} - \frac{1}{c^2}\right)(a - b)^2 + \left(\frac{2}{bc} - \frac{1}{a^2}\right)(b - c)^2 + \left(\frac{2}{ca} - \frac{1}{b^2}\right)(c - a)^2 \geq 0.$$

Therefore, $S_a = \frac{2}{bc} - \frac{1}{a^2}$, $S_b = \frac{2}{ca} - \frac{1}{b^2}$, and $S_c = \frac{2}{ab} - \frac{1}{c^2}$. Assume $a \geq b \geq c$ without loss of generality, and since $b + c \geq a$, it is easy to see that $S_a \geq 0$, $S_b \geq 0$, and $S_a \geq S_b \geq S_c$. By condition (v), it is sufficient to prove $b^2 S_b + c^2 S_c \geq 0$, which is equivalent to $b^3 + c^3 \geq abc$, and this is obviously true.

However, when dealing with cyclic but non-symmetric inequalities, the SOS method may not easily construct a satisfactory form. In such cases, consider transforming the expression involving a, b, c into

$$S = f(a, b, c) = M(a - b)^2 + N(c - a)(c - b),$$

where M, N are expressions in terms of a, b, c. If M, N are non-negative real numbers, and c is the maximum or minimum among a, b, c, then $S \geq 0$. For instance, Schur's inequality can be represented as

$$\sum_{\text{cyc}} a(a - b)(a - c) = (a + b - c)(a - b)^2 + c(c - a)(c - b) \geq 0.$$

(12) *The pqr method*

The *pqr* method, also known as the elementary symmetric polynomial method, defines $p = a+b+c$, $q = ab+bc+ca$, and $r = abc$. Then any symmetric polynomial in three variables concerning a, b, c can be expressed in terms of p, q, r, and it can be shown that this representation is unique. For non-negative real numbers a, b, c common forms of trivariate inequalities in the *pqr* format include:

(i) $a + b + c \geq 3\sqrt[3]{abc}$: $p \geq 3\sqrt[3]{r}$;
(ii) $(a + b + c)^2 \geq 3(ab + bc + ca)$: $p^2 \geq 3q$;
(iii) $a(a - b)(a - c)+b(b - c)(b - a)+c(c - a)(c - b) \geq 0$: $p^3-4pq+9r \geq 0$.

This allows us to use the relationships among p, q, r to simplify the problems.

Example 4.25 (Chinese High School Mathematics League 2014, Problem 1). Let real numbers a, b, c satisfy $a + b + c = 1$ and $abc > 0$. Prove that

$$ab + bc + ca < \frac{\sqrt{abc}}{2} + \frac{1}{4}.$$

Proof. Without loss of generality, assume $a \geq b \geq c$. If $a > 0 > b \geq c$, then

$$ab + bc + ca = bc + (1 - b - c)(b + c) \leq \frac{(b + c)^2}{4} + (b + c) - (b + c)^2 < 0,$$

implying the original inequality holds.

Suppose a, b, c are all positive real numbers, since $a + b + c = 1$, i.e., $p = 1$, by the *pqr* form of Schur's inequality, $q \leq \frac{1}{4} + \frac{9}{4}r$, and $\frac{9}{4}r < \frac{\sqrt{r}}{2}$ is equivalent to $r < \frac{4}{81}$.

By the *pqr* form of the AM-GM inequality, $r \leq \frac{1}{27}$, and hence

$$q \leq \frac{1}{4} + \frac{9}{4}r < \frac{1}{4} + \frac{\sqrt{r}}{2}.$$

(13) *Local inequality method*

For symmetric sum-type inequalities, if it's challenging to approach the problem globally, one can start locally, utilizing local properties to complete the overall proof.

Example 4.26. Let x, y, z, w be positive real numbers. Prove that

$$\frac{x^3 + y^3 + z^3}{x + y + z} + \frac{y^3 + z^3 + w^3}{y + z + w} + \frac{z^3 + w^3 + x^3}{z + w + x} + \frac{w^3 + x^3 + y^3}{w + x + y}$$
$$\geq x^2 + y^2 + z^2 + w^2.$$

Proof.   Consider the local inequality $\frac{x^3 + y^3 + z^3}{x + y + z} \geq \frac{x^2 + y^2 + z^2}{3}$. Indeed, by the Cauchy–Schwarz inequality,

$$(x + y + z)(x^3 + y^3 + z^3) \geq (x^2 + y^2 + z^2)^2$$

$$\geq (x^2 + y^2 + z^2)\frac{(x + y + z)^2}{3}.$$

Similarly, the remaining three expressions can be derived, and adding these four inequalities together suffices.

Sometimes the "local" part can be a single term or several terms.

Example 4.27 (Chinese Team Selection Test 2005, Problem 12). Let a, b, c, d be positive real numbers such that $abcd = 1$. Prove that

$$\frac{1}{(1 + a)^2} + \frac{1}{(1 + b)^2} + \frac{1}{(1 + c)^2} + \frac{1}{(1 + d)^2} \geq 13.$$

Proof.   Since $\frac{1}{1+x} + \frac{1}{1+y} = 1$ when $xy = 1$, $1 = \frac{1}{1+ab} + \frac{1}{1+cd}$.
Consider the local inequality $\frac{1}{(1+a)^2} + \frac{1}{(1+b)^2} \geq \frac{1}{1+ab}$. It can be shown that

$$\frac{1}{(1 + a)^2} + \frac{1}{(1 + b)^2} - \frac{1}{1 + ab} = \frac{ab(a - b)^2 + (ab - 1)^2}{(1 + a)^2(1 + b)^2(1 + ab)} \geq 0.$$

Similarly, $\frac{1}{(1+c)^2} + \frac{1}{(1+d)^2} \geq \frac{1}{1+cd}$, and summing these two inequalities completes the proof.

4.2 Problems and Solutions

4.2.1 *Solving inequalities*

Problem 4.1 (IMO 2-2, proposed by Hungary). For what values of the variable x is the following inequality

$$\frac{4x^2}{(1 - \sqrt{1 + 2x})^2} < 2x + 9$$

valid?

Solution 1. For the inequality to be valid, it is necessary that

$$\begin{cases} 1 + 2x \geq 0, \\ 1 - \sqrt{1 + 2x} \neq 0, \end{cases}$$

and thus $x \geq -\frac{1}{2}$ and $x \neq 0$. The original inequality is equivalent to

$$\frac{4x^2(1 + \sqrt{1 + 2x})^2}{(1 - \sqrt{1 + 2x})^2(1 + \sqrt{1 + 2x})^2} < 2x + 9,$$

$$\frac{4x^2(1 + \sqrt{1 + 2x})^2}{(1 - (1 + 2x))^2} < 2x + 9,$$

$$(1 + \sqrt{1 + 2x})^2 < 2x + 9,$$

which implies $2\sqrt{1 + 2x} < 7$, yielding $x < \frac{45}{8}$. Therefore, the given inequality holds when $-\frac{1}{2} \leq x < 0$ or $0 < x < \frac{45}{8}$.

Solution 2. From

$$\begin{cases} 1 + 2x \geq 0, \\ 1 - \sqrt{1 + 2x} \neq 0, \end{cases}$$

we have $x \geq -\frac{1}{2}$ and $x \neq 0$. Eliminating the denominator gives

$$4x^2 < (2x + 9)(1 - \sqrt{1 + 2x})^2,$$

which leads to $(2x + 9)\sqrt{1 + 2x} < 11x + 9$. Squaring both sides and simplifying yield $x^2(8x - 45) < 0$. Solving this inequality, we find $x < \frac{45}{8}$. Therefore, the given inequality holds when $-\frac{1}{2} \leq x < 0$ or $0 < x < \frac{45}{8}$.

【Score Situation】This particular problem saw the following distribution of scores among contestants: 2 contestants scored 6 points, 1 contestant scored 5 points, no contestant scored 4 points, 2 contestants scored 3 points, 1 contestant scored 2 points, 1 contestant scored 1 point, and 3 contestants scored 0 point. The average score of this problem is 2.600, indicating that it had a certain level of difficulty.

Among the top five teams in the team scores, the Czechoslovakia team achieved a total score of 257 points, the Hungary team achieved a total score of 248 points, the Romania team achieved a total score of 248 points, the Bulgaria team achieved a total score of 175 points, and the German Democratic Republic team achieved a total score of 38 points.

The gold medal cutoff for this IMO was set at 40 points (with 4 contestants earning gold medals), the silver medal cutoff was 37 points (with 4 contestants earning silver medals), and the bronze medal cutoff was 33 points (with 4 contestants earning bronze medals).

In this IMO, no contestant achieved a perfect score of 44 points.

Problem 4.2 (IMO 4-2, proposed by Hungary). Determine all real numbers x which satisfy the inequality:

$$\sqrt{3-x} - \sqrt{x+1} > \frac{1}{2}.$$

Solution. From

$$\begin{cases} 3 - x \geq 0, \\ x + 1 \geq 0, \end{cases}$$

we obtain $-1 \leq x \leq 3$. When $-1 \leq x \leq 3$, the value of $\sqrt{3-x} - \sqrt{x+1}$ decreases as x increases. Solve the equation

$$\sqrt{3-x} - \sqrt{x+1} = \tfrac{1}{2}, \tag{1}$$

$$\tfrac{7}{4} - 2x = \sqrt{x+1},$$

$$64x^2 - 128x + 33 = 0,$$

yielding $x = \frac{8 \pm \sqrt{31}}{8}$. After verification, $x = \frac{8-\sqrt{31}}{8}$ is a solution to (1). That is, when $x = \frac{8-\sqrt{31}}{8}$, the value of $\sqrt{3-x} - \sqrt{x+1}$ equals $\frac{1}{2}$. Therefore, the solution to the inequality is $-1 \leq x < \frac{8-\sqrt{31}}{8}$.

【Score Situation】 This particular problem saw the following distribution of scores among contestants: 7 contestants scored 6 points, 3 contestants scored 5 points, 2 contestants scored 4 points, 3 contestants scored 3 points, 1 contestant scored 2 points, 1 contestant scored 1 point, and no contestant scored 0 point. The average score of this problem is 4.529, indicating that it was simple.

Among the top five teams in the team scores, the Hungary team achieved a total score of 289 points, the Soviet Union team achieved a total score of 263 points, the Romania team achieved a total score of 257 points, the Czechoslovakia team achieved a total score of 212 points, and the Poland team achieved a total score of 212 points.

The gold medal cutoff for this IMO was set at 41 points (with 4 contestants earning gold medals), the silver medal cutoff was 34 points (with 12 contestants earning silver medals), and the bronze medal cutoff was 29 points (with 15 contestants earning bronze medals).

In this IMO, only one contestant achieved a perfect score of 46 points, namely Iosif Bernstein from the Soviet Union.

Problem 4.3 (IMO 7-1, proposed by Yugoslavia). Determine all values x in the interval $0 \leq x \leq 2\pi$ which satisfy the inequality

$$2\cos x \leq |\sqrt{1 + \sin 2x} - \sqrt{1 - \sin 2x}| \leq \sqrt{2}.$$

Solution. It is evident that

$$
\begin{aligned}
&|\sqrt{1 + \sin 2x} - \sqrt{1 - \sin 2x}| \\
&= |\sqrt{(\sin x + \cos x)^2} - \sqrt{(\sin x - \cos x)^2}| \\
&= ||\sin x + \cos x| - |\sin x - \cos x|| \\
&= \begin{cases} 2|\cos x|, & \frac{\pi}{4} \leq x \leq \frac{3}{4}\pi \text{ or } \frac{5}{4}\pi \leq x \leq \frac{7}{4}\pi, \\ 2|\sin x|, & \text{in other cases.} \end{cases}
\end{aligned}
$$

When $\frac{3}{4}\pi < x < \frac{5}{4}\pi$, we obtain $2\cos x < 0 \leq 2|\sin x|$. When $0 \leq x < \frac{\pi}{4}$ or $\frac{7}{4}\pi < x \leq 2\pi$, there holds $2\cos x > 2|\sin x| \geq 0$.

The right inequality,

$$|\sqrt{1 + \sin 2x} - \sqrt{1 - \sin 2x}| \leq \max\{\sqrt{1 + \sin 2x}, \sqrt{1 - \sin 2x}\} \leq \sqrt{2},$$

is obvious. Therefore, the solution to the inequality is $\frac{\pi}{4} \leq x \leq \frac{7\pi}{4}$.

Note. This problem is the same as the 6th problem in 1967 British Mathematical Olympiad Round 1.

【Score Situation】 This particular problem saw the following distribution of scores among contestants: 37 contestants scored 4 points, 10 contestants scored 3 points, 14 contestants scored 2 points, 13 contestants scored 1 point, and 6 contestants scored 0 point. The average score of this problem is 2.738, indicating that it had a certain level of difficulty.

Among the top five teams in the team scores, the scores of this problem are as follows: the Soviet Union team scored 25 points (with a total team score of 281 points), the Hungary team scored 29 points (with a total team score of 244 points), the Romania team scored 27 points (with a total team score of 222 points), the Poland team scored 28 points (with a total team score of 178 points), and the German Democratic Republic team scored 23 points (with a total team score of 175 points).

The gold medal cutoff for this IMO was set at 38 points (with 8 contestants earning gold medals), the silver medal cutoff was 30 points (with 12 contestants earning silver medals), and the bronze medal cutoff was 20 points (with 17 contestants earning bronze medals).

In this IMO, only two contestants achieved a perfect score of 40 points, namely László Lovász from Hungary and Pavel Bleher from the Soviet Union.

Problem 4.4 (IMO 14-4, proposed by the Netherlands). Find all solutions $(x_1, x_2, x_3, x_4, x_5)$ of the system of inequalities

$$\begin{cases} (x_1^2 - x_3 x_5)(x_2^2 - x_3 x_5) \leq 0, \\ (x_2^2 - x_4 x_1)(x_3^2 - x_4 x_1) \leq 0, \\ (x_3^2 - x_5 x_2)(x_4^2 - x_5 x_2) \leq 0, \\ (x_4^2 - x_1 x_3)(x_5^2 - x_1 x_3) \leq 0, \\ (x_5^2 - x_2 x_4)(x_1^2 - x_2 x_4) \leq 0, \end{cases}$$

where x_1, x_2, x_3, x_4, x_5 are positive real numbers.

Solution 1. Obviously $x_1 = x_2 = x_3 = x_4 = x_5$ is a solution to the system, where $x_i (i = 1, 2, 3, 4, 5)$ are positive real numbers. On the other hand, let $x_{5+i} = x_i$, and then the system of inequalities can be written as

$$(x_i^2 - x_{i+2} x_{i+4})(x_{i+1}^2 - x_{i+2} x_{i+4}) \leq 0, \quad i \in \{1, 2, 3, 4, 5\}.$$

Summing up these five inequalities, we get

$$\sum_{i=1}^{5} (x_i^2 - x_{i+2} x_{i+4})(x_{i+1}^2 - x_{i+2} x_{i+4})$$

$$= \sum_{i=1}^{5} (x_i^2 x_{i+1}^2 - x_i^2 x_{i+2} x_{i+4} - x_{i+1}^2 x_{i+2} x_{i+4} + x_{i+2}^2 x_{i+4}^2)$$

$$= \frac{1}{2} \sum_{i=1}^{5} (x_{i+4}^2 x_i^2 + x_{i+1}^2 x_{i+2}^2 + x_i^2 x_{i+2}^2 + x_{i+4}^2 x_{i+1}^2$$

$$- 2x_i^2 x_{i+2} x_{i+4} - 2x_{i+1}^2 x_{i+2} x_{i+4})$$

$$= \frac{1}{2} \sum_{i=1}^{5} (x_i^2 (x_{i+2}^2 - 2x_{i+2} x_{i+4} + x_{i+4}^2)$$

$$+ x_{i+1}^2 (x_{i+2}^2 - 2x_{i+2} x_{i+4} + x_{i+4}^2))$$

$$= \frac{1}{2} \sum_{i=1}^{5} (x_i^2 + x_{i+1}^2)(x_{i+2} - x_{i+4})^2$$

$$\leq 0.$$

Since $(x_i^2 + x_{i+1}^2)(x_{i+2} - x_{i+4})^2 \geq 0$ for $i \in \{1, 2, 3, 4, 5\}$, the above inequality holds when $x_1 = x_2 = x_3 = x_4 = x_5$.

Therefore, the solution to the system is $x_1 = x_2 = x_3 = x_4 = x_5$.

Solution 2. Let $(x_1, x_2, x_3, x_4, x_5)$ be a solution to the system of inequalities. Consider the case where

$$x_1 = \min\{x_1, x_2, x_3, x_4, x_5\} \quad \text{and} \quad x_2 = \min\{x_2, x_3, x_4, x_5\}.$$

Since $x_1^2 \leq x_3 x_5$ and $x_2^2 \leq x_3 x_5$, along with $(x_1^2 - x_3 x_5)(x_2^2 - x_3 x_5) \leq 0$, we obtain

$$x_1 \leq x_2 = x_3 = x_5 \leq x_4.$$

Therefore, $x_4^2 \geq x_1 x_3$ and $x_5^2 \geq x_1 x_3$. From $(x_5^2 - x_2 x_4)(x_1^2 - x_2 x_4) \leq 0$, it follows that

$$x_1 = x_2 = x_3 = x_4 = x_5. \tag{1}$$

Similar reasoning applies to other cases, confirming the validity of (1). On the other hand, for any positive real number x, it is evident that (x, x, x, x, x) satisfies the given system of inequalities. Therefore, the solution to the system is $x_1 = x_2 = x_3 = x_4 = x_5$.

【Score Situation】 This particular problem saw the following distribution of scores among contestants: 21 contestants scored 7 points, no contestant scored 6 points, 2 contestants scored 5 points, no contestant scored 4 points, no contestant scored 3 points, no contestant scored 2 points, 5 contestants scored 1 point, and 5 contestants scored 0 point. The average score of this problem is 4.909, indicating that it was simple.

Among the top five teams in the team scores, the scores of this problem are as follows: the Soviet Union team scored 48 points (with a total team score of 270 points), the Hungary team scored 52 points (with a total team score of 263 points), the German Democratic Republic team scored 43 points (with a total team score of 239 points), the Romania team scored 34 points (with a total team score of 208 points), and the United Kingdom team scored 30 points (with a total team score of 179 points).

The gold medal cutoff for this IMO was set at 40 points (with 8 contestants earning gold medals), the silver medal cutoff was 30 points (with 16 contestants earning silver medals), and the bronze medal cutoff was 19 points (with 30 contestants earning bronze medals).

In this IMO, a total of eight contestants achieved a perfect score of 40 points.

Problem 4.5 (IMO 29-4, proposed by Ireland). Show that the set of real numbers x which satisfy the inequality

$$\sum_{k=1}^{70} \frac{k}{x-k} \geq \frac{5}{4}$$

is a union of disjoint intervals, the sum of whose lengths is 1988.

Solution. Consider the function $y = \frac{1}{x-1} + \frac{2}{x-2} + \frac{3}{x-3} + \cdots + \frac{70}{x-70}$ and its graph intersecting with the line $y = \frac{5}{4}$. Let the solutions of the equation

$$\frac{1}{x-1} + \frac{2}{x-2} + \frac{3}{x-3} + \cdots + \frac{70}{x-70} = \frac{5}{4}$$

be denoted by $x_1 < x_2 < \cdots < x_{70}$. The solution set for the inequality is given by

$$S = (1, x_1) \cup (2, x_2) \cup \cdots \cup (70, x_{70}).$$

Thus, the sum of the lengths of all solution intervals is

$$L = (x_1 - 1) + (x_2 - 2) + \cdots + (x_{70} - 70)$$
$$= x_1 + x_2 + \cdots + x_{70} - (1 + 2 + \cdots + 70).$$

The graphs of the function $y = \frac{1}{x-1} + \frac{2}{x-2} + \frac{3}{x-3} + \cdots + \frac{70}{x-70}$ and the line $y = \frac{5}{4}$ are illustrated in Figure 4.1.

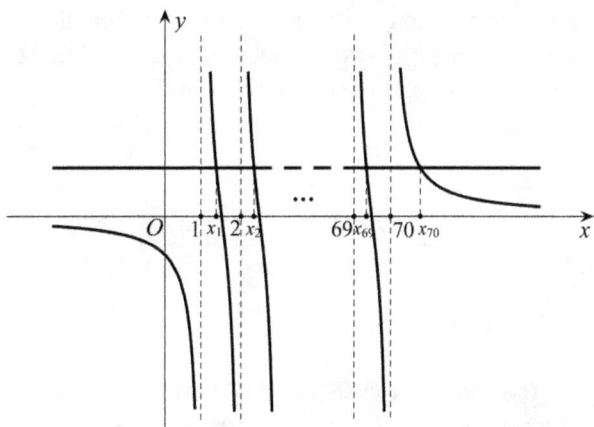

Figure 4.1 The Graphs of the Function $y = \frac{1}{x-1} + \frac{2}{x-2} + \frac{3}{x-3} + \cdots + \frac{70}{x-70}$ **and the Line** $y = \frac{5}{4}$

Calculate $x_1 + x_2 + \cdots + x_{70}$, and it is evident that the equation

$$\frac{1}{x-1} + \frac{2}{x-2} + \frac{3}{x-3} + \cdots + \frac{70}{x-70} = \frac{5}{4} \tag{1}$$

is equivalent to

$$(x - x_1)(x - x_2) \cdots (x - x_{70}) = 0. \tag{2}$$

Equation (1) can be expressed as

$$(x-2)(x-3) \cdots (x-70) + 2(x-1)(x-3) \cdots (x-70) + \cdots$$

$$+ 70(x-1)(x-2) \cdots (x-69)$$

$$= \frac{5}{4}(x-1)(x-2)(x-3) \cdots (x-70),$$

which simplifies to $x^{70} - \frac{9}{5}(1 + 2 + \cdots + 70)x^{69} + \cdots = 0$.

Equation (2) can be expressed as

$$x^{70} - (x_1 + x_2 + \cdots + x_{70})x^{69} + \cdots = 0.$$

Therefore,

$$x_1 + x_2 + \cdots + x_{70} = \frac{9}{5}(1 + 2 + \cdots + 70),$$

yielding

$$L = \frac{9}{5}(1 + 2 + \cdots + 70) - (1 + 2 + \cdots + 70) = 1988.$$

Note. Although the solution to this problem is not complicated, it encompasses a considerable range of knowledge, including the Fundamental Theorem of Algebra and the relationship between roots and coefficients.

For any positive integer m, the total length of the solution set to

$$\frac{1}{x-1} + \frac{2}{x-2} + \cdots + \frac{m}{x-m} \geq r \ (r > 0)$$

is $\frac{m(m+1)}{2r}$.

【Score Situation】 This particular problem saw the following distribution of scores among contestants: 66 contestants scored 7 points, 6 contestants scored 6 points, 3 contestants scored 5 points, 3 contestants scored 4 points, 17 contestants scored 3 points, 12 contestants scored 2 points, 25 contestants scored 1 point, and 136 contestants scored 0 point. The average score of this problem is 2.332, indicating that it had a certain level of difficulty.

Among the top five teams in the team scores, the scores of this problem are as follows: the Soviet Union team scored 38 points (with a total team score of 217 points), the China team scored 42 points (with a total team score of 201 points), the Romania team scored 42 points (with a total team score of 201 points), the Germany team scored 31 points (with a total team score of 174 points), and the Vietnam team scored 40 points (with a total team score of 166 points).

The gold medal cutoff for this IMO was set at 32 points (with 17 contestants earning gold medals), the silver medal cutoff was 23 points (with 48 contestants earning silver medals), and the bronze medal cutoff was 14 points (with 65 contestants earning bronze medals).

In this IMO, a total of five contestants achieved a perfect score of 42 points.

4.2.2 *Proving inequalities*

Problem 4.6 (IMO 3-2, proposed by Poland). Let a, b, c be the side lengths of $\triangle ABC$, and S its area. Prove that

$$a^2 + b^2 + c^2 \geq 4\sqrt{3}S.$$

In what case does the equality hold?

Proof 1. By the law of cosine and the area formulas,

$$a^2 + b^2 + c^2 - 4\sqrt{3}S$$
$$= 2a^2 + 2c^2 - 2ac\cos B - 2\sqrt{3}ac\sin B$$
$$= 2\left(a^2 + c^2 - 2ac\sin\left(B + \frac{\pi}{6}\right)\right)$$
$$\geq 2(a^2 + c^2 - 2ac)$$
$$= 2(a - c)^2 \geq 0.$$

Thus, the inequality holds, and the equality holds if and only if $a = b = c$.

Proof 2. By Heron's formula, where p is the semi-perimeter of $\triangle ABC$, the original inequality is equivalent to

$$(a^2 + b^2 + c^2)^2 \geq 48p(p - a)(p - b)(p - c)$$
$$\Leftrightarrow (a^2 + b^2 + c^2)^2 \geq 3(a + b + c)(a + b - c)(b + c - a)(c + a - b)$$
$$\Leftrightarrow (a^2 + b^2 + c^2)^2 \geq 3((b + c)^2 - a^2)(a^2 - (b - c)^2)$$
$$\Leftrightarrow (a^2 + b^2 + c^2)^2 \geq 6(a^2b^2 + b^2c^2 + c^2a^2) - 3(a^4 + b^4 + c^4)$$
$$\Leftrightarrow a^4 + b^4 + c^4 \geq a^2b^2 + b^2c^2 + c^2a^2$$
$$\Leftrightarrow (a^2 - b^2)^2 + (b^2 - c^2)^2 + (c^2 - a^2)^2 \geq 0.$$

Thus, the inequality holds, and the equality holds if and only if $a = b = c$.

Note. This problem has a variety of solutions, such as using the AM-GM inequality, or discussing cases based on the size of angle A. Additionally, this inequality is known as Weitzenböck's inequality, and it has several generalizations:

Generalization 1. Let a, b, c be the side lengths of $\triangle ABC$, and S its area. Then

$$a^2 + b^2 + c^2 \geq 4\sqrt{3}S + (a - b)^2 + (b - c)^2 + (c - a)^2.$$

Proof. Assume without loss of generality that $a \geq b \geq c$. Then

$$(a - c)^2 = ((a - b) + (b - c))^2$$

$$= (a - b)^2 + (b - c)^2 + 2(a - b)(b - c)$$

$$\geq (a - b)^2 + (b - c)^2.$$

Therefore,

$$a^2 + b^2 + c^2 - 4\sqrt{3}S = 2a^2 + 2c^2 - 2ac\cos B - 2\sqrt{3}ac\sin B$$

$$= 2\left(a^2 + c^2 - 2ac\sin\left(B + \frac{\pi}{6}\right)\right)$$

$$\geq 2(a^2 + c^2 - 2ac) = 2(a - c)^2$$

$$\geq (a - b)^2 + (b - c)^2 + (c - a)^2.$$

This inequality is called the Finsler–Hadwiger inequality, named after Paul Finsler and Hugo Hadwiger. We can also prove that:

(i) $a^2 + b^2 + c^2 \geq 4\sqrt{3}S + 2(a - b)^2$;

(ii) $a^2 + b^2 + c^2 \geq 4\sqrt{3}S + \frac{(a^2 - b^2)^2}{2c^2}$.

Generalization 2. Let a, b, c be the side lengths of $\triangle ABC$, and S its area. For $\alpha, \beta, \gamma \in \mathbf{R}_+$,

$$\alpha a^2 + \beta b^2 + \gamma c^2 \geq 4S\sqrt{\alpha\beta + \beta\gamma + \gamma\alpha},$$

with equality if and only if $\frac{a^2}{\beta + \gamma} = \frac{b^2}{\gamma + \alpha} = \frac{c^2}{\alpha + \beta}$.

This is a weighted generalization of Weitzenböck's inequality. Setting $\alpha = \beta = \gamma$ yields Weitzenböck's inequality. The proof requires the following lemmas:

Lemma 1. $16S^2 = 2(a^2b^2 + b^2c^2 + c^2a^2) - (a^4 + b^4 + c^4)$.

Lemma 2. For $\alpha, \beta, \gamma \in \mathbf{R}_+$ and $x, y, z \in \mathbf{R}$,

$$(\alpha x + \beta y + \gamma z)^2 \geq (\alpha\beta + \beta\gamma + \gamma\alpha)(2xy + 2yz + 2zx - x^2 - y^2 - z^2),$$

with equality if and only if $\frac{x}{\beta + \gamma} = \frac{y}{\gamma + \alpha} = \frac{z}{\alpha + \beta}$.

Proof of Lemma 2. The inequality follows from

$$(\alpha x + \beta y + \gamma z)^2 - (\alpha\beta + \beta\gamma + \gamma\alpha)(2xy + 2yz + 2zx - x^2 - y^2 - z^2)$$

$$= (\alpha + \beta)(\gamma + \alpha)x^2 + (\beta + \gamma)(\alpha + \beta)y^2$$

$$+(\gamma + \alpha)(\beta + \gamma)z^2 - 2\gamma(\alpha + \beta)xy$$

$$- 2\beta(\gamma + \alpha)zx - 2\alpha(\beta + \gamma)yz$$

$$= (\alpha + \beta)(\beta + \gamma)(\gamma + \alpha)\left(\gamma\left(\frac{x}{\beta + \gamma} - \frac{y}{\gamma + \alpha}\right)^2\right.$$

$$\left. + \beta\left(\frac{z}{\alpha + \beta} - \frac{x}{\beta + \gamma}\right)^2 + \alpha\left(\frac{y}{\gamma + \alpha} - \frac{z}{\alpha + \beta}\right)^2\right)$$

$$\geq 0.$$

Using Lemmas 1 and 2, and setting $x = a^2$, $y = b^2$, and $z = c^2$, we get Generalization 2.

Generalization 3. Let a, b, c and a', b', c' be the side lengths of $\triangle ABC$ and $\triangle A'B'C'$, respectively, with areas S and S'. Then

$$a'^2(b^2 + c^2 - a^2) + b'^2(c^2 + a^2 - b^2) + c'^2(a^2 + b^2 - c^2) \geq 16SS',$$

with equality if and only if the two triangles are similar.

This inequality, initially found by J. Neuberg in 1897, was rediscovered and proved by the American geometer D. Pedoe in 1942. It also attracted many Chinese scholars with its elegant symmetry. Notably, in 1981, Professors Lu Yang and Jingzhong Zhang established a similar inequality for simplices in higher-dimensional spaces.

Furthermore, there are several similar problems:

- **(British Mathematical Olympiad 2006, 1st Round, Problem 5).** For positive real numbers a, b, c, prove that

$$(a^2 + b^2)^2 \geq (a + b + c)(a + b - c)(b + c - a)(c + a - b).$$

- **(William Lowell Putnam Mathematical Competition 1998, A6).** Let A, B, C denote distinct points with integer coordinates in $\mathbf{R}^2$. Prove that if $(|AB| + |BC|)^2 < 8 \cdot [ABC] + 1$, then A, B, C are three vertices of a square.
  Here, $|XY|$ is the length of segment XY and $[ABC]$ is the area of $\triangle ABC$.

【Score Situation】This particular problem saw the following distribution of scores among contestants: 3 contestants scored 7 points, no contestant scored 6 points, 1 contestant scored 5 points, no contestant scored 4 points, no contestant scored 3 points, 1 contestant scored 2 points, no contestant scored 1 point, and 4 contestants scored 0 point. The average score of this problem is 3.111, indicating that it was relatively straightforward.

Among the top five teams in the team scores, the Hungary team achieved a total score of 270 points, the Poland team achieved a total score of 203 points, the Romania team achieved a total score of 197 points, the Czechoslovakia team achieved a total score of 159 points, and the German Democratic Republic team achieved a total score of 146 points.

The gold medal cutoff for this IMO was set at 37 points (with 3 contestants earning gold medals), the silver medal cutoff was 34 points (with 4 contestants earning silver medals), and the bronze medal cutoff was 30 points (with 4 contestants earning bronze medals).

In this IMO, only one contestant achieved a perfect score of 40 points, namely Béla Bollobás from Hungary.

Problem 4.7 (IMO 6-2, proposed by Hungary). Suppose a, b, c are the side lengths of $\triangle ABC$. Prove that

$$a^2(b + c - a) + b^2(c + a - b) + c^2(a + b - c) \leq 3abc.$$

Proof 1. Let $a = y + z$, $b = z + x$, and $c = x + y$, where $x, y, z > 0$. The original inequality is equivalent to

$$2x(y + z)^2 + 2y(z + x)^2 + 2z(x + y)^2 \leq 3(x + y)(y + z)(z + x),$$
$$6xyz \leq x^2y + y^2z + z^2x + x^2z + y^2x + z^2y. \tag{1}$$

By the AM-GM inequality,

$$
\begin{aligned}
& x^2y + y^2z + z^2x + x^2z + y^2x + z^2y \\
& \geq 6\sqrt[6]{x^2y \cdot y^2z \cdot z^2x \cdot x^2z \cdot y^2x \cdot z^2y} \\
& = 6xyz,
\end{aligned}
$$

which proves (1).

Thus, the original inequality holds, and the equality is valid if and only if $a = b = c$.

Proof 2. Since a, b, c are the side lengths of a triangle,

$$(a-b)^2(a+b-c) \geq 0,$$
$$(b-c)^2(b+c-a) \geq 0,$$
$$(c-a)^2(c+a-b) \geq 0.$$

Summing these three inequalities, we obtain

$$a^2((a+b-c)+(c+a-b)-2b-2c)$$
$$+b^2((a+b-c)+(b+c-a)-2a-2c)$$
$$+c^2((b+c-a)+(c+a-b)-2b-2a)+6abc \geq 0,$$

which simplifies to

$$-2a^2(b+c-a)-2b^2(c+a-b)-2c^2(a+b-c)+6abc \geq 0.$$

Therefore, $a^2(b+c-a)+b^2(c+a-b)+c^2(a+b-c) \leq 3abc$, and the equality holds if and only if $a=b=c$.

Note. The substitution used in Proof 1 is often called the incircle substitution. As illustrated in Figure 4.2, consider the incircle of $\triangle ABC$, with D, E, F being the tangency points on the three sides, respectively. Let $AE = AF = x$, $BD = BF = y$, and $CD = CE = z$. Then $a = y + z$, $b = z + x$, and $c = x + y$.

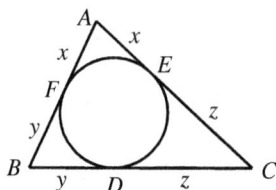

Figure 4.2 The Incircle of $\triangle ABC$

This substitution is very useful in proving inequalities related to the side lengths of a triangle.

Similar to Proof 1, this problem can also be approached by using

$$\frac{1}{8}(x+y)(y+z)(z+x) \geq \sqrt{xy} \cdot \sqrt{yz} \cdot \sqrt{zx} = xyz,$$

which implies $abc \geq (b+c-a)(c+a-b)(a+b-c)$. Expanding this leads to the inequality to be proved.

Furthermore, there are several similar problems:

- **(Austrian–Polish Mathematical Competition 2001, Problem 3).** Let a, b, c be side lengths of a triangle. Prove that

$$2 < \frac{a+b}{c} + \frac{b+c}{a} + \frac{c+a}{b} - \frac{a^3 + b^3 + c^3}{abc} \leq 3.$$

- **(Swedish Mathematical Competition 1982, Problem 2).** Show that $abc \geq (a+b-c)(b+c-a)(c+a-b)$ for positive reals a, b, c.
- **(Proposed by Adrian Andreescu).** Prove that for any non-negative real numbers a, b, c,

$$(a - 2b + 4c)(c - 2a + 4b)(b - 2c + 4a) \leq 27abc.$$

【Score Situation】 This particular problem saw the following distribution of scores among contestants: 7 contestants scored 7 points, no contestant scored 6 points, 1 contestant scored 5 points, no contestant scored 4 points, no contestant scored 3 points, no contestant scored 2 points, 2 contestants scored 1 point, and 7 contestants scored 0 point. The average score of this problem is 3.294, indicating that it was relatively straightforward.

Among the top five teams in the team scores, the scores of this problem are as follows: the Soviet Union team scored 54 points (with a total team score of 269 points), the Hungary team scored 37 points (with a total team score of 253 points), the Romania team scored 35 points (with a total team score of 213 points), the Poland team scored 33 points (with a total team score of 209 points), and the Bulgaria team scored 29 points (with a total team score of 198 points).

The gold medal cutoff for this IMO was set at 38 points (with 7 contestants earning gold medals), the silver medal cutoff was 31 points (with 9 contestants earning silver medals), and the bronze medal cutoff was 27 points (with 19 contestants earning bronze medals).

In this IMO, only one contestant achieved a perfect score of 42 points, namely David Bernstein from the Soviet Union.

Problem 4.8 (IMO 11-6, proposed by the Soviet Union).

Prove that for all real numbers $x_1, x_2, y_1, y_2, z_1, z_2$, with $x_1 > 0$, $x_2 > 0$, $x_1 y_1 - z_1^2 > 0$, and $x_2 y_2 - z_2^2 > 0$, the inequality

$$\frac{8}{(x_1 + x_2)(y_1 + y_2) - (z_1 + z_2)^2} \leq \frac{1}{x_1 y_1 - z_1^2} + \frac{1}{x_2 y_2 - z_2^2}$$

is satisfied. Give necessary and sufficient conditions for equality.

Proof 1. From $x_1y_1 > z_1^2 \geq 0$ and $x_1 > 0$, it follows that $y_1 > 0$. Similarly, $y_2 > 0$. Therefore,

$$x_1y_1z_2^2 + x_2y_2z_1^2 \geq 2z_1z_2\sqrt{x_1x_2y_1y_2},$$

$$(\sqrt{x_1x_2y_1y_2} - z_1z_2)^2 = x_1x_2y_1y_2 - 2z_1z_2\sqrt{x_1x_2y_1y_2} + z_1^2z_2^2$$

$$\geq x_1x_2y_1y_2 + z_1^2z_2^2 - (x_1y_1z_2^2 + x_2y_2z_1^2)$$

$$= (x_1y_1 - z_1^2)(x_2y_2 - z_2^2). \tag{1}$$

Also

$$(x_1 + x_2)(y_1 + y_2) - (z_1 + z_2)^2$$

$$= (x_1y_1 - z_1^2) + (x_2y_2 - z_2^2) + (x_1y_2 + x_2y_1 - 2z_1z_2)$$

$$\geq (x_1y_1 - z_1^2) + (x_2y_2 - z_2^2) + 2(\sqrt{x_1x_2y_1y_2} - z_1z_2)$$

$$\geq (x_1y_1 - z_1^2) + (x_2y_2 - z_2^2) + 2\sqrt{(x_1y_1 - z_1^2)(x_2y_2 - z_2^2)} \text{ (Using (1))}$$

$$\geq 4\sqrt{(x_1y_1 - z_1^2)(x_2y_2 - z_2^2)},$$

which implies

$$\frac{8}{(x_1 + x_2)(y_1 + y_2) - (z_1 + z_2)^2} \leq \frac{2}{\sqrt{(x_1y_1 - z_1^2)(x_2y_2 - z_2^2)}}$$

$$\leq \frac{1}{x_1y_1 - z_1^2} + \frac{1}{x_2y_2 - z_2^2}.$$

The equality holds if and only if $x_1 = x_2$, $y_1 = y_2$, and $z_1 = z_2$.

Proof 2. Let $A_1 = x_1y_1 - z_1^2$ and $A_2 = x_2y_2 - z_2^2$, where $A_1, A_2 > 0$. The original inequality becomes

$$\left(\frac{1}{A_1} + \frac{1}{A_2}\right) \cdot (A_1 + A_2 + x_1y_2 + x_2y_1 - 2z_1z_2) \geq 8.$$

Also

$$\left(\frac{1}{A_1} + \frac{1}{A_2}\right) \cdot (A_1 + A_2 + x_1y_2 + x_2y_1 - 2z_1z_2)$$

$$\geq \left(\frac{1}{A_1} + \frac{1}{A_2}\right) \cdot (A_1 + A_2 + 2\sqrt{x_1y_2x_2y_1} - 2z_1z_2)$$

$$= \left(\frac{1}{A_1} + \frac{1}{A_2}\right) \cdot \left(A_1 + A_2 + 2\sqrt{(A_1 + z_1^2)(A_2 + z_2^2)} - 2z_1z_2\right)$$

$$\geq \left(\frac{1}{A_1} + \frac{1}{A_2}\right) \cdot \left(A_1 + A_2 + 2(\sqrt{A_1A_2} + z_1z_2) - 2z_1z_2\right)$$

(using the Cauchy–Schwarz inequality)

$$= \left(\frac{1}{A_1} + \frac{1}{A_2}\right) \cdot \left(A_1 + A_2 + 2\sqrt{A_1A_2}\right)$$

$$= \left(\frac{1}{A_1} + \frac{1}{A_2}\right) \cdot (A_1 + A_2) + 2\frac{A_1 + A_2}{\sqrt{A_1A_2}}$$

$$\geq \left(\sqrt{\frac{1}{A_1}}\sqrt{A_1} + \sqrt{\frac{1}{A_2}}\sqrt{A_2}\right)^2 + 2\frac{2\sqrt{A_1A_2}}{\sqrt{A_1A_2}}$$

$$= 4 + 2 \cdot 2 = 8.$$

Hence, the original inequality holds. The equality holds if and only if $x_1 = x_2$, $y_1 = y_2$, and $z_1 = z_2$.

Note. The A_1 and A_2 in the problem can be represented as the determinants of the following matrices $\boldsymbol{M}_1$ and $\boldsymbol{M}_2$ respectively, namely

$$\boldsymbol{M}_1 = \begin{pmatrix} x_1 & z_1 \\ z_1 & y_1 \end{pmatrix} \quad \text{and} \quad \boldsymbol{M}_2 = \begin{pmatrix} x_2 & z_2 \\ z_2 & y_2 \end{pmatrix},$$

and the inequality to be proved can be written as

$$\frac{8}{\det(\boldsymbol{M}_1 + \boldsymbol{M}_2)} \leq \frac{1}{\det \boldsymbol{M}_1} + \frac{1}{\det \boldsymbol{M}_2}.$$

If $\boldsymbol{M}_1$ and $\boldsymbol{M}_2$ are $n \times n$ positive definite matrices, then the above inequality still holds.

Furthermore, there are several similar problems:

- **(Austrian–Polish Mathematical Competition 1998, Problem 1).**
  Let x_1, x_2, y_1, y_2 be real numbers such that $x_1^2 + x_2^2 \leq 1$. Prove that

$$(x_1y_1 + x_2y_2 - 1)^2 \geq (x_1^2 + x_2^2 - 1)(y_1^2 + y_2^2 - 1).$$

- **(From the Chinese journal *High-School Mathematics*, 1992(03): 26–27).** Let $a_1, a_2, b_1, b_2, c_1, c_2$ be real numbers such that $a_1a_2 > 0$, $a_1c_1 \geq b_1^2$, and $a_2c_2 \geq b_2^2$. Prove that

$$(a_1 + a_2)(c_1 + c_2) \geq (b_1 + b_2)^2 + 2\sqrt{(a_1c_1 - b_1^2)(a_2c_2 - b_2^2)}.$$

- (**Eötvös Mathematics Competition 1939, Problem 1**). Let $a_1, a_2, b_1, b_2, c_1, c_2$ be real numbers such that $a_1 a_2 > 0$, $a_1 c_1 \geq b_1^2$, and $a_2 c_2 \geq b_2^2$. Prove that

$$(a_1 + a_2)(c_1 + c_2) \geq (b_1 + b_2)^2.$$

【Score Situation】 This particular problem saw the following distribution of scores among contestants: 7 contestants scored 8 points, 2 contestants scored 7 points, 3 contestants scored 6 points, 9 contestants scored 5 points, 13 contestants scored 4 points, 13 contestants scored 3 points, 17 contestants scored 2 points, 15 contestants scored 1 point, and 33 contestants scored 0 point. The average score of this problem is 2.438, indicating that it had a certain level of difficulty.

Among the top five teams in the team scores, the scores of this problem are as follows: the Hungary team scored 30 points (with a total team score of 247 points), the German Democratic Republic team scored 29 points (with a total team score of 240 points), the Soviet Union team scored 29 points (with a total team score of 231 points), the Romania team scored 33 points (with a total team score of 219 points), and the United Kingdom team scored 23 points (with a total team score of 193 points).

The gold medal cutoff for this IMO was set at 40 points (with 3 contestants earning gold medals), the silver medal cutoff was 30 points (with 20 contestants earning silver medals), and the bronze medal cutoff was 24 points (with 21 contestants earning bronze medals).

In this IMO, only three contestants achieved a perfect score of 40 points, namely Tibor Fiala from Hungary, Vladimir Drinfeld from the Soviet Union, and Simon Phillips Norton from the United Kingdom.

Problem 4.9 (IMO 13-1, proposed by Hungary). Prove that the following assertion is true for $n = 3$ and $n = 5$, and that it is false for every other natural number $n > 2$:

If $a_1, a_2, \ldots, a_n$ are arbitrary real numbers, then

$$(a_1 - a_2)(a_1 - a_3) \cdots (a_1 - a_n) + (a_2 - a_1)(a_2 - a_3) \cdots (a_2 - a_n)$$
$$+ \cdots + (a_n - a_1)(a_n - a_2) \cdots (a_n - a_{n-1}) \geq 0.$$

Proof. Without loss of generality, assume $a_1 \leq a_2 \leq \cdots \leq a_n$. Let A_n denote the given expression.

If n is even, let $a_1 < a_2 = a_3 = \cdots = a_n$, then

$$A_n = (-1)^{n-1}(a_2 - a_1)^{n-1} < 0,$$

and therefore, the inequality does not hold.

If $n = 3$, then

$$A_3 = (a_1 - a_2)(a_1 - a_3) + (a_2 - a_1)(a_2 - a_3) + (a_3 - a_1)(a_3 - a_2)$$

$$= a_1^2 + a_2^2 + a_3^2 - a_1 a_2 - a_2 a_3 - a_3 a_1$$

$$= \frac{1}{2}((a_1 - a_2)^2 + (a_2 - a_3)^2 + (a_3 - a_1)^2) \geq 0.$$

If $n = 5$, then

$$A_5 = (a_1 - a_2)(a_1 - a_3)(a_1 - a_4)(a_1 - a_5)$$

$$+ (a_2 - a_1)(a_2 - a_3)(a_2 - a_4)(a_2 - a_5)$$

$$+ (a_3 - a_1)(a_3 - a_2)(a_3 - a_4)(a_3 - a_5)$$

$$+ (a_4 - a_1)(a_4 - a_2)(a_4 - a_3)(a_4 - a_5)$$

$$+ (a_5 - a_1)(a_5 - a_2)(a_5 - a_3)(a_5 - a_4),$$

where the sum of the first two terms is

$$(a_2 - a_1)((a_3 - a_1)(a_4 - a_1)(a_5 - a_1) - (a_3 - a_2)(a_4 - a_2)(a_5 - a_2)) \geq 0,$$

and similarly, the sum of the last two terms is also non-negative. Additionally, the third term is non-negative, so $A_5 \geq 0$.

If n is odd and $n \geq 7$, let $a_1 = a_2 = a_3 < a_4 < a_5 = a_6 = \cdots = a_n$, then

$$A_n = (a_4 - a_1)(a_4 - a_2) \cdots (a_4 - a_n) < 0.$$

In conclusion, $A_n \geq 0$ for any real numbers $a_1, a_2, \ldots, a_n$ if and only if $n = 3$ or $n = 5$.

Note. The homogeneous polynomial of degree 4,

$$A_5 = \prod_{j \neq 1} (a_1 - a_j) + \prod_{j \neq 2} (a_2 - a_j) + \prod_{j \neq 3} (a_3 - a_j) + \prod_{j \neq 4} (a_4 - a_j)$$

$$+ \prod_{j \neq 5} (a_5 - a_j),$$

cannot be written as a sum of squares of quadratic forms in a_1, a_2, a_3, a_4, a_5, and we can prove it by the following lemma.

Lemma. *Let Q be a quadratic form in a_1, a_2, a_3, a_4, a_5, which is zero whenever the a_i satisfy*

$$a_1 = a_2, \quad a_3 = a_4 = a_5$$

or some permutation. Then $Q = 0$ for all values of the a_i.

In fact, the general theory of these forms goes back to Hilbert, who showed that every non-negative quartic form in three variables can be written as a sum of squares, but gave an example of a quartic form in four variables which cannot be so represented. In his famous list of 23 problems, Hilbert posed the following problem:

Suppose that $f \in \mathbf{R}(x_1, x_2, \ldots, x_n)$ is non-negative at all points of $\mathbf{R}^n$ where f is defined. Is f a finite sum of squares of rational functions?

This is indeed always possible and was proved by Emil Artin in 1927.

Furthermore, there is another problem related to Hilbert's 17th Problem:

- **(Chinese Team Selection Test 2017, Problem 29).** For any $x, y, z \in \mathbf{R}$, a three-variable real-coefficient polynomial

$$f(x, y, z) = f_2(x, y)z^2 + 2f_3(x, y)z + f_4(x, y) \geq 0,$$

where $f_k(x, y)$ are homogeneous polynomials of degree k ($k = 2, 3, 4$). If there is a real-coefficient polynomial $r(x, y)$ such that

$$f_2(x, y)f_4(x, y) - f_3^2(x, y) = (r(x, y))^2,$$

prove that there must be two real-coefficient polynomials $g(x, y, z)$ and $h(x, y, z)$ such that

$$f(x, y, z) = g^2(x, y, z) + h^2(x, y, z).$$

【Score Situation】This particular problem saw the following distribution of scores among contestants: 4 contestants scored 5 points, 2 contestants scored 4 points, 3 contestants scored 3 points, 11 contestants scored 2 points, 6 contestants scored 1 point, and 2 contestants scored 0 point. The average score of this problem is 2.321, indicating that it had a certain level of difficulty.

Among the top five teams in the team scores, the scores of this problem are as follows: the Hungary team scored 35 points (with a total team score of 255 points), the Soviet Union team scored 34 points (with a total team score of 205 points), the German Democratic Republic team scored 20 points (with a total team score of 142 points), the Poland team scored 29 points (with a total team score of 118 points), the Romania team scored 22 points (with a total team score of 110 points), and the United Kingdom team scored 21 points (with a total team score of 110 points).

The gold medal cutoff for this IMO was set at 35 points (with 7 contestants earning gold medals), the silver medal cutoff was 23 points (with 12 contestants earning silver medals), and the bronze medal cutoff was 11 points (with 29 contestants earning bronze medals).

In this IMO, only one contestant achieved a perfect score of 42 points, namely Imre Ruzsa from Hungary.

Problem 4.10 (IMO 17-1, proposed by Czechoslovakia). Let x_i, $y_i (i = 1, 2, \ldots, n)$ be real numbers such that

$$x_1 \geq x_2 \geq \cdots \geq x_n \quad \text{and} \quad y_1 \geq y_2 \geq \cdots \geq y_n.$$

Prove that, if $z_1, z_2, \ldots, z_n$ is any permutation of $y_1, y_2, \ldots, y_n$, then

$$\sum_{i=1}^{n} (x_i - y_i)^2 \leq \sum_{i=1}^{n} (x_i - z_i)^2.$$

Proof.   The original inequality is equivalent to

$$\sum_{i=1}^{n} x_i^2 - 2\sum_{i=1}^{n} x_i y_i + \sum_{i=1}^{n} y_i^2 \leq \sum_{i=1}^{n} x_i^2 - 2\sum_{i=1}^{n} x_i z_i + \sum_{i=1}^{n} z_i^2$$

$$\Leftrightarrow \sum_{i=1}^{n} x_i z_i \leq \sum_{i=1}^{n} x_i y_i \quad \left(\sum_{i=1}^{n} y_i^2 = \sum_{i=1}^{n} z_i^2 \right),$$

which is true according to the "mixed order sum" and "same order sum" from the rearrangement inequality. Therefore, the original inequality is valid.

【Score Situation】 This particular problem saw the following distribution of scores among contestants: 76 contestants scored 6 points, 12 contestants scored 5 points, 11 contestants scored 4 points, 5 contestants scored 3 points, 9 contestants scored 2 points, 12 contestants scored 1 point, and 3 contestants scored 0 point. The average score of this problem is 4.727, indicating that it was simple.

Among the top five teams in the team scores, the scores of this problem are as follows: the Hungary team scored 47 points (with a total team score of 258 points), the German Democratic Republic team scored 46 points (with a total team score of 249 points), the United States team scored 43 points (with a total team score of 247 points), the Soviet Union team scored 48 points (with a total team score of 246 points), and the United Kingdom team scored 45 points (with a total team score of 241 points).

The gold medal cutoff for this IMO was set at 38 points (with 8 contestants earning gold medals), the silver medal cutoff was 32 points (with 25 contestants earning silver medals), and the bronze medal cutoff was 23 points (with 36 contestants earning bronze medals).

In this IMO, a total of six contestants achieved a perfect score of 40 points.

Problem 4.11 (IMO 19-4, proposed by the United Kingdom).

Four real constants a, b, A, B are given, and

$$f(\theta) = 1 - a\cos\theta - b\sin\theta - A\cos 2\theta - B\sin 2\theta.$$

Prove that if $f(\theta) \geq 0$ for all real θ, then

$$a^2 + b^2 \leq 2 \quad \text{and} \quad A^2 + B^2 \leq 1.$$

Proof. Since for any real number θ,

$$f(\theta) = 1 - a\cos\theta - b\sin\theta - A\cos 2\theta - B\sin 2\theta \geq 0, \tag{1}$$

so it follows that

$$f\left(\theta + \frac{\pi}{2}\right) = 1 + a\sin\theta - b\cos\theta + A\cos 2\theta + B\sin 2\theta \geq 0, \tag{2}$$

$$f(\theta + \pi) = 1 + a\cos\theta + b\sin\theta - A\cos 2\theta - B\sin 2\theta \geq 0. \tag{3}$$

Adding (1) and (2),

$$2 + (a - b)\sin\theta - (a + b)\cos\theta \geq 0,$$

$$2 + \sqrt{(a - b)^2 + (a + b)^2}\sin(\theta - \varphi_1) \geq 0,$$

where $\cos\varphi_1 = \dfrac{a-b}{\sqrt{(a-b)^2+(a+b)^2}}$ and $\sin\varphi_1 = \dfrac{a+b}{\sqrt{(a-b)^2+(a+b)^2}}$. Therefore,

$$2 - \sqrt{(a-b)^2 + (a+b)^2} \geq 0,$$

which implies $a^2 + b^2 \leq 2$.

By (1) and (3),

$$1 - A\cos 2\theta - B\sin 2\theta \geq 0,$$

$$1 - \sqrt{A^2 + B^2}\cos(2\theta - \varphi_2) \geq 0,$$

where $\cos\varphi_2 = \dfrac{A}{\sqrt{A^2+B^2}}$ and $\sin\varphi_2 = \dfrac{B}{\sqrt{A^2+B^2}}$. Consequently, $1 - \sqrt{A^2+B^2} \geq 0$, which implies $A^2 + B^2 \leq 1$.

Note. The expression of $f(\theta)$ can also be transformed into

$$f(\theta) = 1 - r\cos(\theta - \alpha) - R\cos 2(\theta - \beta),$$

where $r = \sqrt{a^2+b^2}$ and $R = \sqrt{A^2+B^2}$. Alternatively, one can employ proof by contradiction to solve this problem.

In fact, the problem is a special case of the theorem: If

$$f(\theta) = 1 + \sum_{j=1}^{n} (a_j \cos j\theta + b_j \sin j\theta) \geq 0,$$

then $a_j^2 + b_j^2 \leq 2$ for $j = 1, 2, \ldots, n-1$ and $a_n^2 + b_n^2 \leq 1$.

This is an important property of Fourier series of positive functions.

【Score Situation】This particular problem saw the following distribution of scores among contestants: 16 contestants scored 6 points, 1 contestant scored 5 points, 1 contestant scored 4 points, 1 contestant scored 3 points, 2 contestants scored 2 points, 7 contestants scored 1 point, and 9 contestants scored 0 point. The average score of this problem is 3.216, indicating that it was relatively straightforward.

Among the top five teams in the team scores, the United States team achieved a total score of 202 points, the Soviet Union team achieved a total score of 192 points, the Hungary team achieved a total score of 190 points, the United Kingdom team achieved a total score of 190 points, and the Netherlands team achieved a total score of 185 points.

The gold medal cutoff for this IMO was set at 34 points (with 13 contestants earning gold medals), the silver medal cutoff was 24 points (with 29 contestants earning silver medals), and the bronze medal cutoff was 17 points (with 35 contestants earning bronze medals).

In this IMO, a total of five contestants achieved a perfect score of 40 points.

Problem 4.12 (IMO 20-5, proposed by France).

Let $\{a_k\}(k = 1, 2, 3, \ldots, n, \ldots)$ be a sequence of distinct positive integers. Prove that for all positive integers n,

$$a_1 + \frac{a_2}{4} + \frac{a_3}{9} + \cdots + \frac{a_n}{n^2} \geq 1 + \frac{1}{2} + \frac{1}{3} + \cdots + \frac{1}{n}.$$

Proof 1. For any positive integer n, let $b_1, b_2, \ldots, b_n$ be a permutation of $a_1, a_2, \ldots, a_n$ such that $b_1 < b_2 < \cdots < b_n$, implying $b_i \geq i$ for $i = 1, 2, \ldots, n$.

Since $b_1, b_2, \ldots, b_n$ and $1, \frac{1}{2^2}, \ldots, \frac{1}{n^2}$ are in reverse order, by the rearrangement inequality,

$$\sum_{k=1}^{n} \frac{a_k}{k^2} \geq \sum_{k=1}^{n} \frac{b_k}{k^2} \geq \sum_{k=1}^{n} \frac{k}{k^2} = \sum_{k=1}^{n} \frac{1}{k}.$$

Proof 2. Since the positive integers $a_1, a_2, \ldots, a_k$ are distinct,

$$a_1 + a_2 + \cdots + a_k \geq 1 + 2 + \cdots + k.$$

Consequently,

$$\frac{a_1 - 1}{1^2} + \frac{a_2 - 2}{2^2} + \cdots + \frac{a_n - n}{n^2}$$

$$\geq \frac{a_1 + a_2 - (1 + 2)}{2^2} + \frac{a_3 - 3}{3^2} + \cdots + \frac{a_n - n}{n^2}$$

$$\geq \frac{a_1 + a_2 + a_3 - (1 + 2 + 3)}{3^2} + \frac{a_4 - 4}{4^2} + \cdots + \frac{a_n - n}{n^2}$$

$$\cdots\cdots\cdots\cdots\cdots\cdots\cdots\cdots\cdots\cdots$$

$$\geq \frac{a_1 + a_2 + \cdots + a_n - (1 + 2 + \cdots + n)}{n^2} \geq 0,$$

and thus, $a_1 + \frac{a_2}{4} + \frac{a_3}{9} + \cdots + \frac{a_n}{n^2} \geq 1 + \frac{1}{2} + \frac{1}{3} + \cdots + \frac{1}{n}$.

Note. There are several similar problems:

- (**Japan Team Selection Test 2022, Problem 4; IMO Shortlist 2021, Algebra, Problem 3**). Given a positive integer n, find the smallest value of

$$\left\lfloor \frac{a_1}{1} \right\rfloor + \left\lfloor \frac{a_2}{2} \right\rfloor + \cdots + \left\lfloor \frac{a_n}{n} \right\rfloor$$

over all permutations $(a_1, a_2, \ldots, a_n)$ of $(1, 2, \ldots, n)$.

- (**Polish Mathematical Olympiad Finals 2008, Problem 1**). The numbers $1, 2, \ldots, n^2$ are arranged in the cells of an $n \times n$ board in such a way that the numbers $1, 2, \ldots, n$ are in the first row (in this order), $n+1, n+2, \ldots, 2n$ in the second, etc. We choose n cells of the board, no two of which are in the same row or column. Let a_i be the chosen number in the ith row. Prove that

$$\frac{1^2}{a_1} + \frac{2^2}{a_2} + \cdots + \frac{n^2}{a_n} \geq \frac{n+2}{2} - \frac{1}{n^2+1}.$$

- (**Polish Mathematical Olympiad Finals 2007, Problem 6**). A sequence of real numbers $a_0, a_1, a_2, \ldots$ is defined by $a_0 = -1$ and

$$a_n + \frac{a_{n-1}}{2} + \frac{a_{n-2}}{3} + \cdots + \frac{a_1}{n} + \frac{a_0}{n+1} = 0 \quad \text{for } n \geq 1.$$

Show that $a_n > 0$ for $n \geq 1$.

- (**Polish Mathematical Olympiad Finals 1995, Problem 4**). Let n be a positive integer. Find the smallest possible value of

$$x_1 + \frac{x_2^2}{2} + \frac{x_3^3}{3} + \cdots + \frac{x_n^n}{n},$$

where $x_1, x_2, \ldots, x_n$ are positive reals such that $\frac{1}{x_1} + \frac{1}{x_2} + \cdots + \frac{1}{x_n} = n$.

- (**All Soviet Union Mathematical Olympiad 1979**). For every n, a decreasing sequence $\{x_k\}$ satisfies a condition

$$x_1 + \frac{x_4}{2} + \frac{x_9}{3} + \cdots + \frac{x_{n^2}}{n} \leq 1.$$

Prove that for every n, it also satisfies $x_1 + \frac{x_2}{2} + \frac{x_3}{3} + \cdots + \frac{x_n}{n} \leq 3$.

【Score Situation】This particular problem saw the following distribution of scores among contestants: 39 contestants scored 6 points, 1 contestant scored 5 points, no contestant scored 4 points, 2 contestants scored 3 points, no contestant scored 2 points, 3 contestants scored 1 point, and 3 contestants scored 0 point. The average score of this problem is 5.167, indicating that it was simple.

Among the top five teams in the team scores, the scores of this problem are as follows: the Romania team scored 48 points (with a total team score of 237 points), the United States team scored 48 points (with a total team score of 225 points), the United Kingdom team scored 45 points (with a total team score of 201 points), the Vietnam team scored 45 points (with a total team score of 200 points), and the Czechoslovakia team scored 45 points (with a total team score of 195 points).

The gold medal cutoff for this IMO was set at 35 points (with 5 contestants earning gold medals), the silver medal cutoff was 27 points (with 20 contestants earning silver medals), and the bronze medal cutoff was 22 points (with 38 contestants earning bronze medals).

In this IMO, only one contestant achieved a perfect score of 40 points, namely Mark Kleiman from the United States.

Problem 4.13 (IMO 24-6, proposed by the United States). Let a, b, and c be the lengths of the sides of a triangle. Prove that

$$a^2b(a-b) + b^2c(b-c) + c^2a(c-a) \geq 0.$$

Determine when the equality occurs.

Proof 1. Let $a = y + z$, $b = z + x$, and $c = x + y$, where $x, y, z > 0$. The original inequality then becomes:

The left-hand side

$$
\begin{aligned}
&= (y+z)^2(z+x)(y-x) + (z+x)^2(x+y)(z-y) \\
&\quad + (x+y)^2(y+z)(x-z) \\
&= 2(xy^3 + yz^3 + zx^3 - x^2yz - xy^2z - xyz^2) \\
&= 2(xy(y-z)^2 + yz(z-x)^2 + zx(x-y)^2) \\
&\geq 0,
\end{aligned}
$$

with the equality holding when $x = y = z$.

Proof 2. Without loss of generality, assume $a = \max\{a, b, c\}$. The original inequality then becomes:

The left-hand side $= a(b-c)^2(b+c-a) + b(a-b)(a-c)(a+b-c) \geq 0$.

Note. In this problem, the inequality to be proven is not symmetric, so one cannot assume $a \geq b \geq c$ without loss of generality. Additionally, in

Proof 1, the Cauchy–Schwarz inequality can also be employed, as shown below:

$$(xy^3 + yz^3 + zx^3)(z + x + y)$$
$$\geq (y\sqrt{xyz} + z\sqrt{xyz} + x\sqrt{xyz})^2 = xyz(x + y + z)^2,$$

which might be more straightforward than completing the square.

Furthermore, there is a similar problem:

- **(From Oles Dobosevych's paper "On a Method of Proving Symmetric Inequalities," 2009: 3).** Prove that

$$a^3 b + b^3 c + c^3 a \geq abc(a + b + c)$$

for all non-negative real numbers a, b, c.

【Score Situation】This particular problem saw the following distribution of scores among contestants: 6 contestants scored 7 points, no contestant scored 6 points, 1 contestant scored 5 points, 7 contestants scored 4 points, 2 contestants scored 3 points, no contestant scored 2 points, 4 contestants scored 1 point, and 20 contestants scored 0 point. The average score of this problem is 2.125, indicating that it had a certain level of difficulty.

Among the top five teams in the team scores, the Germany team achieved a total score of 212 points, the United States team achieved a total score of 171 points, the Hungary team achieved a total score of 170 points, the Soviet Union team achieved a total score of 169 points, and the Romania team achieved a total score of 161 points.

The gold medal cutoff for this IMO was set at 38 points (with 9 contestants earning gold medals), the silver medal cutoff was 26 points (with 27 contestants earning silver medals), and the bronze medal cutoff was 15 points (with 57 contestants earning bronze medals).

In this IMO, a total of four contestants achieved a perfect score of 42 points.

Problem 4.14 (IMO 25-1, proposed by Germany). Prove that $0 \leq yz + zx + xy - 2xyz \leq \frac{7}{27}$, where x, y, and z are non-negative real numbers for which $x + y + z = 1$.

Proof 1. Without loss of generality, assume $x \geq y \geq z$. Then $z \leq \frac{1}{3}$, and we have

$$xy + yz + zx - 2xyz = xy(1 - 2z) + yz + zx \geq 0,$$

with equality when $x = 1$ and $y = z = 0$.

Let $z = \frac{1}{3} - k$, where $0 \le k \le \frac{1}{3}$. Hence $x + y = \frac{2}{3} + k$. Further,

$$xy + yz + zx - 2xyz = xy(1 - 2z) + z(x + y)$$

$$= xy\left(\frac{1}{3} + 2k\right) + \left(\frac{1}{3} - k\right)\left(\frac{2}{3} + k\right)$$

$$\le \frac{(x + y)^2}{4}\left(\frac{1}{3} + 2k\right) + \left(\frac{1}{3} - k\right)\left(\frac{2}{3} + k\right)$$

$$= \left(\frac{1}{3} + \frac{k}{2}\right)^2\left(\frac{1}{3} + 2k\right) + \left(\frac{1}{3} - k\right)\left(\frac{2}{3} + k\right)$$

$$= \frac{1}{2}k^3 - \frac{1}{4}k^2 + \frac{7}{27}$$

$$= \frac{1}{2}k^2\left(k - \frac{1}{2}\right) + \frac{7}{27} \le \frac{7}{27},$$

where the equality holds when $k = 0$ and $x = y$, i.e., when $x = y = z = \frac{1}{3}$.

Proof 2. Since $0 \le x, y, z \le 1$,

$$xy + yz + zx - 2xyz \ge xy + yz + zx - yz - zx = xy \ge 0.$$

Also,

$$(1 - 2x)(1 - 2y)(1 - 2z) = 1 - 2(x + y + z) + 4(xy + yz + zx) - 8xyz$$

$$= -1 + 4(xy + yz + zx - 2xyz).$$

When $1 - 2x$, $1 - 2y$, and $1 - 2z$ are all non-negative real numbers,

$$(1 - 2x)(1 - 2y)(1 - 2z) \le \left(\frac{(1 - 2x) + (1 - 2y) + (1 - 2z)}{3}\right)^3 = \frac{1}{27}. \quad (1)$$

Since $x + y + z = 1$, at most one of $1 - 2x$, $1 - 2y$, and $1 - 2z$ is negative. In this case, inequality (1) still holds. Therefore,

$$-1 + 4(xy + yz + zx - 2xyz) \le \frac{1}{27},$$

implying $xy + yz + zx - 2xyz \le \frac{7}{27}$.

Note. The original inequality can also be proven by using the adjustment method or *pqr* method. Furthermore, if $x_i (i = 1, 2, \ldots, n)$ are non-negative real numbers satisfying $x_1 + x_2 + \cdots + x_n = 1$, then

$$0 \le \sum_{1 \le i < j \le n} x_i x_j - \frac{n - 1}{n - 2} \sum_{1 \le i < j < k \le n} x_i x_j x_k \le \frac{1}{6}\left(1 - \frac{1}{n}\right)\left(2 + \frac{1}{n}\right).$$

Furthermore, there are several similar problems:

- (**From the book *Problems and Solutions in Mathematical Olympiad High School 2*, 2022: 31**). Let x, y, z be non-negative real numbers such that $x + y + z = 1$. Prove that

$$\frac{1}{2} \le x^2 + y^2 + z^2 + 6xyz \le 1.$$

- (**Chinese Team Selection Test 2019, Problem 13**). Let x, y, z be complex numbers satisfying $|x|^2 + |y|^2 + |z|^2 = 1$. Prove that

$$|x^3 + y^3 + z^3 - 3xyz| \le 1.$$

- (**British Mathematical Olympiad 2010, 2nd Round, Problem 4**). Prove that for all positive real numbers x, y, and z,

$$4(x + y + z)^3 > 27(x^2 y + y^2 z + z^2 x).$$

- (**From Mildorf's paper "Olympiad Inequalities," 2005: 20**). Let $0 \le a, b, c \le \frac{1}{2}$ be real numbers with $a + b + c = 1$. Show that

$$a^3 + b^3 + c^3 + 4abc \le \frac{9}{32}.$$

- (**British Mathematical Olympiad 2002, 1st Round, Problem 3**). Let x, y, z be positive real numbers such that $x^2 + y^2 + z^2 = 1$. Prove that

$$x^2 yz + xy^2 z + xyz^2 \le \frac{1}{3}.$$

- (**United States of America Mathematical Olympiad 2001, Problem 3**). Let a, b, c be non-negative real numbers such that $a^2 + b^2 + c^2 + abc = 4$. Show that

$$0 \le ab + bc + ca - abc \le 2.$$

- (**Canadian Mathematical Olympiad 1999, Problem 5**). Let x, y, and z be non-negative real numbers satisfying $x + y + z = 1$. Show that

$$x^2 y + y^2 z + z^2 x \le \frac{4}{27},$$

and find when the equality occurs.

- **(British Mathematical Olympiad 1999, 2nd Round, Problem 3).** Non-negative real numbers p, q, and r satisfy $p + q + r = 1$. Prove that

$$7(pq + qr + rp) \leq 2 + 9pqr.$$

- **(All Soviet Union Mathematical Olympiad 1989).** Prove that if a, b, c are the lengths of the sides of a triangle and $a + b + c = 1$, then the following inequality holds:

$$a^2 + b^2 + c^2 + 4abc < \frac{1}{2}.$$

【Score Situation】 This particular problem saw the following distribution of scores among contestants: 86 contestants scored 7 points, 10 contestants scored 6 points, 7 contestants scored 5 points, 9 contestants scored 4 points, 23 contestants scored 3 points, 35 contestants scored 2 points, 3 contestants scored 1 point, and 19 contestants scored 0 point. The average score of this problem is 4.557, indicating that it was simple.

Among the top five teams in the team scores, the scores of this problem are as follows: the Soviet Union team scored 42 points (with a total team score of 235 points), the Bulgaria team scored 42 points (with a total team score of 203 points), the Romania team scored 42 points (with a total team score of 199 points), the Hungary team scored 37 points (with a total team score of 195 points), and the United States team scored 42 points (with a total team score of 195 points).

The gold medal cutoff for this IMO was set at 40 points (with 14 contestants earning gold medals), the silver medal cutoff was 26 points (with 35 contestants earning silver medals), and the bronze medal cutoff was 17 points (with 49 contestants earning bronze medals).

In this IMO, a total of eight contestants achieved a perfect score of 42 points.

Problem 4.15 (IMO 28-3, proposed by Germany). Let $x_1, x_2, \ldots, x_n$ be real numbers satisfying $x_1^2 + x_2^2 + \cdots + x_n^2 = 1$. Prove that for every integer $k \geq 2$ there are integers $a_1, a_2, \ldots, a_n$, not all 0, such that $|a_i| \leq k - 1$ for all i and

$$|a_1 x_1 + a_2 x_2 + \cdots + a_n x_n| \leq \frac{(k-1)\sqrt{n}}{k^n - 1}.$$

Proof. Since changing the sign of x_i does not affect the condition $x_1^2 + x_2^2 + \cdots + x_n^2 = 1$, and simultaneously changing the signs of both a_i and x_i does not affect the sum $a_1 x_1 + a_2 x_2 + \cdots + a_n x_n$, we can assume that $x_1, x_2, \ldots, x_n$ are all non-negative real numbers.

Consider the number $e_1 x_1 + e_2 x_2 + \cdots + e_n x_n$, where $e_i \in \{0, 1, 2, \ldots, k-1\}$. If there exist two different arrays $(e_1', e_2', \ldots, e_n')$ and

$(e_1'', e_2'', \ldots, e_n'')$ such that $\sum_{i=1}^n e_i' x_i = \sum_{i=1}^n e_i'' x_i$, let $a_i = e_i' - e_i''$ for $i = 1, 2, \ldots, n$. Then a_i are integers, not all zero, and $|a_i| \le k - 1$, yielding

$$|a_1 x_1 + a_2 x_2 + \cdots + a_n x_n| = 0 < \frac{(k-1)\sqrt{n}}{k^n - 1}.$$

Thus, the original inequality holds.

If $e_1 x_1 + e_2 x_2 + \cdots + e_n x_n$ are all distinct, then the set

$$A = \left\{ \sum_{i=1}^n e_i x_i \, \middle| \, e_i \in \{0, 1, 2, \ldots, k-1\} \right\}$$

contains k^n distinct numbers. By the Cauchy–Schwarz inequality,

$$0 \le \sum_{i=1}^n e_i x_i \le (k-1) \sum_{i=1}^n x_i \le (k-1) \sqrt{\sum_{i=1}^n x_i^2} \cdot \sqrt{n} = (k-1)\sqrt{n},$$

which implies that all k^n numbers in A fall within the interval $[0, (k-1)\sqrt{n}]$.

Dividing this interval into $k^n - 1$ smaller intervals, each with a length of $\frac{(k-1)\sqrt{n}}{k^n - 1}$, we use the Pigeonhole Principle, implying that there must be two distinct numbers $\sum_{i=1}^n e_i' x_i$ and $\sum_{i=1}^n e_i'' x_i$ in the same smaller interval (including endpoints).

Let $a_i = e_i' - e_i''$ for $i = 1, 2, \ldots, n$. Then $|a_i| \le k - 1$ and there is the inequality

$$|a_1 x_1 + a_2 x_2 + \cdots + a_n x_n| \le \frac{(k-1)\sqrt{n}}{k^n - 1}.$$

Thus, the original inequality holds.

Note. There is a similar problem:

- **(Polish Mathematical Olympiad Finals 2009, Problem 4).** Let $x_1, x_2, \ldots, x_n$ be non-negative numbers whose sum is 1. Show that there exist numbers $a_1, a_2, \ldots, a_n$ chosen from amongst 0,1,2,3,4 such that $a_1, a_2, \ldots, a_n$ are different from $2, 2, \ldots, 2$ and

$$2 \le a_1 x_1 + a_2 x_2 + \cdots + a_n x_n \le 2 + \frac{2}{3^n - 1}.$$

【Score Situation】 This particular problem saw the following distribution of scores among contestants: 59 contestants scored 7 points, 6 contestants scored 6 points, 2 contestants scored 5 points, no contestant scored 4 points, 1 contestant scored 3 points, 5 contestants scored 2 points, 39 contestants scored 1 point, and 125 contestants scored 0 point. The average score of this problem is 2.156, indicating that it had a certain level of difficulty.

Among the top five teams in the team scores, the scores of this problem are as follows: the Romania team scored 40 points (with a total team score of 250 points), the Germany team scored 42 points (with a total team score of 248 points), the Soviet Union team scored 28 points (with a total team score of 235 points), the German Democratic Republic team scored 42 points (with a total team score of 231 points), and the United States team scored 35 points (with a total team score of 220 points).

The gold medal cutoff for this IMO was set at 42 points (with 22 contestants earning gold medals), the silver medal cutoff was 32 points (with 42 contestants earning silver medals), and the bronze medal cutoff was 18 points (with 56 contestants earning bronze medals).

In this IMO, a total of 22 contestants achieved a perfect score of 42 points.

Problem 4.16 (IMO 35-1, proposed by France). Let m and n be positive integers. Let $a_1, a_2, \ldots, a_m$ be distinct elements of $\{1, 2, \ldots, n\}$ such that whenever $a_i + a_j \leq n$ for some i, j with $1 \leq i < j \leq m$, there exists k with $1 \leq k \leq m$ such that $a_i + a_j = a_k$. Prove that

$$\frac{a_1 + a_2 + \cdots + a_m}{m} \geq \frac{n+1}{2}.$$

Proof. Without loss of generality, assume $a_1 > a_2 > \cdots > a_m$. For any positive integer i $(1 \leq i \leq m)$,

$$a_i + a_{m+1-i} \geq n + 1. \tag{1}$$

Indeed, if there exists $i \in \{1, 2, \ldots, m\}$ such that (1) is not satisfied, then for this i, we have $a_i + a_{m+1-i} \leq n$. From $a_1 > a_2 > \cdots > a_m > 0$, it follows that

$$a_i < a_i + a_m < a_i + a_{m-1} < \cdots < a_i + a_{m+1-i} \leq n.$$

Thus, $\{a_i + a_m, a_i + a_{m-1}, \ldots, a_i + a_{m+1-i}\} \in \{a_1, a_2, \ldots, a_{i-1}\}$, but

$$|\{a_i + a_m, a_i + a_{m-1}, \ldots, a_i + a_{m+1-i}\}| = i > i - 1 = |\{a_1, a_2, \ldots, a_{i-1}\}|,$$

which leads to a contradiction. Therefore, the inequality (1) holds, and

$$2(a_1 + a_2 + \cdots + a_m)$$
$$= (a_1 + a_m) + (a_2 + a_{m-1}) + \cdots + (a_m + a_1) \geq m(n+1),$$

implying $\frac{a_1 + a_2 + \cdots + a_m}{m} \geq \frac{n+1}{2}$.

【Score Situation】This particular problem saw the following distribution of scores among contestants: 92 contestants scored 7 points, 11 contestants scored 6 points, 13 contestants scored 5 points, 10 contestants scored 4 points, 15 contestants scored 3 points, 33 contestants scored 2 points, 60 contestants scored 1 point, and 151 contestants scored 0 point. The average score of this problem is 2.561, indicating that it had a certain level of difficulty.

Among the top five teams in the team scores, the scores of this problem are as follows: the United States team scored 40 points (with a total team score of 252 points), the China team scored 42 points (with a total team score of 229 points), the Russia team scored 29 points (with a total team score of 224 points), the Bulgaria team scored 36 points (with a total team score of 223 points), and the Hungary team scored 35 points (with a total team score of 221 points).

The gold medal cutoff for this IMO was set at 40 points (with 30 contestants earning gold medals), the silver medal cutoff was 30 points (with 64 contestants earning silver medals), and the bronze medal cutoff was 19 points (with 98 contestants earning bronze medals).

In this IMO, a total of 22 contestants achieved a perfect score of 42 points.

Problem 4.17 (IMO 36-2, proposed by Russia). Let a, b, c be positive real numbers such that $abc = 1$. Prove that

$$\frac{1}{a^3(b+c)} + \frac{1}{b^3(c+a)} + \frac{1}{c^3(a+b)} \geq \frac{3}{2}.$$

Proof 1. Since $abc = 1$, it follows that $\frac{1}{a} + \frac{1}{b} + \frac{1}{c} = bc + ca + ab$. By the Cauchy–Schwarz inequality,

$$\left(\frac{1}{a^3(b+c)} + \frac{1}{b^3(a+c)} + \frac{1}{c^3(a+b)} \right) (a(b+c) + b(a+c) + c(a+b))$$

$$\geq \left(\frac{1}{a} + \frac{1}{b} + \frac{1}{c} \right)^2,$$

and thus, $\left(\frac{1}{a^3(b+c)} + \frac{1}{b^3(a+c)} + \frac{1}{c^3(a+b)} \right) \geq \frac{1}{2}\left(\frac{1}{a} + \frac{1}{b} + \frac{1}{c} \right) \geq \frac{3}{2} \cdot \sqrt[3]{\frac{1}{abc}} = \frac{3}{2}$.

Proof 2. Without loss of generality, assume $a \leq b \leq c$. Then,

$$ab \leq ac \leq bc, \qquad \frac{bc}{a(b+c)} \geq \frac{ca}{b(c+a)} \geq \frac{ab}{c(a+b)}.$$

By the rearrangement inequality,

$$\frac{1}{a^3(b+c)} + \frac{1}{b^3(c+a)} + \frac{1}{c^3(a+b)}$$

$$= \frac{(abc)^2}{a^3(b+c)} + \frac{(abc)^2}{b^3(c+a)} + \frac{(abc)^2}{c^3(a+b)}$$

$$= \frac{bc}{a(b+c)} \cdot bc + \frac{ca}{b(c+a)} \cdot ca + \frac{ab}{c(a+b)} \cdot ab$$

$$\geq \frac{bc}{a(b+c)} \cdot ca + \frac{ca}{b(c+a)} \cdot ab + \frac{ab}{c(a+b)} \cdot bc,$$

and also

$$\frac{bc}{a(b+c)} \cdot bc + \frac{ca}{b(c+a)} \cdot ca + \frac{ab}{c(a+b)} \cdot ab$$

$$\geq \frac{bc}{a(b+c)} \cdot ab + \frac{ca}{b(c+a)} \cdot bc + \frac{ab}{c(a+b)} \cdot ca.$$

Summing up these two inequalities, we get

$$2\left(\frac{1}{a^3(b+c)} + \frac{1}{b^3(c+a)} + \frac{1}{c^3(a+b)}\right) \geq \frac{1}{a} + \frac{1}{b} + \frac{1}{c} \geq 3\sqrt[3]{\frac{1}{a} \cdot \frac{1}{b} \cdot \frac{1}{c}} = 3.$$

Hence $\frac{1}{a^3(b+c)} + \frac{1}{b^3(c+a)} + \frac{1}{c^3(a+b)} \geq \frac{3}{2}$.

Proof 3. By the AM-GM inequality,

$$\frac{1}{a^3(b+c)} + \frac{b+c}{4bc} \geq 2\sqrt{\frac{b+c}{4a^3bc(b+c)}} = \frac{1}{a}.$$

Similarly, $\frac{1}{b^3(c+a)} + \frac{c+a}{4ca} \geq \frac{1}{b}$ and $\frac{1}{c^3(a+b)} + \frac{a+b}{4ab} \geq \frac{1}{c}$. Thus,

$$\frac{1}{a^3(b+c)} + \frac{1}{b^3(a+c)} + \frac{1}{c^3(a+b)} + \frac{1}{4}\left(\frac{b+c}{bc} + \frac{c+a}{ca} + \frac{a+b}{ab}\right)$$

$$\geq \frac{1}{a} + \frac{1}{b} + \frac{1}{c},$$

implying

$$\frac{1}{a^3(b+c)} + \frac{1}{b^3(a+c)} + \frac{1}{c^3(a+b)}$$

$$\geq \frac{1}{2}\left(\frac{1}{a} + \frac{1}{b} + \frac{1}{c}\right) \geq \frac{1}{2} \cdot 3\sqrt[3]{\frac{1}{a} \cdot \frac{1}{b} \cdot \frac{1}{c}} = \frac{3}{2}.$$

【Score Situation】 This particular problem saw the following distribution of scores among contestants: 90 contestants scored 7 points, 5 contestants scored 6 points, 4 contestants scored 5 points, 1 contestant scored 4 points, 1 contestant scored 3 points, 5 contestants scored 2 points, 7 contestants scored 1 point, and 299 contestants scored 0 point. The average score of this problem is 1.709, indicating that it was relatively challenging.

Among the top five teams in the team scores, the scores of this problem are as follows: the China team scored 42 points (with a total team score of 236 points), the Romania team scored 35 points (with a total team score of 230 points), the Russia team scored 42 points (with a total team score of 227 points), the Vietnam team scored 42 points (with a total team score of 220 points), and the Hungary team scored 21 points (with a total team score of 210 points).

The gold medal cutoff for this IMO was set at 37 points (with 30 contestants earning gold medals), the silver medal cutoff was 29 points (with 71 contestants earning silver medals), and the bronze medal cutoff was 19 points (with 100 contestants earning bronze medals).

In this IMO, a total of 14 contestants achieved a perfect score of 42 points.

Problem 4.18 (IMO 38-3, proposed by Russia). Let $x_1, x_2, \ldots, x_n$ be real numbers satisfying the conditions

$$|x_1 + x_2 + \cdots + x_n| = 1 \quad \text{and} \quad |x_i| \leq \frac{n+1}{2} \quad \text{for } i = 1, 2, \ldots, n.$$

Show that there exists a permutation $y_1, y_2, \ldots, y_n$ of $x_1, x_2, \ldots, x_n$ such that

$$|y_1 + 2y_2 + \cdots + ny_n| \leq \frac{n+1}{2}.$$

Proof. For any permutation $\pi = (y_1, y_2, \ldots, y_n)$ of $x_1, x_2, \ldots, x_n$, let $S(\pi)$ denote the value of the sum $y_1 + 2y_2 + \cdots + ny_n$.

Let $r = \frac{n+1}{2}$, $\pi_0 = (x_1, x_2, \ldots, x_n)$, and $\overline{\pi} = (x_n, x_{n-1}, \ldots, x_1)$. If $|S(\pi_0)| \leq r$ or $|S(\overline{\pi})| \leq r$, then the original inequality holds. Otherwise, $|S(\pi_0)| > r$ and $|S(\overline{\pi})| > r$. In this case,

$$|S(\pi_0) + S(\overline{\pi})| = |(n+1)(x_1 + x_2 + \cdots + x_n)| = 2r.$$

Since the absolute values of $S(\pi_0)$ and $S(\overline{\pi})$ both exceed r, their signs must be opposite, with one greater than r and the other less than $-r$.

It is observed that starting from π_0, one can reach any permutation by performing several exchanges of adjacent elements in the permutation π_0. Thus, there exists a sequence of permutations $\pi_0, \pi_1, \ldots, \pi_m$ such that $\pi_m = \overline{\pi}$, and for each $i \in \{0, 1, 2, \ldots, m-1\}$, the permutation π_{i+1} is obtained by swapping two adjacent elements in π_i.

This implies that if $\pi_i = (a_1, a_2, \ldots, a_n)$ and $\pi_{i+1} = (b_1, b_2, \ldots, b_n)$, then there exists some index $k \in \{1, 2, \ldots, n-1\}$ such that $b_k = a_{k+1}$, $b_{k+1} = a_k$, and $b_j = a_j$ for $j \neq k, k+1$. Since $|x_i| \leq \frac{n+1}{2}$ for $i = 1, 2, \ldots, n$, it follows that

$$|S(\pi_{i+1}) - S(\pi_i)| = |kb_k + (k+1)b_{k+1} - ka_k - (k+1)a_{k+1}|$$

$$= |a_k - a_{k+1}|$$

$$\leq |a_k| + |a_{k+1}|$$

$$\leq 2r.$$

This indicates that in the sequence $S(\pi_0), S(\pi_1), \ldots, S(\pi_m)$, the distance between any two consecutive numbers on the number line does not exceed $2r$. Furthermore, $S(\pi_0)$ and $S(\pi_m)$ are on opposite sides of the interval $[-r, r]$, so there must be at least one $S(\pi_i)$ within this interval, implying the existence of a permutation π_i such that $|S(\pi_i)| \leq r$. This concludes the proof.

Note. There are several similar problems:

- **(Germany Team Selection Test 2022, AIMO6, Problem 3).** Determine all integers $n \geq 2$ with the following property: every n distinct integers whose sum is not divisible by n can be arranged in some order $a_1, a_2, \ldots, a_n$ such that n divides $a_1 + 2a_2 + 3a_3 + \cdots + na_n$.
- **(Polish Mathematical Olympiad 2020, 2nd Round, Problem 5).** Let $p > 2$ be a prime number and S be a set of $p+1$ integers. Prove that there exist distinct numbers $a_1, a_2, \ldots, a_{p-1} \in S$ such that

$$a_1 + 2a_2 + 3a_3 + \cdots + (p-1)a_{p-1}$$

is divisible by p.
- **(William Lowell Putnam Mathematical Competition 2017, B6).** Find the number of ordered 64-tuples $(x_0, x_1, \ldots, x_{63})$ such that $x_0, x_1, \ldots, x_{63}$ are distinct elements of $\{1, 2, \ldots, 2017\}$ and

$$x_0 + x_1 + 2x_2 + 3x_3 + \cdots + 63x_{63}$$

is divisible by 2017.

【Score Situation】This particular problem saw the following distribution of scores among contestants: 88 contestants scored 7 points, 6 contestants scored 6 points, 2 contestants scored 5 points, 4 contestants scored 4 points, 22 contestants scored 3 points, 5 contestants scored 2 points, 64 contestants scored 1 point, and 269 contestants scored 0 point. The average score of this problem is 1.778, indicating that it was relatively challenging.

Among the top five teams in the team scores, the scores of this problem are as follows: the China team scored 38 points (with a total team score of 223 points), the Hungary team scored 35 points (with a total team score of 219 points), the Iran team scored 42 points (with a total team score of 217 points), the United States team scored 32 points (with a total team score of 202 points), and the Russia team scored 29 points (with a total team score of 202 points).

The gold medal cutoff for this IMO was set at 35 points (with 39 contestants earning gold medals), the silver medal cutoff was 25 points (with 70 contestants earning silver medals), and the bronze medal cutoff was 15 points (with 122 contestants earning bronze medals).

In this IMO, a total of four contestants achieved a perfect score of 42 points.

Problem 4.19 (IMO 41-2, proposed by the United States).
Suppose a, b, c are positive reals with product 1. Prove that

$$\left(a - 1 + \frac{1}{b}\right)\left(b - 1 + \frac{1}{c}\right)\left(c - 1 + \frac{1}{a}\right) \leq 1.$$

Proof 1. From $abc = 1$, let $a = \frac{x}{y}$, $b = \frac{y}{z}$, and $c = \frac{z}{x}$, where $x, y, z \in \mathbf{R}_+$. The original inequality becomes

$$(x - y + z) \cdot (y - z + x) \cdot (z - x + y) \leq xyz.$$

If two of $x - y + z$, $y - z + x$, and $z - x + y$ are negative, without loss of generality, assume $x - y + z < 0$ and $y - z + x < 0$, then

$$2x = (x - y + z) + (y - z + x) < 0,$$

leading to a contradiction.

If only one of $x - y + z$, $y - z + x$, and $z - x + y$ is negative, then the inequality obviously holds.

If $x - y + z$, $y - z + x$, and $z - x + y$ are all positve, then by the AM-GM inequality,

$$\sqrt{(x - y + z) \cdot (y - z + x)} \leq \frac{(x - y + z) + (y - z + x)}{2} = x,$$

$$\sqrt{(y - z + x) \cdot (z - x + y)} \leq \frac{(y - z + x) + (z - x + y)}{2} = y,$$

$$\sqrt{(z - x + y) \cdot (x - y + z)} \leq \frac{(z - x + y) + (x - y + z)}{2} = z.$$

Multiplying these three inequalities yields

$$(x - y + z) \cdot (y - z + x) \cdot (z - x + y) \leq xyz,$$

and thus, the original inequality holds.

Proof 2. Since $b - 1 + \frac{1}{c} = b(1 - \frac{1}{b} + \frac{1}{bc}) = b(1 + a - \frac{1}{b})$,

$$\left(a - 1 + \frac{1}{b}\right) \cdot \left(b - 1 + \frac{1}{c}\right)$$

$$= b\left(a - 1 + \frac{1}{b}\right)\left(a + 1 - \frac{1}{b}\right) = b\left(a^2 - \left(1 - \frac{1}{b}\right)^2\right) \le ba^2.$$

Similarly, $(b - 1 + \frac{1}{c}) \cdot (c - 1 + \frac{1}{a}) \le cb^2$, $(c - 1 + \frac{1}{a}) \cdot (a - 1 + \frac{1}{b}) \le ac^2$.

If $a - 1 + \frac{1}{b}$, $b - 1 + \frac{1}{c}$, and $c - 1 + \frac{1}{a}$ are not all positive, assume without loss of generality that $a - 1 + \frac{1}{b} \le 0$, then $a \le 1 - \frac{1}{b} < 1$, and $b \ge 1$. Thus

$$b - 1 + \frac{1}{c} > 0, \quad c - 1 + \frac{1}{a} > 0,$$

indicating that the original inequality holds in this case.

If $a - 1 + \frac{1}{b}$, $b - 1 + \frac{1}{c}$, and $c - 1 + \frac{1}{a}$ are all positive, then from the above three inequalities,

$$\left(a - 1 + \frac{1}{b}\right)^2 \cdot \left(b - 1 + \frac{1}{c}\right)^2 \cdot \left(c - 1 + \frac{1}{a}\right)^2 \le a^3 b^3 c^3 = 1,$$

which implies $(a - 1 + \frac{1}{b}) \cdot (b - 1 + \frac{1}{c}) \cdot (c - 1 + \frac{1}{a}) \le 1$.

Proof 3. Since $abc = 1$,

$$\frac{1}{b}\left(b - 1 + \frac{1}{c}\right) + a\left(c - 1 + \frac{1}{a}\right) = 2,$$

$$\frac{1}{c}\left(c - 1 + \frac{1}{a}\right) + b\left(a - 1 + \frac{1}{b}\right) = 2,$$

$$\frac{1}{a}\left(a - 1 + \frac{1}{b}\right) + c\left(b - 1 + \frac{1}{c}\right) = 2.$$

Let $u = a - 1 + \frac{1}{b}$, $v = b - 1 + \frac{1}{c}$, and $w = c - 1 + \frac{1}{a}$. If u, v, w are not all positive, assume without loss of generality that $u \le 0$, then $a \le 1 - \frac{1}{b} < 1$, and $b \ge 1$. Hence,

$$v = b - 1 + \frac{1}{c} > 0, \quad w = c - 1 + \frac{1}{a} > 0,$$

implying $uvw \le 0 < 1$.

If u, v, w are all positive, then by the AM-GM inequality,

$$2 = \frac{1}{a}u + cv \geq 2\sqrt{\frac{c}{a}uv},$$

$$2 = \frac{1}{b}v + aw \geq 2\sqrt{\frac{a}{b}vw},$$

$$2 = \frac{1}{c}w + bu \geq 2\sqrt{\frac{b}{c}wu},$$

so $uv \leq \frac{a}{c}$, $vw \leq \frac{b}{a}$, and $wu \leq \frac{c}{b}$. Consequently, $(uvw)^2 \leq \frac{a}{c} \cdot \frac{b}{a} \cdot \frac{c}{b} = 1$, and $uvw \leq 1$.

In conclusion, the original inequality holds.

Note. In Proof 1, the inequality $(x - y + z) \cdot (y - z + x) \cdot (z - x + y) \leq xyz$ is equivalent to

$$x^3 + y^3 + z^3 + 3xyz \geq xy(x + y) + yz(y + z) + zx(z + x),$$

which is known as Schur's inequality.

【Score Situation】This particular problem saw the following distribution of scores among contestants: 108 contestants scored 7 points, 12 contestants scored 6 points, 18 contestants scored 5 points, 6 contestants scored 4 points, 13 contestants scored 3 points, 41 contestants scored 2 points, 213 contestants scored 1 point, and 50 contestants scored 0 point. The average score of this problem is 2.768, indicating that it had a certain level of difficulty.

Among the top five teams in the team scores, the scores of this problem are as follows: the China team scored 42 points (with a total team score of 218 points), the Russia team scored 36 points (with a total team score of 215 points), the United States team scored 40 points (with a total team score of 184 points), the South Korea team scored 40 points (with a total team score of 172 points), the Vietnam team scored 38 points (with a total team score of 169 points), and the Bulgaria team scored 35 points (with a total team score of 169 points).

The gold medal cutoff for this IMO was set at 30 points (with 39 contestants earning gold medals), the silver medal cutoff was 21 points (with 71 contestants earning silver medals), and the bronze medal cutoff was 11 points (with 119 contestants earning bronze medals).

In this IMO, a total of four contestants achieved a perfect score of 42 points.

Problem 4.20 (IMO 42-2, proposed by South Korea). Prove that

$$\frac{a}{\sqrt{a^2 + 8bc}} + \frac{b}{\sqrt{b^2 + 8ca}} + \frac{c}{\sqrt{c^2 + 8ab}} \geq 1$$

for all positive real numbers a, b, and c.

Proof 1. We introduce a parameter λ and set it such that

$$\frac{a}{\sqrt{a^2 + 8bc}} \geq \frac{a^\lambda}{a^\lambda + b^\lambda + c^\lambda}.$$

By the AM-GM inequality,

$$a^2(a^\lambda + b^\lambda + c^\lambda)^2 \geq a^2\left(a^\lambda + 2b^{\frac{\lambda}{2}}c^{\frac{\lambda}{2}}\right)^2 = a^2\left(a^{2\lambda} + 4a^\lambda b^{\frac{\lambda}{2}}c^{\frac{\lambda}{2}} + 4b^\lambda c^\lambda\right)$$

$$\geq a^2\left(a^{2\lambda} + 8a^{\frac{\lambda}{2}}b^{\frac{3\lambda}{4}}c^{\frac{3\lambda}{4}}\right)$$

$$= a^{2\lambda+2} + 8a^{\frac{\lambda}{2}+2}b^{\frac{3\lambda}{4}}c^{\frac{3\lambda}{4}}.$$

Compared with $a^{2\lambda}(a^2 + 8bc)$, we find that when $\lambda = \frac{4}{3}$,

$$a^{2\lambda+2} + 8a^{\frac{\lambda}{2}+2}b^{\frac{3\lambda}{4}}c^{\frac{3\lambda}{4}} = a^{2\lambda}(a^2 + 8bc).$$

Thus, $\frac{a}{\sqrt{a^2+8bc}} \geq \frac{a^{\frac{4}{3}}}{a^{\frac{4}{3}}+b^{\frac{4}{3}}+c^{\frac{4}{3}}}$. Similarly,

$$\frac{b}{\sqrt{b^2 + 8ca}} \geq \frac{b^{\frac{4}{3}}}{a^{\frac{4}{3}} + b^{\frac{4}{3}} + c^{\frac{4}{3}}} \quad \text{and} \quad \frac{c}{\sqrt{c^2 + 8bc}} \geq \frac{c^{\frac{4}{3}}}{a^{\frac{4}{3}} + b^{\frac{4}{3}} + c^{\frac{4}{3}}}.$$

Adding these three inequalities, we obtain

$$\frac{a}{\sqrt{a^2 + 8bc}} + \frac{b}{\sqrt{b^2 + 8ca}} + \frac{c}{\sqrt{c^2 + 8ab}} \geq 1.$$

Proof 2. Using Hölder's inequality, we have

$$\left(\sum_{\text{cyc}} \frac{a}{\sqrt{a^2 + 8bc}}\right)\left(\sum_{\text{cyc}} \frac{a}{\sqrt{a^2 + 8bc}}\right)\left(\sum_{\text{cyc}} a(a^2 + 8bc)\right) \geq \left(\sum_{\text{cyc}} a\right)^3.$$

Thus, we only need to prove $(\sum_{\text{cyc}} a)^3 \geq \sum_{\text{cyc}} a(a^2 + 8bc)$, which is equivalent to proving $\sum_{\text{cyc}} a(b - c)^2 \geq 0$.

Therefore, the original inequality holds.

Proof 3. Without loss of generality, let $a+b+c=1$ and $f(x)=\frac{1}{\sqrt{x}}$. The function $f(x)$ is convex, so by Jensen's inequality,

$$af(a^2+8bc)+bf(b^2+8ca)+cf(c^2+8ab)$$

$$\geq f(a(a^2+8bc)+b(b^2+8ca)+c(c^2+8ab)).$$

Thus, we only need to prove $a(a^2+8bc)+b(b^2+8ca)+c(c^2+8ab)\leq 1$, which is equivalent to

$$a(a^2+8bc)+b(b^2+8ca)+c(c^2+8ab)\leq (a+b+c)^3$$

$$\Leftrightarrow \quad a(b-c)^2+b(c-a)^2+c(a-b)^2\geq 0.$$

Therefore, the original inequality holds.

Note. In Proof 1, the left side of the inequality we wished to prove is a fraction, and we transformed the denominator into a symmetric algebraic expression in terms of a,b,c through the use of local inequalities. By the method of undetermined coefficients, we then determined these coefficients based on the inequality that needed to be proved. This is one of the common methods used in proving inequalities.

Additionally, we can also employ proof by contradiction to solve this problem. Assume

$$x=\frac{a}{\sqrt{a^2+8bc}}, \quad y=\frac{b}{\sqrt{b^2+8ca}}, \quad z=\frac{c}{\sqrt{c^2+8ab}},$$

and then use $(\frac{1}{x^2}-1)(\frac{1}{y^2}-1)(\frac{1}{z^2}-1)=512$ to derive a contradiction.

Next, we can prove a more general proposition:

Let $a_1,a_2,\ldots,a_n$ be positive real numbers with $a_1a_2\cdots a_n=1$. Then

$$\frac{1}{\sqrt{1+(n^2-1)a_1}}+\frac{1}{\sqrt{1+(n^2-1)a_2}}+\cdots+\frac{1}{\sqrt{1+(n^2-1)a_n}}\geq 1.$$

Let $x_i=\frac{1}{\sqrt{1+(n^2-1)a_i}}$ for $i\in\{1,2,\ldots,n\}$. We prove that if $x_1+x_2+\cdots+x_n<1$, then $a_1a_2\cdots a_n>1$. Denote $x_1x_2\cdots x_n=P$. Since $a_i=\frac{1-x_i^2}{(n^2-1)x_i^2}$,

$$\prod_{i=1}^{n}(1-x_i^2)=(n^2-1)^nP^2\prod_{i=1}^{n}a_i.$$

From $x_1 + x_2 + \cdots + x_n < 1$ and the AM-GM inequality, for $j \in \{1, 2, \ldots, n\}$,

$$1 - x_j > x_1 + \cdots + x_{j-1} + x_{j+1} + \cdots + x_n \geq (n-1)\sqrt[n-1]{\frac{P}{x_j}},$$

$$1 + x_j > x_j + x_1 + x_2 + \cdots + x_n \geq (n+1)\sqrt[n+1]{x_j P}.$$

Thus,

$$1 - x_j^2 > (n^2 - 1)x_j^{-\frac{2}{n^2-1}} P^{\frac{2n}{n^2-1}},$$

$$\prod_{i=1}^{n}(1 - x_i^2) > (n^2 - 1)^n P^2.$$

Hence, $(n^2 - 1)^n P^2 \prod_{i=1}^{n} a_i = \prod_{i=1}^{n}(1 - x_i^2) > (n^2 - 1)^n P^2$, implying $a_1 a_2 \cdots a_n > 1$.

Furthermore, there are several related propositions:

(i) Let $a, b, c, d \in \mathbf{R}_+$. Then

$$\frac{a^{\frac{3}{2}}}{\sqrt{a^3 + 15bcd}} + \frac{b^{\frac{3}{2}}}{\sqrt{b^3 + 15cda}} + \frac{c^{\frac{3}{2}}}{\sqrt{c^3 + 15dab}} + \frac{d^{\frac{3}{2}}}{\sqrt{d^3 + 15abc}} \geq 1.$$

(ii) Let $x_i \in \mathbf{R}_+ (i = 1, 2, \ldots, p)$ with $\prod_{i=1}^{p} x_i = 1$ and $\lambda \geq p^n - 1$. Then

$$\sum_{i=1}^{p} \frac{1}{\sqrt[n]{1 + \lambda x_i}} \geq \frac{p}{\sqrt[n]{1 + \lambda}}.$$

(iii) Let $a, b, c \in \mathbf{R}_+$ and $\lambda \geq 8$. Then

$$\frac{a}{\sqrt{a^2 + \lambda bc}} + \frac{b}{\sqrt{b^2 + \lambda ca}} + \frac{c}{\sqrt{c^2 + \lambda ab}} \geq \frac{3}{\sqrt{1 + \lambda}}.$$

There are several similar problems:

- **(Chinese Team Selection Test 2018, Problem 11).** Given positive integers n and k satisfying $n \geq 4k$, find the smallest real number $\lambda = \lambda(n, k)$ such that for any positive real numbers $a_1, a_2, \ldots, a_n$,

$$\sum_{i=1}^{n} \frac{a_i}{\sqrt{a_i^2 + a_{i+1}^2 + \cdots + a_{i+k}^2}} \leq \lambda,$$

where $a_{n+j} = a_j$ for $j = 1, 2, \ldots, k$.

- **(From the book *Secrets in Inequalities*, 2007: 50, proposed by Pham Kim Hung).** Let a, b, c be positive real numbers with $a+b+c = 1$. Prove that

$$\frac{a}{\sqrt[3]{a + 2b}} + \frac{b}{\sqrt[3]{b + 2c}} + \frac{c}{\sqrt[3]{c + 2a}} \geq 1.$$

- **(From the book *Secrets in Inequalities*, 2007: 60).** Let a, b, c be positive real numbers. Prove that

$$\sqrt{a^2 + 8bc} + \sqrt{b^2 + 8ca} + \sqrt{c^2 + 8ab} \leq 3(a + b + c).$$

- **(Chinese Team Selection Test 2006, Problem 11).** Let a, b, c be positive real numbers such that $a + b + c = 1$. Prove that

$$\frac{ab}{\sqrt{ab + bc}} + \frac{bc}{\sqrt{bc + ca}} + \frac{ca}{\sqrt{ca + ab}} \leq \frac{\sqrt{2}}{2}.$$

- **(Mathematical Olympiad Program 2002).** Let a, b, c be positive reals. Prove that

$$\left(\frac{2a}{b + c}\right)^{\frac{2}{3}} + \left(\frac{2b}{c + a}\right)^{\frac{2}{3}} + \left(\frac{2c}{a + b}\right)^{\frac{2}{3}} \geq 3.$$

【Score Situation】This particular problem saw the following distribution of scores among contestants: 77 contestants scored 7 points, 3 contestants scored 6 points, 7 contestants scored 5 points, 11 contestants scored 4 points, 9 contestants scored 3 points, 15 contestants scored 2 points, 40 contestants scored 1 point, and 311 contestants scored 0 point. The average score of this problem is 1.550, indicating that it was relatively challenging.

Among the top five teams in the team scores, the scores of this problem are as follows: the China team scored 40 points (with a total team score of 225 points), the United States team scored 31 points (with a total team score of 196 points), the Russia team scored 28 points (with a total team score of 196 points), the South Korea team scored 42 points (with a total team score of 185 points), and the Bulgaria team scored 19 points (with a total team score of 185 points).

The gold medal cutoff for this IMO was set at 30 points (with 39 contestants earning gold medals), the silver medal cutoff was 20 points (with 81 contestants earning silver medals), and the bronze medal cutoff was 11 points (with 122 contestants earning bronze medals).

In this IMO, a total of four contestants achieved a perfect score of 42 points.

Problem 4.21 (IMO 44-5, proposed by Ireland). Given $n > 2$ and reals $x_1 \le x_2 \le \cdots \le x_n$, show that

$$\left(\sum_{i=1}^{n}\sum_{j=1}^{n}|x_i - x_j|\right)^2 \le \frac{2(n^2 - 1)}{3}\sum_{i=1}^{n}\sum_{j=1}^{n}(x_i - x_j)^2.$$

Also show that the equality is valid if and only if the sequence is an arithmetic progression.

Proof.   (i) Since subtracting the same number from all x_i does not change the inequality, without loss of generality, we assume $\sum_{i=1}^{n} x_i = 0$. From the given condition,

$$\sum_{i=1}^{n}\sum_{j=1}^{n}|x_i - x_j| = 2\sum_{i<j}(x_j - x_i) = 2\sum_{i=1}^{n}(2i - n - 1)x_i,$$

and by the Cauchy–Schwarz inequality,

$$\left(\sum_{i=1}^{n}\sum_{j=1}^{n}|x_i - x_j|\right)^2 = 4\left(\sum_{i=1}^{n}(2i - n - 1)x_i\right)^2$$

$$\le 4\sum_{i=1}^{n}(2i - n - 1)^2\sum_{i=1}^{n}x_i^2$$

$$= \frac{4n(n+1)(n-1)}{3}\sum_{i=1}^{n}x_i^2.$$

Furthermore,

$$\sum_{i=1}^{n}\sum_{j=1}^{n}(x_i - x_j)^2 = n\sum_{i=1}^{n}x_i^2 - 2\sum_{i=1}^{n}x_i\sum_{j=1}^{n}x_j + n\sum_{j=1}^{n}x_j^2 = 2n\sum_{i=1}^{n}x_i^2.$$

Thus, $(\sum_{i=1}^{n}\sum_{j=1}^{n}|x_i - x_j|)^2 \le \frac{2(n^2-1)}{3}\sum_{i=1}^{n}\sum_{j=1}^{n}(x_i - x_j)^2$.

(ii) From the conditions for the equality in the Cauchy–Schwarz inequality, the equality holds if and only if there exists a real number k such that $x_i = k(2i - n - 1)$, i.e., $x_i = 2ki - k(n + 1)$. Thus, $x_1, x_2, \ldots, x_n$ form an arithmetic progression.

If $x_1, x_2, \ldots, x_n$ form an arithmetic sequence with common difference d, then $x_i = \frac{d}{2}(2i - n - 1) + \frac{x_1 + x_n}{2}$. Subtracting $\frac{x_1 + x_n}{2}$ from each x_i yields

$$x_i = \frac{d}{2}(2i - n - 1), \quad \text{and} \quad \sum_{i=1}^{n}x_i = 0.$$

In this case, the equality holds in the given inequality.

Note. There are several similar problems:

- (From the book ***Problems and Solutions in Mathematical Olympiad High School 2, 2022: 17***). Let $x_1, x_2, \ldots, x_n$ be real numbers. Prove that

$$\left(\sum_{1 \leq i < j \leq n} |x_i - x_j| \right)^2 \geq (n-1) \sum_{1 \leq i < j \leq n} |x_i - x_j|^2.$$

- (**Iran Team Selection Test 2006, Problem 4**). Let $x_1, x_2, \ldots, x_n$ be real numbers. Prove that

$$\sum_{i=1}^{n} \sum_{j=1}^{n} |x_i + x_j| \geq n \sum_{i=1}^{n} |x_i|.$$

- (**Chinese Team Selection Test 2006, Problem 3**). Given n real numbers $a_1, a_2, \ldots, a_n$, prove that there exist real numbers $b_1, b_2, \ldots, b_n$ satisfying:

  (a) for any $1 \leq i \leq n$, the number $a_i - b_i$ is a positive integer;

  (b) $\sum_{1 \leq i < j \leq n} (b_i - b_j)^2 \leq \frac{n^2 - 1}{12}$.

- (**Canadian Mathematical Olympiad 1974, Problem 4**). Let n be a fixed positive integer. To any choice of n real numbers satisfying

$$0 \leq x_i \leq 1, \quad i = 1, 2, \ldots, n,$$

there corresponds the sum

$$\sum_{1 \leq i < j \leq n} |x_i - x_j| = |x_1 - x_2| + |x_1 - x_3| + \cdots + |x_1 - x_n| + |x_2 - x_3|$$

$$+ |x_2 - x_4| + \cdots + |x_2 - x_n| + |x_3 - x_4|$$

$$+ |x_3 - x_5| + \cdots + |x_3 - x_n| + \cdots + |x_{n-1} - x_n|.$$

$$(*)$$

Let $S(n)$ denote the largest possible value of the sum $(*)$. Find $S(n)$.

【Score Situation】This particular problem saw the following distribution of scores among contestants: 67 contestants scored 7 points, 4 contestants scored 6 points, 1 contestant scored 5 points, 5 contestants scored 4 points, 6 contestants scored 3 points, 20 contestants scored 2 points, 161 contestants scored 1 point, and 193 contestants scored 0 point. The average score of this problem is 1.613, indicating that it was relatively challenging.

Among the top five teams in the team scores, the scores of this problem are as follows: the Bulgaria team scored 37 points (with a total team score of 227 points), the China team scored 42 points (with a total team score of 211 points), the United States team scored 36 points (with a total team score of 188 points), the Vietnam team scored 36 points (with a total team score of 172 points), and the Russia team scored 30 points (with a total team score of 167 points).

The gold medal cutoff for this IMO was set at 29 points (with 37 contestants earning gold medals), the silver medal cutoff was 19 points (with 69 contestants earning silver medals), and the bronze medal cutoff was 13 points (with 104 contestants earning bronze medals).

In this IMO, only three contestants achieved a perfect score of 42 points, namely Bảo Lê Hùng Việt and Trọng Cảnh Nguyễn from Vietnam, and Yunhao Fu from China.

Problem 4.22 (IMO 45-4, proposed by South Korea). Let $n \geq 3$ be an integer. Let $t_1, t_2, \ldots, t_n$ be positive real numbers such that

$$n^2 + 1 > (t_1 + t_2 + \cdots + t_n)\left(\frac{1}{t_1} + \frac{1}{t_2} + \cdots + \frac{1}{t_n}\right).$$

Show that t_i, t_j, t_k are side lengths of a triangle for all i, j, k with $1 \leq i < j < k \leq n$.

Proof 1. We employ proof by contradiction. Suppose there are three numbers among $t_1, t_2, \ldots, t_n$ that cannot form the side lengths of a triangle. Without loss of generality, assume these are t_1, t_2, t_3 and $t_1 + t_2 \leq t_3$. Since

$$(t_1 + t_2 + \cdots + t_n)\left(\frac{1}{t_1} + \frac{1}{t_2} + \cdots + \frac{1}{t_n}\right)$$

$$= \sum_{1 \leq i < j \leq n}\left(\frac{t_i}{t_j} + \frac{t_j}{t_i}\right) + n$$

$$= \frac{t_1}{t_3} + \frac{t_3}{t_1} + \frac{t_2}{t_3} + \frac{t_3}{t_2} + \sum_{\substack{1 \leq i < j \leq n \\ (i,j) \notin \{(1,3),(2,3)\}}}\left(\frac{t_i}{t_j} + \frac{t_j}{t_i}\right) + n$$

$$\geq \frac{t_1 + t_2}{t_3} + t_3\left(\frac{1}{t_1} + \frac{1}{t_2}\right) + \sum_{\substack{1 \leq i < j \leq n \\ (i,j) \notin \{(1,3),(2,3)\}}} 2 + n$$

$$\geq \frac{t_1 + t_2}{t_3} + \frac{4t_3}{t_1 + t_2} + 2(C_n^2 - 2) + n$$

$$= \frac{t_1 + t_2}{t_3} + \frac{4t_3}{t_1 + t_2} + n^2 - 4. \tag{1}$$

Let $x = \frac{t_3}{t_1 + t_2}$. Then $x \geq 1$, and

$$4x + \frac{1}{x} - 5 = \frac{(x-1)(4x-1)}{x} \geq 0.$$

From (1),

$$(t_1 + t_2 + \cdots + t_n)\left(\frac{1}{t_1} + \frac{1}{t_2} + \cdots + \frac{1}{t_n}\right) \geq 5 + n^2 - 4 = n^2 + 1,$$

which leads to a contradiction. Therefore, the original proposition holds.

Proof 2. We employ mathematical induction to prove that if there exist t_i, t_j, t_k such that $t_i + t_j \leq t_k$, then

$$(t_1 + t_2 + \cdots + t_n)\left(\frac{1}{t_1} + \frac{1}{t_2} + \cdots + \frac{1}{t_n}\right) \geq n^2 + 1.$$

For $n = 3$, assume $t_1 + t_2 \leq t_3$, and it is easy to prove that

$$(t_1 + t_2 + t_3)\left(\frac{1}{t_1} + \frac{1}{t_2} + \frac{1}{t_3}\right) \geq 10.$$

Suppose when $n = l \geq 3$ and $t_1 + t_2 \leq t_3$,

$$(t_1 + t_2 + \cdots + t_l)\left(\frac{1}{t_1} + \frac{1}{t_2} + \cdots + \frac{1}{t_l}\right) \geq l^2 + 1.$$

For $n = l + 1$,

$$(t_1 + t_2 + \cdots + t_{l+1})\left(\frac{1}{t_1} + \frac{1}{t_2} + \cdots + \frac{1}{t_{l+1}}\right)$$

$$= (t_1 + \cdots + t_l)\left(\frac{1}{t_1} + \cdots + \frac{1}{t_l}\right) + t_{l+1}\left(\frac{1}{t_1} + \cdots + \frac{1}{t_l}\right)$$

$$+ \frac{1}{t_{l+1}}(t_1 + \cdots + t_l) + 1$$

$$\geq l^2 + 1 + 2\sqrt{t_{l+1}\left(\frac{1}{t_1} + \cdots + \frac{1}{t_l}\right)\frac{1}{t_{l+1}}(t_1 + \cdots + t_l)} + 1$$

$$\geq l^2 + 1 + 2\sqrt{l^2 + 1} + 1$$

$$> l^2 + 1 + 2l + 1$$

$$= (l + 1)^2 + 1.$$

Therefore, for any integer $n \geq 3$, t_i, t_j, t_k are side lengths of a triangle for all i, j, k with $1 \leq i < j < k \leq n$.

Note. In Proof 1, from the Cauchy–Schwarz inequality,

$$(t_1 + t_2)\left(\frac{1}{t_1} + \frac{1}{t_2}\right) \geq 4.$$

Thus, $\frac{1}{t_1} + \frac{1}{t_2} \geq \frac{4}{t_1+t_2}$.

Furthermore, we can also prove that

$$(t_1 + t_2 + \cdots + t_n)\left(\frac{1}{t_1} + \frac{1}{t_2} + \cdots + \frac{1}{t_n}\right) \leq (n + \sqrt{10} - 3)^2.$$

There are several similar problems:

- Problem 4.30 (IMO 64-4) in this chapter.
- (**Chinese Girls' Mathematical Olympiad 2023, Problem 6**). Let x_i $(i = 1, 2, \ldots, 22)$ be reals such that $x_i \in [2^{i-1}, 2^i]$. Find the maximum possible value of

$$(x_1 + x_2 + \cdots + x_{22})\left(\frac{1}{x_1} + \frac{1}{x_2} + \cdots + \frac{1}{x_{22}}\right).$$

- (**Chinese Mathematical Olympiad 2016, Problem 6**). Given an integer $n \geq 2$ and positive numbers $a < b$, let $x_1, x_2, \ldots, x_n \in [a, b]$. Find the maximum value of

$$\frac{\frac{x_1^2}{x_2} + \frac{x_2^2}{x_3} + \cdots + \frac{x_{n-1}^2}{x_n} + \frac{x_n^2}{x_1}}{x_1 + x_2 + \cdots + x_{n-1} + x_n}.$$

- (**Chinese Mathematical Olympiad 2014, Problem 1**). Given a real number $r \in (0, 1)$, prove that if n complex numbers $z_1, z_2, \ldots, z_n$ satisfy $|z_k - 1| \leq r$ for $k = 1, 2, \ldots, n$, then

$$|z_1 + z_2 + \cdots + z_n| \cdot \left|\frac{1}{z_1} + \frac{1}{z_2} + \cdots + \frac{1}{z_n}\right| \geq n^2(1 - r^2).$$

- (**United States of America Mathematical Olympiad 2012, Problem 1**). Find all integers $n \geq 3$ such that among any n positive real numbers $a_1, a_2, \ldots, a_n$ with

$$\max\{a_1, a_2, \ldots, a_n\} \leq n \cdot \min\{a_1, a_2, \ldots, a_n\},$$

there exist three that are the side lengths of an acute triangle.

- (**United States of America Mathematical Olympiad 2009, Problem 4**). For $n \geq 2$, let $a_1, a_2, \ldots, a_n$ be positive real numbers such that

$$(a_1 + a_2 + \cdots + a_n)\left(\frac{1}{a_1} + \frac{1}{a_2} + \cdots + \frac{1}{a_n}\right) \leq \left(n + \frac{1}{2}\right)^2.$$

  Prove that $\max\{a_1, a_2, \ldots, a_n\} \leq 4\min\{a_1, a_2, \ldots, a_n\}$.
- (**From the book *Secrets in Inequalities*, 2007: 77**). Suppose that $p < q$ are positive constants and $a_1, a_2, \ldots, a_n \in [p, q]$. Prove that

$$(a_1 + a_2 + \cdots + a_n)\left(\frac{1}{a_1} + \frac{1}{a_2} + \cdots + \frac{1}{a_n}\right) \leq n^2 + \frac{k_n(p-q)^2}{4pq},$$

  where $k_n = n^2$ if n is even and $n^2 - 1$ if n is odd.
- (**Vietnam Team Selection Test 2006, Problem 4**). Prove that for all real numbers $a, b, c \in [1, 2]$,

$$(a + b + c)\left(\frac{1}{a} + \frac{1}{b} + \frac{1}{c}\right) \geq 6\left(\frac{a}{b+c} + \frac{b}{c+a} + \frac{c}{a+b}\right).$$

- (**British Mathematical Olympiad 2005, 2nd Round, Problem 3**). Let a, b, c be positive real numbers. Prove that

$$\left(\frac{a}{b} + \frac{b}{c} + \frac{c}{a}\right)^2 \geq (a + b + c)\left(\frac{1}{a} + \frac{1}{b} + \frac{1}{c}\right).$$

- (**Chinese Mathematical Olympiad 1988, Problem 4**).
  (a) Let a, b, c be positive real numbers satisfying

$$(a^2 + b^2 + c^2)^2 > 2(a^4 + b^4 + c^4).$$

  Prove that a, b, c can be the lengths of three sides of a triangle respectively.
  (b) Let $a_1, a_2, \ldots, a_n$ be $n(n > 3)$ positive real numbers satisfying

$$(a_1^2 + a_2^2 + \cdots + a_n^2)^2 > (n-1)(a_1^4 + a_2^4 + \cdots + a_n^4).$$

  Prove that any three of $a_1, a_2, \ldots, a_n$ can be the lengths of three sides of a triangle respectively.
- (**All Soviet Union Mathematical Olympiad 1978**). Given $0 < a \leq x_1 \leq x_2 \leq \cdots \leq x_n \leq b$, prove that

$$(x_1 + x_2 + \cdots + x_n)\left(\frac{1}{x_1} + \frac{1}{x_2} + \cdots + \frac{1}{x_n}\right) \leq \frac{(a+b)^2}{4ab}n^2.$$

- **(United States of America Mathematical Olympiad 1977, Problem 5).** Let a, b, c, d, e be positive numbers bounded by p and q, i.e., they lie in $[p, q]$ with $0 < p$. Prove that

$$(a + b + c + d + e)\left(\frac{1}{a} + \frac{1}{b} + \frac{1}{c} + \frac{1}{d} + \frac{1}{e}\right) \leq 25 + 6\left(\sqrt{\frac{p}{q}} - \sqrt{\frac{q}{p}}\right)^2$$

and determine when there is the equality.

【Score Situation】This particular problem saw the following distribution of scores among contestants: 250 contestants scored 7 points, 6 contestants scored 6 points, 6 contestants scored 5 points, 8 contestants scored 4 points, 16 contestants scored 3 points, 27 contestants scored 2 points, 33 contestants scored 1 point, and 140 contestants scored 0 point. The average score of this problem is 4.080, indicating that it was simple.

Among the top five teams in the team scores, the scores of this problem are as follows: the China team scored 42 points (with a total team score of 220 points), the United States team scored 42 points (with a total team score of 212 points), the Russia team scored 42 points (with a total team score of 205 points), the Vietnam team scored 42 points (with a total team score of 196 points), and the Bulgaria team scored 42 points (with a total team score of 194 points).

The gold medal cutoff for this IMO was set at 32 points (with 45 contestants earning gold medals), the silver medal cutoff was 24 points (with 78 contestants earning silver medals), and the bronze medal cutoff was 16 points (with 120 contestants earning bronze medals).

In this IMO, a total of four contestants achieved a perfect score of 42 points.

Problem 4.23 (IMO 46-3, proposed by South Korea). Let x, y, z be three positive reals such that $xyz \geq 1$. Prove that

$$\frac{x^5 - x^2}{x^5 + y^2 + z^2} + \frac{y^5 - y^2}{x^2 + y^5 + z^2} + \frac{z^5 - z^2}{x^2 + y^2 + z^5} \geq 0.$$

Proof 1. The original inequality can be transformed into

$$\frac{x^2 + y^2 + z^2}{x^5 + y^2 + z^2} + \frac{x^2 + y^2 + z^2}{y^5 + z^2 + x^2} + \frac{x^2 + y^2 + z^2}{z^5 + x^2 + y^2} \leq 3.$$

Using the Cauchy–Schwarz inequality and the given condition $xyz \geq 1$, we have

$$(x^5 + y^2 + z^2)(yz + y^2 + z^2) \geq (x^2(xyz)^{\frac{1}{2}} + y^2 + z^2)^2$$
$$\geq (x^2 + y^2 + z^2)^2.$$

Therefore, $\frac{x^2+y^2+z^2}{x^5+y^2+z^2} \leq \frac{yz+y^2+z^2}{x^2+y^2+z^2}$. Similarly,

$$\frac{x^2+y^2+z^2}{y^5+z^2+x^2} \leq \frac{zx+z^2+x^2}{x^2+y^2+z^2}, \quad \frac{x^2+y^2+z^2}{z^5+x^2+y^2} \leq \frac{xy+x^2+y^2}{x^2+y^2+z^2}.$$

Summing up these three inequalities and using $x^2+y^2+z^2 \geq xy+yz+zx$, we obtain

$$\frac{x^2+y^2+z^2}{x^5+y^2+z^2} + \frac{x^2+y^2+z^2}{y^5+z^2+x^2} + \frac{x^2+y^2+z^2}{z^5+x^2+y^2} \leq 2 + \frac{xy+yz+zx}{x^2+y^2+z^2} \leq 3.$$

Proof 2. It suffices to prove:

$$\sum_{\text{cyc}} \frac{x^5}{x^5+y^2+z^2} \geq 1 \geq \sum_{\text{cyc}} \frac{x^2}{x^5+y^2+z^2}.$$

Given $xyz \geq 1$, we have

$$\sum_{\text{cyc}} \frac{x^5}{x^5+y^2+z^2} \geq \sum_{\text{cyc}} \frac{x^5}{x^5+xyz(y^2+z^2)}$$

$$= \sum_{\text{cyc}} \frac{x^4}{x^4+y^3z+yz^3}.$$

Since $(y-z)(y^3-z^3) \geq 0$, it follows that $y^4+z^4 \geq y^3z+yz^3$. Hence,

$$\sum_{\text{cyc}} \frac{x^4}{x^4+y^3z+yz^3} \geq \sum_{\text{cyc}} \frac{x^4}{x^4+y^4+z^4} = 1.$$

Furthermore, since

$$x^5+xyz(y^2+z^2) \leq xyz(x^5+y^2+z^2),$$

$$\sum_{\text{cyc}} \frac{x^2}{x^5+y^2+z^2} \leq \sum_{\text{cyc}} \frac{x^2 \cdot xyz}{x^5+xyz(y^2+z^2)}$$

$$= \sum_{\text{cyc}} \frac{x^2yz}{x^4+yz(y^2+z^2)}.$$

By the AM-GM inequality,

$$x^4+x^4+y^3z+yz^3 \geq 4x^2yz,$$

$$x^4+y^3z+y^3z+y^2z^2 \geq 4xy^2z,$$

$$x^4+yz^3+yz^3+y^2z^2 \geq 4xyz^2,$$

$$y^3z+yz^3 \geq 2y^2z^2,$$

and summing up these four inequalities yields

$$x^4 + yz(y^2 + z^2) \geq x^2yz + xy^2z + xyz^2.$$

Hence,

$$\sum_{cyc} \frac{x^2yz}{x^4 + yz(y^2 + z^2)} \leq \sum_{cyc} \frac{x^2yz}{x^2yz + xy^2z + xyz^2} = 1.$$

Thus, the original inequality holds.

Proof 3. Since

$$\frac{x^5 - x^2}{x^5 + y^2 + z^2} - \frac{x^5 - x^2}{x^3(x^2 + y^2 + z^2)} = \frac{x^2(x^3 - 1)^2(y^2 + z^2)}{x^3(x^5 + y^2 + z^2)(x^2 + y^2 + z^2)} \geq 0,$$

we have

$$\sum_{cyc} \frac{x^5 - x^2}{x^5 + y^2 + z^2} \geq \sum_{cyc} \frac{x^5 - x^2}{x^3(x^2 + y^2 + z^2)} = \frac{1}{x^2 + y^2 + z^2} \sum_{cyc} \left(x^2 - \frac{1}{x} \right).$$

Given $xyz \geq 1$, we obtain $\frac{1}{x} + \frac{1}{y} + \frac{1}{z} \leq yz + zx + xy$. Therefore,

$$\sum_{cyc} \left(x^2 - \frac{1}{x} \right) \geq \sum_{cyc} x^2 - \sum_{cyc} yz \geq 0.$$

Thus, the original inequality holds.

Note. Proof 3 was provided by the Moldovan contestant Iurie Boreico, who received a special prize for this solution. Furthermore, there are several related propositions:

(i) Let n be a positive integer and $x_1, x_2, \ldots, x_n$ be positive real numbers satisfying $x_1 x_2 \ldots x_n \geq 1$. Then

$$\sum_{cyc} \frac{x_1^{2n-1} - x_1^{n-1}}{x_1^{2n-1} + x_2^{n-1} + \cdots + x_n^{n-1}} \geq 0,$$

where $\sum_{cyc}$ denotes the cyclic sum over $x_1, x_2, \ldots, x_n$, and the equality holds if and only if $x_1 = x_2 = \cdots = x_n = 1$.

(ii) Let x, y, z be positive real numbers satisfying $yz + zx + xy \geq 3$. Then

$$\sum_{cyc} \frac{x^4 - x^2}{x^4 + y^2 + z^2} \geq 0,$$

where $\sum_{cyc}$ denotes the cyclic sum over x, y, z, and the equality holds if and only if $x = y = z = 1$.

【Score Situation】 This particular problem saw the following distribution of scores among contestants: 55 contestants scored 7 points, 9 contestants scored 6 points, no contestant scored 5 points, no contestant scored 4 points, no contestant scored 3 points, 3 contestants scored 2 points, 23 contestants scored 1 point, and 423 contestants scored 0 point. The average score of this problem is 0.912, indicating that it was extremely difficult.

Among the top five teams in the team scores, the scores of this problem are as follows: the China team scored 35 points (with a total team score of 235 points), the United States team scored 35 points (with a total team score of 213 points), the Russia team scored 20 points (with a total team score of 212 points), the Iran team scored 22 points (with a total team score of 201 points), and the South Korea team scored 27 points (with a total team score of 200 points).

The gold medal cutoff for this IMO was set at 35 points (with 42 contestants earning gold medals), the silver medal cutoff was 23 points (with 79 contestants earning silver medals), and the bronze medal cutoff was 12 points (with 128 contestants earning bronze medals).

In this IMO, a total of 16 contestants achieved a perfect score of 42 points.

Problem 4.24 (IMO 48-1, proposed by New Zealand). Real numbers $a_1, a_2, \ldots, a_n$ are given. For each i $(1 \leq i \leq n)$, define

$$d_i = \max\{a_j : 1 \leq j \leq i\} - \min\{a_j : i \leq j \leq n\}$$

and let

$$d = \max\{d_i : 1 \leq i \leq n\}.$$

(a) Prove that, for any real numbers $x_1 \leq x_2 \leq \cdots \leq x_n$,

$$\max\{|x_i - a_i| : 1 \leq i \leq n\} \geq \frac{d}{2}. \qquad (*)$$

(b) Show that there are real numbers $x_1 \leq x_2 \leq \cdots \leq x_n$ such that the equality holds in $(*)$.

Proof 1. (a) Let $d = d_g (1 \leq g \leq n)$ and define

$$a_p = \max\{a_j : 1 \leq j \leq g\} \quad \text{and} \quad a_r = \min\{a_j : g \leq j \leq n\}.$$

Then $1 \leq p \leq g \leq r \leq n$, and $d = a_p - a_r$. For any real numbers $x_1 \leq x_2 \leq \cdots \leq x_n$, note that

$$(a_p - x_p) + (x_r - a_r) = (a_p - a_r) + (x_r - x_p) \geq a_p - a_r = d,$$

implying either $a_p - x_p \geq \frac{d}{2}$ or $x_r - a_r \geq \frac{d}{2}$. Thus,

$$\max\{|x_i - a_i| : 1 \leq i \leq n\} \geq \max\{|x_p - a_p|, |x_r - a_r|\}$$

$$\geq \max\{a_p - x_p, \ x_r - a_r\} \geq \frac{d}{2}.$$

(b) Define a sequence $\{x_k\}$ as follows:

$$x_1 = a_1 - \frac{d}{2}, \quad x_k = \max\left\{x_{k-1}, a_k - \frac{d}{2}\right\}, \quad 2 \le k \le n.$$

It is evident that the sequence $\{x_k\}$ is non-decreasing by definition, and $x_k - a_k \ge -\frac{d}{2}$ for all $1 \le k \le n$. Next, we prove that

$$x_k - a_k \le \frac{d}{2} \quad \text{for all } 1 \le k \le n. \tag{1}$$

For any $1 \le k \le n$, let $l \le k$ be the smallest index such that $x_k = x_l$. Then either $l = 1$ or $l \ge 2$ and $x_l > x_{l-1}$. In both cases,

$$x_k = x_l = a_l - \frac{d}{2}. \tag{2}$$

Since $a_l - a_k \le \max\{a_j : 1 \le j \le k\} - \min\{a_j : k \le j \le n\} = d_k \le d$, by (2),

$$x_k - a_k = a_l - a_k - \frac{d}{2} \le d - \frac{d}{2} = \frac{d}{2},$$

which is (1). This shows that $-\frac{d}{2} \le x_k - a_k \le \frac{d}{2}$ for all $1 \le k \le n$. Hence,

$$\max\{|x_i - a_i| : 1 \le i \le n\} \le \frac{d}{2}.$$

By (a), the sequence $\{x_k\}$ indeed makes the equality in (*).

Proof 2. (a) The same as Proof 1.

(b) For each $i(1 \le i \le n)$, let

$$M_i = \max\{a_j : 1 \le j \le i\} \quad \text{and} \quad m_i = \min\{a_j : i \le j \le n\}.$$

Then

$$M_i = \max\{a_j : 1 \le j \le i\} \le \max\{a_1, \ldots, a_i, a_{i+1}\} = M_{i+1},$$

$$m_i = \min\{a_j : i \le j \le n\} \le \min\{a_{i+1}, a_{i+2}, \ldots, a_n\} = m_{i+1},$$

indicating both sequences $\{M_i\}$ and $\{m_i\}$ are non-decreasing, and by their definition, $m_i \le a_i \le M_i$.

Let $x_i = \frac{M_i + m_i}{2}$, and from $d_i = M_i - m_i$,

$$-\frac{d_i}{2} = \frac{m_i - M_i}{2} = x_i - M_i \le x_i - a_i \le x_i - m_i = \frac{M_i - m_i}{2} = \frac{d_i}{2}.$$

Therefore,

$$\max\{|x_i - a_i| : 1 \le i \le n\} \le \max\left\{\frac{d_i}{2} : 1 \le i \le n\right\} = \frac{d}{2}.$$

By (a), the sequence $\{x_k\}$ indeed makes the equality in (*).

Note. There are several similar problems:

- **(Turkey Team Selection Test 2002, Problem 3).** Given a positive integer n and real numbers $a_1, a_2, \ldots, a_n$, show that there exist integers m and k such that $|\sum_{i=1}^{m} a_i - \sum_{i=m+1}^{n} a_i| \le |a_k|$.
- Let $a_1, a_2, \ldots, a_n$ be real numbers. Determine the real numbers $x_1, x_2, \ldots, x_n$, $x_1 \le x_2 \le \cdots \le x_n$, such that $\max_{1 \le k \le n} |a_k - x_k|$ is minimized.

【Score Situation】This particular problem saw the following distribution of scores among contestants: 161 contestants scored 7 points, 21 contestants scored 6 points, 18 contestants scored 5 points, 18 contestants scored 4 points, 105 contestants scored 3 points, 8 contestants scored 2 points, 13 contestants scored 1 point, and 176 contestants scored 0 point. The average score of this problem is 3.383, indicating that it was relatively straightforward.

Among the top five teams in the team scores, the scores of this problem are as follows: the Russia team scored 42 points (with a total team score of 184 points), the China team scored 36 points (with a total team score of 181 points), the Vietnam team scored 41 points (with a total team score of 168 points), the South Korea team scored 38 points (with a total team score of 168 points), and the United States team scored 41 points (with a total team score of 155 points).

The gold medal cutoff for this IMO was set at 29 points (with 39 contestants earning gold medals), the silver medal cutoff was 21 points (with 83 contestants earning silver medals), and the bronze medal cutoff was 14 points (with 131 contestants earning bronze medals).

In this IMO, no contestant achieved a perfect score of 42 points.

Problem 4.25 (IMO 49-2, proposed by Austria). (a) Prove that

$$\frac{x^2}{(x-1)^2} + \frac{y^2}{(y-1)^2} + \frac{z^2}{(z-1)^2} \ge 1$$

for all real numbers x, y, z, each different from 1, and satisfying $xyz = 1$.

(b) Prove that the equality holds above for infinitely many triples of rational numbers x, y, z, each different from 1, and satisfying $xyz = 1$.

Proof 1. (a) Let $a = \frac{x}{x-1}$, $b = \frac{y}{y-1}$, and $c = \frac{z}{z-1}$. Then $x = \frac{a}{a-1}$, $y = \frac{b}{b-1}$, and $z = \frac{c}{c-1}$. From $xyz = 1$, we have $abc = (a-1)(b-1)(c-1)$, i.e.,

$$a + b + c - 1 = ab + bc + ca.$$

Therefore,

$$a^2 + b^2 + c^2 = (a+b+c)^2 - 2(ab + bc + ca)$$
$$= (a+b+c)^2 - 2(a+b+c-1)$$
$$= (a+b+c-1)^2 + 1 \geq 1.$$

Hence, $\frac{x^2}{(x-1)^2} + \frac{y^2}{(y-1)^2} + \frac{z^2}{(z-1)^2} \geq 1$.

(b) Let $(x, y, z) = (-\frac{k}{(k-1)^2}, k - k^2, \frac{k-1}{k^2})$, where k is a positive integer. Then (x, y, z) is a triple of rational numbers, and x, y, z are all not equal to 1. For different positive integers k, the triples of rational numbers (x, y, z) are distinct. In this case,

$$\frac{x^2}{(x-1)^2} + \frac{y^2}{(y-1)^2} + \frac{z^2}{(z-1)^2}$$
$$= \frac{k^2}{(k^2-k+1)^2} + \frac{(k-k^2)^2}{(k^2-k+1)^2} + \frac{(k-1)^2}{(k^2-k+1)^2}$$
$$= \frac{k^4 - 2k^3 + 3k^2 - 2k + 1}{(k^2-k+1)^2} = 1.$$

Thus, the proposition holds.

Proof 2. (a) From $xyz = 1$, let $p = x$, $q = 1$, and $r = \frac{1}{y}$. Then $x = \frac{p}{q}$, $y = \frac{q}{r}$, and $z = \frac{1}{xy} = \frac{r}{p}$, where p, q, r are distinct. Therefore,

$$\frac{x^2}{(x-1)^2} + \frac{y^2}{(y-1)^2} + \frac{z^2}{(z-1)^2} \geq 1$$

$$\Leftrightarrow \frac{p^2}{(p-q)^2} + \frac{q^2}{(q-r)^2} + \frac{r^2}{(r-p)^2} \geq 1. \tag{1}$$

Let $a = \frac{p}{p-q}$, $b = \frac{q}{q-r}$, and $c = \frac{r}{r-p}$. Then the inequality (1) becomes $\sum a^2 \geq 1$. Since

$$\frac{a-1}{a} = \frac{q}{p}, \quad \frac{b-1}{b} = \frac{r}{q}, \quad \text{and} \quad \frac{c-1}{c} = \frac{p}{r},$$

it follows that $\frac{a-1}{a} \cdot \frac{b-1}{b} \cdot \frac{c-1}{c} = 1$, i.e.,

$$1 - \sum_{\text{cyc}} a + \sum_{\text{cyc}} ab = 0. \tag{2}$$

From (2), $1 - \sum_{\text{cyc}} a^2 = -(a+b+c-1)^2 \leq 0$. Thus, $\sum_{\text{cyc}} a^2 \geq 1$, and the inequality (1) holds.

(b) Let $b = \frac{t^2+t}{t^2+t+1}$, $c = \frac{t+1}{t^2+t+1}$, and $a = -\frac{bc}{b+c}$, where t can take any rational number except 0 and -1. Changing t results in an infinite number of rational triples (a, b, c), where a, b, c are all not equal to 1 and $\sum_{\text{cyc}} a = \sum_{\text{cyc}} a^2 = 1$.

Therefore, $(x, y, z) = (\frac{a}{a-1}, \frac{b}{b-1}, \frac{c}{c-1})$ makes (b) hold.

Proof 3. (a) Let $a = \sqrt[3]{x}$, $b = \sqrt[3]{y}$, and $c = \sqrt[3]{z}$. Then $\frac{a^2}{bc} = \frac{\sqrt[3]{x^2}}{\sqrt[3]{yz}} = \frac{x}{\sqrt[3]{xyz}} = x$, and similarly $\frac{b^2}{ca} = y$ and $\frac{c^2}{ab} = z$. Thus, the inequality to be proved becomes

$$\frac{a^4}{(a^2 - bc)^2} + \frac{b^4}{(b^2 - ca)^2} + \frac{c^4}{(c^2 - ab)^2} \geq 1.$$

By the Cauchy–Schwarz inequality,

$$\frac{a^4}{(a^2 - bc)^2} + \frac{b^4}{(b^2 - ca)^2} + \frac{c^4}{(c^2 - ab)^2}$$

$$\geq \frac{(a^2 + b^2 + c^2)^2}{(a^2 - bc)^2 + (b^2 - ca)^2 + (c^2 - ab)^2},$$

and further,

$$(a^2 + b^2 + c^2)^2 - ((a^2 - bc)^2 + (b^2 - ca)^2 + (c^2 - ab)^2)$$

$$= (ab + bc + ca)^2 \geq 0.$$

Thus, $\frac{(a^2+b^2+c^2)^2}{(a^2-bc)^2+(b^2-ca)^2+(c^2-ab)^2} \geq 1$.

Consequently, the original inequality is true.

(b) The same as Proof 1 or Proof 2.

Note. There are several similar problems:

- **(Iran Team Selection Test 2018, Problem 2).** Determine the least real number k such that the inequality

$$\left(\frac{2a}{a-b}\right)^2 + \left(\frac{2b}{b-c}\right)^2 + \left(\frac{2c}{c-a}\right)^2 + k \geq 4\left(\frac{2a}{a-b} + \frac{2b}{b-c} + \frac{2c}{c-a}\right)$$

is satisfied for all real numbers a, b, c.

- **(Canadian Mathematical Olympiad 2017, Problem 1).** Let a, b, and c be non-negative real numbers, no two of which are equal. Prove that

$$\frac{a^2}{(b-c)^2} + \frac{b^2}{(c-a)^2} + \frac{c^2}{(a-b)^2} > 2.$$

- **(German Mathematical Olympiad 2010, Problem 2).** Let a, b, c be distinct real numbers. Show that

$$\left(\frac{2a-b}{a-b}\right)^2 + \left(\frac{2b-c}{b-c}\right)^2 + \left(\frac{2c-a}{c-a}\right)^2 \geq 5.$$

- **(From the book *Secrets in Inequalities*, 2007: 151).** Let x, y, z be distinct real numbers. Prove that

$$\frac{x^2}{(x-y)^2} + \frac{y^2}{(y-z)^2} + \frac{z^2}{(z-x)^2} \geq 1.$$

- **(United Kingdom Team Selection Test 2005, NST2, Problem 3).** Let $n \geq 3$ be an integer and $a_1, a_2, \ldots, a_n$ be positive real numbers such that $a_1 a_2 \ldots a_n = 1$. Prove that

$$\frac{a_1 + 3}{(a_1 + 1)^2} + \frac{a_2 + 3}{(a_2 + 1)^2} + \cdots + \frac{a_n + 3}{(a_n + 1)^2} \geq 3.$$

【Score Situation】This particular problem saw the following distribution of scores among contestants: 94 contestants scored 7 points, 7 contestants scored 6 points, 53 contestants scored 5 points, 35 contestants scored 4 points, 3 contestants scored 3 points, 24 contestants scored 2 points, 209 contestants scored 1 point, and 110 contestants scored 0 point. The average score of this problem is 2.563, indicating that it had a certain level of difficulty.

Among the top five teams in the team scores, the scores of this problem are as follows: the China team scored 42 points (with a total team score of 217 points), the Russia team scored 33 points (with a total team score of 199 points), the United States team scored 30 points (with a total team score of 190 points), the South Korea team scored 42 points (with a total team score of 188 points), and the Iran team scored 37 points (with a total team score of 181 points).

The gold medal cutoff for this IMO was set at 31 points (with 47 contestants earning gold medals), the silver medal cutoff was 22 points (with 100 contestants earning silver medals), and the bronze medal cutoff was 15 points (with 120 contestants earning bronze medals).

In this IMO, only three contestants achieved a perfect score of 42 points, namely Xiaosheng Mu and Dongyi Wei from China, and Alex Zhai from the United States.

Problem 4.26 (IMO 53-2, proposed by Australia). Let $n \geq 3$ be an integer, and let $a_2, a_3, \ldots, a_n$ be positive real numbers such that $a_2 a_3 \ldots a_n = 1$. Prove that

$$(1 + a_2)^2 (1 + a_3)^3 \cdots (1 + a_n)^n > n^n.$$

Proof.   By the AM-GM inequality, for $k \in \{2, 3, \ldots, n\}$,

$$(1 + a_k)^k = \left(\underbrace{\frac{1}{k-1} + \frac{1}{k-1} + \cdots + \frac{1}{k-1}}_{k-1 \text{ times}} + a_k \right)^k \geq k^k \left(\frac{1}{k-1} \right)^{k-1} a_k.$$

Thus,

$$(1 + a_2)^2 (1 + a_3)^3 \cdots (1 + a_n)^n$$

$$\geq 2^2 a_2 \times \frac{3^3 a_3}{2^2} \times \frac{4^4 a_4}{3^3} \times \cdots \times \frac{n^n a_n}{(n-1)^{n-1}}$$

$$= n^n,$$

and the equality holds when $a_k = \frac{1}{k-1}$ for $k \in \{2, 3, \ldots, n\}$.

However, this contradicts the condition $a_2 a_3 \cdots a_n = 1$. Hence,

$$(1 + a_2)^2 (1 + a_3)^3 \cdots (1 + a_n)^n > n^n.$$

Note. There are several similar problems:

- Let $n \geq 3$ be an integer and $a_2, a_3, \ldots, a_n$ be positive real numbers such that $a_2 a_3 \cdots a_n = 1$. Prove that

$$(1 + a_2)^2 (1 + a_3)^3 \cdots (1 + a_n)^n > \frac{1}{4^{n-1}} n^n (n-1)^{n-1}.$$

- (**Chinese Beiyue University League Independent Recruitment Test 2014, Problem 10**). Let $x_1, x_2, \ldots, x_n$ be positive real numbers with $x_1 x_2 \cdots x_n = 1$. Prove that

$$(\sqrt{2} + x_1)(\sqrt{2} + x_2) \cdots (\sqrt{2} + x_n) \geq (\sqrt{2} + 1)^n.$$

- (**All-Russian Mathematical Olympiad 2007, Regional Round, Grade 11, Problem 8**). Let $x_1, x_2, \ldots, x_n$ be positive real numbers. Prove that

$$(1 + x_1)(1 + x_1 + x_2) \cdots (1 + x_1 + x_2 + \cdots + x_n)$$
$$\geq \sqrt{(n+1)^{n+1}} \sqrt{x_1 x_2 \cdots x_n}.$$

- (**From Mildorf's paper "Olympiad Inequalities," 2005: 14**). Let $x_1, x_2, \ldots, x_n$ be positive reals such that

$$\frac{1}{x_1 + 1998} + \frac{1}{x_2 + 1998} + \cdots + \frac{1}{x_n + 1998} = \frac{1}{1998}.$$

Prove that $\frac{\sqrt[n]{x_1 x_2 \cdots x_n}}{n-1} \geq 1998$.

- (**Asian Pacific Mathematics Olympiad 2002, Problem 1**). Let $a_1, a_2, \ldots, a_n$ be a sequence of non-negative integers, where n is a positive integer. Let $A_n = \frac{a_1 + a_2 + \cdots + a_n}{n}$. Prove that

$$a_1! a_2! \cdots a_n! \geq (\lfloor A_n \rfloor !)^n,$$

where $\lfloor A_n \rfloor$ is the greatest integer less than or equal to A_n, and $a! = 1 \times 2 \times \cdots \times a$ for $a \geq 1$ (and $0! = 1$).

- (**Open Mathematical Olympiad of 239 Presidential Physics and Mathematics Lyceum in Saint-Petersburg 2000, Grade 10-11, Problem 3**). Let $a_1, a_2, \ldots, a_n$ be positive real numbers. Prove that

$$(\sqrt{2})^n (a_1 + a_2)(a_2 + a_3) \cdots (a_n + a_1)$$
$$\leq (a_1 + a_2 + a_3)(a_2 + a_3 + a_4) \cdots (a_n + a_1 + a_2).$$

- (**Turkey Team Selection Test 1992, Problem 3**). Let $x_1, x_2, \ldots, x_{n+1}$ be positive real numbers satisfying

$$\frac{1}{1 + x_1} + \frac{1}{1 + x_2} + \cdots + \frac{1}{1 + x_{n+1}} = 1.$$

Prove that $x_1 x_2 \cdots x_{n+1} \geq n^{n+1}$.

【Score Situation】 This particular problem saw the following distribution of scores among contestants: 171 contestants scored 7 points, 7 contestants scored 6 points, 2 contestants scored 5 points, 8 contestants scored 4 points, 5 contestants scored 3 points, 8 contestants scored 2 points, 83 contestants scored 1 point, and 263 contestants scored 0 point. The average score of this problem is 2.550, indicating that it had a certain level of difficulty.

Among the top five teams in the team scores, the scores of this problem are as follows: the South Korea team scored 42 points (with a total team score of 209 points), the China team scored 40 points (with a total team score of 195 points), the United States team scored 40 points (with a total team score of 194 points), the Russia team scored 35 points (with a total team score of 177 points), the Thailand team scored 42 points (with a total team score of 159 points), and the Canada team scored 32 points (with a total team score of 159 points).

The gold medal cutoff for this IMO was set at 28 points (with 51 contestants earning gold medals), the silver medal cutoff was 21 points (with 88 contestants earning silver medals), and the bronze medal cutoff was 14 points (with 137 contestants earning bronze medals).

In this IMO, only one contestant achieved a perfect score of 42 points, namely Jeck Lim from Singapore.

Problem 4.27 (IMO 55-1, proposed by Austria). Let $a_0 < a_1 < a_2 < \cdots$ be an infinite sequence of positive integers. Prove that there exists a unique integer $n \geq 1$ such that

$$a_n < \frac{a_0 + a_1 + \cdots + a_n}{n} \leq a_{n+1}. \tag{1}$$

Proof. For $n = 1, 2, \ldots$, define

$$d_n = (a_0 + a_1 + \cdots + a_n) - n a_n.$$

Then the first inequality in (1) holds if and only if $d_n > 0$. Furthermore,

$$na_{n+1} - (a_0 + a_1 + \cdots + a_n)$$
$$= (n+1)a_{n+1} - (a_0 + a_1 + \cdots + a_n + a_{n+1}) = -d_{n+1},$$

implying that the second inequality in (1) holds if and only if $d_{n+1} \leq 0$.

By definition, the sequence $d_1, d_2, \ldots$ consists of integers, and

$$d_1 = (a_0 + a_1) - 1 \cdot a_1 = a_0 > 0,$$

while

$$d_{n+1} - d_n = ((a_0 + a_1 + \cdots + a_n) - na_{n+1}) - ((a_0 + a_1 + \cdots + a_n) - na_n)$$
$$= n(a_n - a_{n+1}) < 0,$$

implying $d_{n+1} < d_n$.

Therefore, the sequence $d_1, d_2, \ldots$ is a strictly decreasing integer sequence with a positive first term. Hence, there exists a unique n such that $d_n > 0 \geq d_{n+1}$.

Note. This problem is novel but not difficult to solve. Essentially, it is a discrete form of the Intermediate Value Theorem.

Furthermore, there are several similar problems:

- **(British Mathematical Olympiad 2023, 2nd Round, Problem 3).** For an integer $n \geq 3$, we say that $A = (a_1, a_2, \ldots, a_n)$ is an n-list if every a_k is an integer satisfying $1 \leq a_k \leq n$. For each $k = 1, 2, \ldots, n-1$, let M_k be the minimal possible non-zero value of $\left| \frac{a_1 + a_2 + \cdots + a_{k+1}}{k+1} - \frac{a_1 + a_2 + \cdots + a_k}{k} \right|$ across all n-lists. We say that an n-list A is *ideal* if

$$\left| \frac{a_1 + a_2 + \cdots + a_{k+1}}{k+1} - \frac{a_1 + a_2 + \cdots + a_k}{k} \right|$$
$$= M_k \quad \text{for each } k = 1, 2, \ldots, n-1.$$

Find the number of *ideal* n-lists.

- **(All-Russian Mathematical Olympiad 2000, Final Round, Grade 11, Problem 4).** Let $a_1, a_2, \ldots, a_n$ be a sequence of non-negative integers. For $k = 1, 2, \ldots, n$, denote

$$m_k = \max_{1 \leq l \leq k} \frac{a_{k-l+1} + a_{k-l+2} + \cdots + a_k}{l}.$$

Prove that for every $\alpha > 0$ the number of values of k for which $m_k > \alpha$ is less than $\frac{a_1 + a_2 + \cdots + a_n}{\alpha}$.

- **(All Soviet Union Mathematical Olympiad 1978).** Given $a_1, a_2, \ldots, a_n$, define $b_k = \frac{a_1 + a_2 + \cdots + a_k}{k}$ for $1 \leq k \leq n$. Let

$$C = (a_1 - b_1)^2 + (a_2 - b_2)^2 + \cdots + (a_n - b_n)^2,$$
$$D = (a_1 - b_n)^2 + (a_2 - b_n)^2 + \cdots + (a_n - b_n)^2.$$

Prove that $C \leq D \leq 2C$.

【Score Situation】This particular problem saw the following distribution of scores among contestants: 370 contestants scored 7 points, 23 contestants scored 6 points, 18 contestants scored 5 points, 15 contestants scored 4 points, 22 contestants scored 3 points, 14 contestants scored 2 points, 23 contestants scored 1 point, and 75 contestants scored 0 point. The average score of this problem is 5.348, indicating that it was simple.

Among the top five teams in the team scores, the scores of this problem are as follows: the China team scored 42 points (with a total team score of 201 points), the United States team scored 42 points (with a total team score of 193 points), the Chinese Taiwan team scored 42 points (with a total team score of 192 points), the Russia team scored 42 points (with a total team score of 191 points), and the Japan team scored 38 points (with a total team score of 177 points).

The gold medal cutoff for this IMO was set at 29 points (with 49 contestants earning gold medals), the silver medal cutoff was 22 points (with 113 contestants earning silver medals), and the bronze medal cutoff was 16 points (with 133 contestants earning bronze medals).

In this IMO, only three contestants achieved a perfect score of 42 points, namely Jiyang Gao from China, Po-Sheng Wu from Chinese Taiwan, and Alexander Gunning from Australia.

Problem 4.28 (IMO 61-2, proposed by Belgium). Four real numbers a, b, c, d are such that $a \geq b \geq c \geq d > 0$ and $a + b + c + d = 1$. Prove that

$$(a + 2b + 3c + 4d)a^a b^b c^c d^d < 1.$$

Proof 1. Since $a + b + c + d = 1$, by the weighted AM-GM inequality,

$$a^a b^b c^c d^d \leq a \cdot a + b \cdot b + c \cdot c + d \cdot d = a^2 + b^2 + c^2 + d^2.$$

Thus, it suffices to prove that

$$(a + 2b + 3c + 4d)(a^2 + b^2 + c^2 + d^2) < 1.$$

Given $a \geq b \geq c \geq d > 0$,

$$
\begin{aligned}
(a + b + c + d)^3 &= \sum a^3 + 3 \sum a^2 b + 6 \sum abc \\
&> a^2(a + 3b + 3c + 3d) + b^2(3a + b + 3c + 3d) \\
&\quad + c^2(3a + 3b + c + 3d) + d^2(3a + 3b + 3c + d) \\
&\geq (a + 2b + 3c + 4d)(a^2 + b^2 + c^2 + d^2),
\end{aligned}
$$

implying $(a + 2b + 3c + 4d)(a^2 + b^2 + c^2 + d^2) < 1$.

Hence, the original inequality is valid.

Proof 2. If $a \leq \frac{1}{2}$, then

$$(a + 2b + 3c + 4d)a^a b^b c^c d^d \leq (a + 3b + 3c + 3d)a^{a+b+c+d}$$

$$= (3 - 2a)a$$

$$= 1 - (1 - a)(1 - 2a)$$

$$\leq 1.$$

The equality holds if and only if $a = b = c = d = \frac{1}{2}$, which is a contradiction.

If $a > \frac{1}{2}$, then,

$$(a + 2b + 3c + 4d)a^a b^b c^c d^d \leq (3 - 2a)a^a (1 - a)^{1-a}.$$

By the weighted AM-GM inequality,

$$\frac{a}{a + (1 - a)} \cdot a + \frac{1 - a}{a + (1 - a)} \cdot (1 - a) \geq a^a (1 - a)^{1-a},$$

and thus,

$$(3 - 2a)a^a (1 - a)^{1-a} \leq (3 - 2a)(a^2 + (1 - a)^2)$$

$$= (3 - 2a)(1 - 2a + 2a^2)$$

$$= 1 + 2(1 - a)^2 (1 - 2a)$$

$$< 1.$$

Hence, the original inequality is true.

Note. The key to Proof 1 is to use the weighted AM-GM inequality to transform $a^a b^b c^c d^d$ into $a^2 + b^2 + c^2 + d^2$, and then to prove a four-variable cubic inequality.

【Score Situation】 This particular problem saw the following distribution of scores among contestants: 138 contestants scored 7 points, 9 contestants scored 6 points, 7 contestants scored 5 points, 4 contestants scored 4 points, 9 contestants scored 3 points, 129 contestants scored 2 points, 29 contestants scored 1 point, and 291 contestants scored 0 point. The average score of this problem is 2.248, indicating that it had a certain level of difficulty.

Among the top five teams in the team scores, the scores of this problem are as follows: the China team scored 38 points (with a total team score of 215 points), the Russia team scored 31 points (with a total team score of 185 points), the United States team scored 36 points (with a total team score of 183 points), the South Korea team scored 32 points (with a total team score of 175 points), and the Thailand team scored 37 points (with a total team score of 174 points).

The gold medal cutoff for this IMO was set at 31 points (with 49 contestants earning gold medals), the silver medal cutoff was 24 points (with 112 contestants earning silver medals), and the bronze medal cutoff was 15 points (with 155 contestants earning bronze medals).

In this IMO, only one contestant achieved a perfect score of 42 points, namely Jinmin Li from China.

Problem 4.29 (IMO 62-2, proposed by Canada). Show that the inequality

$$\sum_{i=1}^{n}\sum_{j=1}^{n}\sqrt{|x_i - x_j|} \leq \sum_{i=1}^{n}\sum_{j=1}^{n}\sqrt{|x_i + x_j|} \tag{*}$$

holds for all real numbers $x_1, x_2, \ldots, x_n$.

Proof. Apply mathematical induction on n. For $n = 1$ and $n = 2$, the inequality (*) holds obviously.

Assume the inequality (*) holds for fewer than n real numbers. If there is some $x_i = 0$, then the terms containing x_i on both sides of inequality (*) are equal. By removing x_i from $\{x_1, x_2, \ldots, x_n\}$, we reduce it to the inductive hypothesis for $n - 1$ real numbers.

Similarly, if $x_i = -x_j \neq 0$, then all terms on both sides of inequality (*) involving x_i or x_j are

$$2\sqrt{2|x_i|} + \sum_{\substack{k = 1, \ldots, n \\ k \neq i, j}} (\sqrt{|x_k - x_i|} + \sqrt{|x_k + x_i|}),$$

and by removing x_i and x_j from $\{x_1, x_2, \ldots, x_n\}$, we reduce it to the inductive hypothesis for $n - 2$ real numbers.

Assume $x_i + x_j \neq 0$ for all i, $j \in \{1, \ldots, n\}$(allowing $i = j$). Apply a translation to all variables $x_i \mapsto x_i + a$ such that after the translation, there exist two variables whose sum is 0. Note that this translation does not change the value of the left side of inequaltiy (*).

If there exists a positive real number a such that $x_i + x_j + a = 0$ for some i, $j \in \{1, \ldots, n\}$, let a_+ be the smallest such a; otherwise, let $a_+ = +\infty$. If there exists a negative real number a such that $x_i + x_j + a = 0$ for some $i, j \in \{1, \ldots, n\}$, let a_- be the largest such a; otherwise, let $a_- = -\infty$.

Since the functions $\sqrt{x}$ and $\sqrt{-x}$ are concave in their domains, for any $i, j \in \{1, \ldots, n\}$, $\sqrt{x_i + x_j + a}$ is a concave function of a in the interval $[a_-, a_+]$. Moreover, the sum $\sum_{i,j=1}^{n} \sqrt{x_i + x_j + a}$ is also a concave function of a in the interval $[a_-, a_+]$. Therefore, there exists a number b (either a_+ or a_-, but $b \neq \pm\infty$) such that

$$\sum_{i,j=1}^{n} \sqrt{|x_i + x_j|} \geq \sum_{i,j=1}^{n} \sqrt{|x_i + x_j + b|}.$$

(Note that $a_+ = +\infty$ and or $a_- = -\infty$ cannot be valid at the same time. If $a_+ = +\infty$, then $x_i + x_j > 0$ for $i, j \in \{1, 2, \ldots, n\}$ and $b = a_-$. If $a_- = -\infty$, then $x_i + x_j < 0$ for $i, j \in \{1, 2, \ldots, n\}$ and $b = a_+$.)

Let $y_i = x_i + \frac{b}{2}$. Then there exist i, j such that $y_i + y_j = 0$. Consequently,

$$\sum_{i=1}^{n} \sum_{j=1}^{n} \sqrt{|x_i - x_j|} = \sum_{i=1}^{n} \sum_{j=1}^{n} \sqrt{|y_i - y_j|} \leq \sum_{i=1}^{n} \sum_{j=1}^{n} \sqrt{|y_i + y_j|}$$

$$\leq \sum_{i=1}^{n} \sum_{j=1}^{n} \sqrt{|x_i + x_j|},$$

where the first inequality follows from the previous transformation to the inductive hypothesis. This completes the inductive proof.

Note. If f is a non-decreasing concave function defined on $[0, +\infty)$ with $f(0) = 0$, then for any real numbers $x_1, x_2, \ldots, x_n$,

$$\sum_{i=1}^{n} \sum_{j=1}^{n} f(|x_i - x_j|) \leq \sum_{i=1}^{n} \sum_{j=1}^{n} f(|x_i + x_j|).$$

Bin Zhao provided the proof in his paper "A Generation of the 2021 IMO Problem 2 (2021 年 IMO 第二题的一个推广)," published on the New Star Math Website (http://www.nsmath.cn/jszl).

Furthermore, there are several similar problems:

- **(From Junpu Hu's paper "Two Generations of a Problem (一个问题的两个推广)," published on the New Star Math Website (http://www.nsmath.cn/xszl)).** Given a positive integer n and n distinct real numbers $x_1, x_2, \ldots, x_n$, find the maximum value of $\lambda = \lambda(n)$ such that

$$\sum_{1 \leq i < j \leq n} \left| \frac{x_i + x_j}{x_i - x_j} \right| \geq \lambda(n).$$

- **(From Junpu Hu's paper "Two Generalizations of a Problem (一个问题的两个推广)," published on the New Star Math**

Website (http://www.nsmath.cn/xszl)). Given a positive integer n and n distinct real numbers $x_1, x_2, \ldots, x_n$, find the maximum value of $\lambda = \lambda(n)$ such that

$$\sum_{i=1}^{n} \left| \frac{x_i + x_{i+1}}{x_i - x_{i+1}} \right| \geq \lambda(n).$$

- **(From a New Star Math Camp Test, China).** Given $n \geq 2$, find the minimum value of $\lambda = \lambda(n)$ such that

$$\sum_{i=1}^{n} \frac{a_i - b_i}{a_i + b_i} \leq \lambda$$

for any positive real numbers $a_1, a_2, \ldots, a_n, b_1, b_2, \ldots, b_n$ that satisfy $\sum_{i=1}^{n} a_i = \sum_{i=1}^{n} b_i$.

- **(International Zhautykov Olympiad in Mathematics 2019, Problem 2).** Find the largest real number C such that for all distinct positive real numbers $a_1, a_2, \ldots, a_{2019}$, the following inequality holds:

$$\frac{a_1}{|a_2 - a_3|} + \frac{a_2}{|a_3 - a_4|} + \cdots + \frac{a_{2018}}{|a_{2019} - a_1|} + \frac{a_{2019}}{|a_1 - a_2|} > C.$$

【Score Situation】 This particular problem saw the following distribution of scores among contestants: 16 contestants scored 7 points, 2 contestants scored 6 points, 1 contestant scored 5 points, 3 contestants scored 4 points, 2 contestants scored 3 points, 12 contestants scored 2 points, 61 contestants scored 1 point, and 522 contestants scored 0 point. The average score of this problem is 0.375, indicating that it was extremely difficult.

Among the top five teams in the team scores, the scores of this problem are as follows: the China team scored 16 points (with a total team score of 208 points), the Russia team scored 17 points (with a total team score of 183 points), the South Korea team scored 9 points (with a total team score of 172 points), the United States team scored 2 points (with a total team score of 165 points), and the Canada team scored 8 points (with a total team score of 151 points).

The gold medal cutoff for this IMO was set at 24 points (with 52 contestants earning gold medals), the silver medal cutoff was 19 points (with 103 contestants earning silver medals), and the bronze medal cutoff was 12 points (with 148 contestants earning bronze medals).

In this IMO, only one contestant achieved a perfect score of 42 points, namely Yichuan Wang from China.

Problem 4.30 (IMO 64-4, proposed by the Netherlands). Let $x_1, x_2, \ldots, x_{2023}$ be different positive real numbers such that

$$a_n = \sqrt{(x_1 + x_2 + \cdots + x_n)\left(\frac{1}{x_1} + \frac{1}{x_2} + \cdots + \frac{1}{x_n}\right)}$$

is an integer for every $n = 1, 2, \ldots, 2023$. Prove that $a_{2023} \geq 3034$.

Proof 1. By the Cauchy–Schwarz inequality,

$$a_n = \sqrt{\left(\sum_{i=1}^{n-1} x_i + x_n\right)\left(\sum_{i=1}^{n-1} \frac{1}{x_i} + \frac{1}{x_n}\right)} \geq \sqrt{\left(\sum_{i=1}^{n-1} x_i\right)\left(\sum_{i=1}^{n-1} \frac{1}{x_i}\right)}$$

$$+ \sqrt{x_n \cdot \frac{1}{x_n}} = a_{n-1} + 1,$$

and the equality holds if and only if

$$x_n^2 = \frac{\sum_{i=1}^{n-1} x_i}{\sum_{i=1}^{n-1} \frac{1}{x_i}} = \left(\sum_{i=1}^{n-1} x_i\right)^2 \cdot \frac{1}{a_{n-1}^2} \Leftrightarrow x_n = \frac{\sum_{i=1}^{n-1} x_i}{a_{n-1}}.$$

Next, we prove that $a_{n+1} - a_{n-1} \geq 3$ for all $n \geq 2$. If it is not true, then

$$a_{n+1} - a_n = a_n - a_{n-1} = 1,$$

implying

$$x_n = \frac{\sum_{i=1}^{n-1} x_i}{a_{n-1}}, \quad x_{n+1} = \frac{\sum_{i=1}^{n-1} x_i + x_n}{a_n} = \frac{a_{n-1} x_n + x_n}{a_{n-1} + 1} = x_n.$$

However, this contradicts the assumption that $x_1, x_2, \ldots, x_{2023}$ are all distinct. Therefore, $a_{n+1} - a_{n-1} \geq 3$, and consequently

$$a_{2023} = a_1 + \sum_{i=1}^{1001} (a_{2i+1} - a_{2i-1}) \geq 1 + 1001 \times 3 = 3034.$$

Proof 2. It is sufficient to prove that $a_{n+2} \geq a_n + 3$ for every n in the set $\{1, 2, \ldots, 2021\}$. Then

$$a_{2023} \geq a_{2021} + 3 \geq a_{2019} + 6 \geq \cdots \geq a_1 + 3033 = 3034.$$

By the Cauchy–Schwarz inequality,

$$a_{n+2} = \sqrt{\left(\sum_{i=1}^{n} x_i + x_{n+1} + x_{n+2}\right)\left(\sum_{i=1}^{n}\frac{1}{x_i} + \frac{1}{x_{n+1}} + \frac{1}{x_{n+2}}\right)}$$

$$\geq \sqrt{\left(\sum_{i=1}^{n} x_i\right)\left(\sum_{i=1}^{n}\frac{1}{x_i}\right)} + \sqrt{(x_{n+1} + x_{n+2})\left(\frac{1}{x_{n+1}} + \frac{1}{x_{n+2}}\right)}$$

$$= a_n + \sqrt{(x_{n+1} + x_{n+2})\left(\frac{1}{x_{n+1}} + \frac{1}{x_{n+2}}\right)}.$$

Since $x_1, x_2, \ldots, x_{2023}$ are all distinct and $a_1, a_2, \ldots, a_{2023}$ are integers, $(x_{n+1} + x_{n+2})(\frac{1}{x_{n+1}} + \frac{1}{x_{n+2}}) > 4$, so $a_{n+2} > a_n + 2$, implying $a_{n+2} \geq a_n + 3$ for every n in the set $\{1, 2, \ldots, 2021\}$.

Therefore, $a_{2023} \geq 3034$.

Note. There are several similar problems, which are presented in Problem 4.22.

【Score Situation】 This particular problem saw the following distribution of scores among contestants: 384 contestants scored 7 points, 3 contestants scored 6 points, 1 contestant scored 5 points, 4 contestants scored 4 points, 8 contestants scored 3 points, 32 contestants scored 2 points, 100 contestants scored 1 point, and 86 contestants scored 0 point. The average score of this problem is 4.717, indicating that it was simple.

Among the top five teams in the team scores, the scores of this problem are as follows: the China team scored 42 points (with a total team score of 240 points), the United States team scored 42 points (with a total team score of 222 points), the South Korea team scored 42 points (with a total team score of 215 points), the Romania team scored 42 points (with a total team score of 208 points), and the Canada team scored 42 points (with a total team score of 183 points).

The gold medal cutoff for this IMO was set at 32 points (with 54 contestants earning gold medals), the silver medal cutoff was 25 points (with 90 contestants earning silver medals), and the bronze medal cutoff was 18 points (with 170 contestants earning bronze medals).

In this IMO, a total of five contestants achieved a perfect score of 42 points.

4.2.3 *Determining value ranges*

Problem 4.31 (IMO 16-5, proposed by the Netherlands). Determine all possible values of

$$S = \frac{a}{d+a+b} + \frac{b}{a+b+c} + \frac{c}{b+c+d} + \frac{d}{c+d+a},$$

where a, b, c, d are arbitrary positive numbers.

Solution 1. Since

$$\frac{a}{d+a+b} + \frac{b}{a+b+c} < \frac{a}{a+b} + \frac{b}{a+b} = 1,$$

$$\frac{c}{b+c+d} + \frac{d}{c+d+a} < \frac{c}{c+d} + \frac{d}{c+d} = 1,$$

it follows that $S < 2$. Furthermore,

$$S > \frac{a}{a+b+c+d} + \frac{b}{a+b+c+d} + \frac{c}{a+b+c+d} + \frac{d}{a+b+c+d} = 1,$$

implying $1 < S < 2$.

For any value $1 + t (0 < t < 1)$ in the interval $(1, 2)$, let $a = c = 1$. Then $S = 1 + t$ if

$$t = S - 1 = \frac{2}{1+b+d} - \frac{2}{b+2} + \frac{d}{d+2} = \frac{2}{1+b+d} + \frac{b}{b+2} - \frac{2}{d+2}. \quad (1)$$

Choose $b = b_0$ sufficiently small such that $\frac{2}{(b_0+1)(b_0+2)} > t > \frac{b_0}{b_0+2}$. Then the function of d,

$$f(d) = (1 + b_0 + d)(d + 2)\left(\frac{2}{1+b_0+d} - \frac{2}{b_0+2} + \frac{d}{d+2} - t\right)$$

$$= (1 + b_0 + d)(d + 2)\left(\frac{2}{1+b_0+d} + \frac{b_0}{b_0+2} - \frac{2}{d+2} - t\right),$$

has $f(0) = 2(\frac{2}{b_0+2} - t(b_0 + 1)) > 0$, and $f(d)$ becomes negative as $d \to +\infty$.

Therefore, there must exist d_0 such that $f(d_0) = 0$, i.e., there are b_0, d_0 making (1) hold.

Thus, the range of S is the open interval $(1, 2)$.

Solution 2. It is evident that the sum S is homogeneous in a, b, c, d, so we can assume

$$x = a + c, \quad y = b + d, \quad \text{and} \quad x + y = 1.$$

Consider the sum $S_1 = \frac{a}{d+a+b} + \frac{c}{b+c+d} = \frac{a}{1-c} + \frac{c}{1-a} = \frac{2ac+x-x^2}{ac+1-x}$.

If x is fixed, then the product ac can take any real number from 0 to $(\frac{a+c}{2})^2 = \frac{x^2}{4}$, i.e., $0 < ac \le \frac{x^2}{4}$. Furthermore,

$$S_1 = \frac{2ac + x - x^2}{ac + 1 - x} = 2 + \frac{3x - 2 - x^2}{ac + 1 - x},$$

which is monotonic in ac. Thus, the range of S_1 is $(x, \frac{2x}{2-x}]$.

Similarly, if y is fixed, then the range of $S_2 = \frac{b}{a+b+c} + \frac{d}{c+d+a}$ is $(y, \frac{2y}{2-y}]$.

Consequently, the range of the sum S is $(x + y, \frac{2x}{2-x} + \frac{2y}{2-y}]$, i.e., $(1, \frac{4-4xy}{2+xy}]$.

Noting that the range of xy is $(0, \frac{1}{4}]$, we find the range of $\frac{4-4xy}{2+xy}$ to be $[\frac{4}{3}, 2)$. Therefore, the range of the sum S is $(1, 2)$.

Note. There are several similar problems:

- Let a, b, c, d be positive real numbers satisfying $ab + bc + cd + da = 1$. Prove that

$$\frac{a^3}{b+c+d} + \frac{b^3}{c+d+a} + \frac{c^3}{d+a+b} + \frac{d^3}{a+b+c} \geq \frac{1}{3}.$$

- Let a, b, c, d be positive real numbers. Prove that

$$\frac{a}{b+2c+3d} + \frac{b}{c+2d+3a} + \frac{c}{d+2a+3b} + \frac{d}{a+2b+3c} \geq \frac{2}{3}.$$

- (**From the book** *Secrets in Inequalities*, **2007: 37, proposed by Pham Kim Hung**). Let a, b, c, d be positive real numbers. Prove that

$$\frac{a}{b^2+c^2+d^2} + \frac{b}{c^2+d^2+a^2} + \frac{c}{d^2+a^2+b^2} + \frac{d}{a^2+b^2+c^2} \geq \frac{4}{a+b+c+d}.$$

- (**From the book** *Secrets in Inequalities*, **2007: 38**). Let $a_1, a_2, \ldots, a_n$ be positive real numbers. Prove that

$$\frac{a_1}{a_2^2 + a_3^2 + \cdots + a_n^2} + \frac{a_2}{a_1^2 + a_3^2 + \cdots + a_n^2} + \cdots + \frac{a_n}{a_1^2 + a_2^2 + \cdots + a_{n-1}^2}$$
$$\geq \frac{4}{a_1 + a_2 + \cdots + a_n}.$$

- (**Nesbitt's inequality in six variables**). Prove that for all positive real numbers a, b, c, d, e, f,

$$\frac{a}{b+c} + \frac{b}{c+d} + \frac{c}{d+e} + \frac{d}{e+f} + \frac{e}{f+a} + \frac{f}{a+b} \geq 3.$$

- (**Turkey Team Selection Test 1997, Problem 6**). Let $n \geq 2$ be an integer and $x_1, x_2, \ldots, x_n$ be positive real numbers such that $x_1^2 + x_2^2 + \cdots + x_n^2 = 1$. Determine the smallest possible value of

$$\frac{x_1^5}{x_2 + x_3 + \cdots + x_n} + \frac{x_2^5}{x_3 + \cdots + x_n + x_1} + \cdots + \frac{x_n^5}{x_1 + x_2 + \cdots + x_{n-1}}.$$

- (**From Mildorf's paper "Olympiad Inequalities," 2005: 34**). Let $n \geq 2$ be a positive integer and $k \geq \frac{n-1}{n}$ be a real number. Show that for all positive reals $a_1, a_2, \ldots, a_n$,

$$\left(\frac{(n-1)a_1}{a_2 + a_3 + \cdots + a_n} \right)^k + \left(\frac{(n-1)a_2}{a_3 + \cdots + a_n + a_1} \right)^k + \cdots$$
$$+ \left(\frac{(n-1)a_n}{a_1 + a_2 + \cdots + a_{n-1}} \right)^k \geq n.$$

【Score Situation】 This particular problem saw the following distribution of scores among contestants: 27 contestants scored 7 points, 8 contestants scored 6 points, 15 contestants scored 5 points, 13 contestants scored 4 points, 8 contestants scored 3 points, 12 contestants scored 2 points, 22 contestants scored 1 point, and 35 contestants scored 0 point. The average score of this problem is 3.100, indicating that it was relatively straightforward.

Among the top five teams in the team scores, the scores of this problem are as follows: the Soviet Union team scored 44 points (with a total team score of 256 points), the United States team scored 39 points (with a total team score of 243 points), the Hungary team scored 33 points (with a total team score of 237 points), the German Democratic Republic team scored 28 points (with a total team score of 236 points), and the Yugoslavia team scored 22 points (with a total team score of 216 points).

The gold medal cutoff for this IMO was set at 38 points (with 10 contestants earning gold medals), the silver medal cutoff was 30 points (with 24 contestants earning silver medals), and the bronze medal cutoff was 23 points (with 37 contestants earning bronze medals).

In this IMO, a total of six contestants achieved a perfect score of 40 points.

Problem 4.32 (IMO 40-2, proposed by Poland). Let n be a fixed integer with $n \geq 2$.

(a) Determine the least constant C such that the inequality

$$\sum_{1 \leq i < j \leq n} x_i x_j (x_i^2 + x_j^2) \leq C \left(\sum_{i=1}^n x_i \right)^4$$

holds for all real numbers $x_1, x_2, \ldots, x_n \geq 0$.

(b) For this constant C, determine when the equality holds.

Solution 1. Since the inequality is symmetric and homogeneous, we may assume $x_1 \geq x_2 \geq \cdots \geq x_n \geq 0$ and $\sum_{i=1}^n x_i = 1$. Thus, it suffices to consider the maximum value of

$$F(x_1, x_2, \ldots, x_n) = \sum_{1 \le i < j \le n} x_i x_j (x_i^2 + x_j^2).$$

Let $x_{k+1}(k \ge 2)$ be the last non-zero number among $x_1, x_2, \ldots, x_n$ and adjust

$$x = (x_1, x_2, \ldots, x_{k-1}, x_k, x_{k+1}, 0, \ldots, 0)$$

to

$$x' = (x_1, x_2, \ldots, x_{k-1}, x_k + x_{k+1}, 0, 0, \ldots, 0).$$

Then

$$F(x') - F(x) = x_k x_{k+1} \left(3(x_k + x_{k+1}) \sum_{i=1}^{k-1} x_i - x_k^2 - x_{k+1}^2 \right)$$

$$= x_k x_{k+1} (3(x_k + x_{k+1})(1 - x_k - x_{k+1}) - x_k^2 - x_{k+1}^2)$$

$$= x_k x_{k+1} ((x_k + x_{k+1})(3 - 4(x_k + x_{k+1})) + 2x_k x_{k+1}).$$

Since $1 \ge x_1 + x_k + x_{k+1} \ge \frac{1}{2}(x_k + x_{k+1}) + (x_k + x_{k+1})$, $x_k + x_{k+1} \le \frac{2}{3} < \frac{3}{4}$.

Hence, $F(x') - F(x) > 0$. In other words, the function value of F strictly increases when x is transformed into x'.

For any $x = (x_1, x_2, \ldots, x_n)$, after several adjustments, one can eventually obtain

$$F(x) \le F(a, b, 0, \ldots, 0) = ab(a^2 + b^2)$$

$$= \frac{1}{2}(2ab)(1 - 2ab) \le \frac{1}{8}.$$

Therefore, the minimum value of C is $\frac{1}{8}$ and the equality holds if and only if two of the x_i are equal and the others are all equal to 0.

Solution 2. When $x_1 = x_2 = 1$ and $x_3 = \cdots = x_n = 0$, we have

$$C \ge \frac{1}{16} \times 1 \times 1 \times (1^2 + 1^2) = \frac{1}{8}.$$

Next, we prove that the inequality

$$\sum_{1 \le i < j \le n} x_i x_j (x_i^2 + x_j^2) \le \frac{1}{8} \left(\sum_{i=1}^n x_i \right)^4$$

is true for all non-negative real numbers $x_1, x_2, \ldots, x_n$.

Since

$$\left(\sum_{i=1}^{n} x_i\right)^4 = \left(\sum_{k=1}^{n} x_k^2 + 2\sum_{1\le i<j\le n} x_i x_j\right)^2$$

$$\ge 4\left(\sum_{k=1}^{n} x_k^2\right)\left(2\sum_{1\le i<j\le n} x_i x_j\right)$$

$$= 8\sum_{1\le i<j\le n}\left(x_i x_j \sum_{k=1}^{n} x_k^2\right)$$

$$\ge 8\sum_{1\le i<j\le n} x_i x_j (x_i^2 + x_j^2),$$

it follows that

$$\sum_{1\le i<j\le n} x_i x_j (x_i^2 + x_j^2) \le \frac{1}{8}\left(\sum_{i=1}^{n} x_i\right)^4.$$

Thus, the minimum value of C is $\frac{1}{8}$, and the equality is true if and only if two of the x_i are equal and the others are all equal to 0.

【Score Situation】 This particular problem saw the following distribution of scores among contestants: 59 contestants scored 7 points, 11 contestants scored 6 points, 4 contestants scored 5 points, 5 contestants scored 4 points, 10 contestants scored 3 points, 37 contestants scored 2 points, 129 contestants scored 1 point, and 195 contestants scored 0 point. The average score of this problem is 1.671, indicating that it was relatively challenging.

Among the top five teams in the team scores, the scores of this problem are as follows: the China team scored 41 points (with a total team score of 182 points), the Russia team scored 37 points (with a total team score of 182 points), the Vietnam team scored 37 points (with a total team score of 177 points), the Romania team scored 42 points (with a total team score of 173 points), and the Bulgaria team scored 23 points (with a total team score of 170 points).

The gold medal cutoff for this IMO was set at 28 points (with 38 contestants earning gold medals), the silver medal cutoff was 19 points (with 70 contestants earning silver medals), and the bronze medal cutoff was 12 points (with 118 contestants earning bronze medals).

In this IMO, no contestant achieved a perfect score of 42 points.

Problem 4.33 (IMO 47-3, proposed by Ireland). Determine the least real number M such that the inequality

$$|ab(a^2 - b^2) + bc(b^2 - c^2) + ca(c^2 - a^2)| \le M(a^2 + b^2 + c^2)^2$$

holds for all real numbers a, b, and c.

Solution. Consider $P(t) = tb(t^2 - b^2) + bc(b^2 - c^2) + ct(c^2 - t^2)$. It is evident that $P(b) = P(c) = P(-c - b) = 0$, and thus,

$$|ab(a^2 - b^2) + bc(b^2 - c^2) + ca(c^2 - a^2)|$$
$$= |P(a)| = |(b - c)(a - b)(a - c)(a + b + c)|.$$

The original inequality is equivalent to

$$|(b - c)(a - b)(a - c)(a + b + c)| \le M(a^2 + b^2 + c^2)^2.$$

Assuming without loss of generality that $a \le b \le c$, we have

$$|(a - b)(b - c)| = (b - a)(c - b) \le \left(\frac{(b - a) + (c - b)}{2}\right)^2 = \frac{(c - a)^2}{4},$$

with the equality if and only if $b - a = c - b$, i.e., $2b = a + c$. Also,

$$\left(\frac{(c - b) + (b - a)}{2}\right)^2 \le \frac{(c - b)^2 + (b - a)^2}{2},$$

which is equivalent to

$$3(c - a)^2 \le 2((b - a)^2 + (c - b)^2 + (c - a)^2),$$

with the equality if and only if $2b = a + c$. Thus,

$$|(b - c)(a - b)(a - c)(a + b + c)|$$
$$\le \frac{1}{4}|(c - a)^3(a + b + c)|$$
$$= \frac{1}{4}\sqrt{(c - a)^6(a + b + c)^2}$$
$$\le \frac{1}{4}\sqrt{\left(\frac{2((b - a)^2 + (c - b)^2 + (c - a)^2)}{3}\right)^3 (a + b + c)^2}$$
$$= \frac{\sqrt{2}}{2}\left(\sqrt[4]{\left(\frac{(b - a)^2 + (c - b)^2 + (c - a)^2}{3}\right)^3 (a + b + c)^2}\right)^2.$$

By the AM-GM inequality,

$$|(b-c)(a-b)(a-c)(a+b+c)|$$

$$\leq \frac{\sqrt{2}}{2}\left(\frac{(b-a)^2+(c-b)^2+(c-a)^2+(a+b+c)^2}{4}\right)^2$$

$$= \frac{9\sqrt{2}}{32}(a^2+b^2+c^2)^2.$$

Thus, $M = \frac{9\sqrt{2}}{32}$, with the equality if and only if $2b = a + c$ and

$$\frac{(b-a)^2+(c-b)^2+(c-a)^2}{3} = (a+b+c)^2,$$

i.e., $2b = a + c$ and $(c - a)^2 = 18b^2$.

Taking $b = 1$, we obtain $a = 1 - \frac{3}{2}\sqrt{2}$ and $c = 1 + \frac{3}{2}\sqrt{2}$. Hence, when

$$(a, b, c) = \left(1 - \frac{3}{2}\sqrt{2}, 1, 1 + \frac{3}{2}\sqrt{2}\right),$$

the equality in the original inequality is achieved, confirming that $M = \frac{9\sqrt{2}}{32}$.

Note. There are several similar problems:

- **(Turkey Mathematical Olympiad 2013, 2nd Round, Problem 5).**
  Find the maximum value of M such that

  $$a^3 + b^3 + c^3 - 3abc \geq M(ab^2 + bc^2 + ca^2 - 3abc)$$

  for all positive real numbers a, b, c.
- **(Chinese Southeast Mathematical Olympiad 2006, Problem 6).**
  Find the minimum value of real numbers m such that

  $$m(a^3 + b^3 + c^3) \geq 6(a^2 + b^2 + c^2) + 1$$

  for all positive real numbers a, b, c with $a + b + c = 1$.

【Score Situation】 This particular problem saw the following distribution of scores among contestants: 28 contestants scored 7 points, 2 contestants scored 6 points, 2 contestants scored 5 points, 1 contestant scored 4 points, 1 contestant scored 3 points, 4 contestants scored 2 points, 95 contestants scored 1 point, and 365 contestants scored 0 point. The average score of this problem is 0.659, indicating that it was extremely difficult.

Among the top five teams in the team scores, the scores of this problem are as follows: the China team scored 35 points (with a total team score of 214 points), the Russia team scored 11 points (with a total team score of 174 points), the South Korea team scored 30 points (with a total team score of 170 points), the Germany team scored 9 points (with a total team score of 157 points), and the United States team scored 7 points (with a total team score of 154 points).

The gold medal cutoff for this IMO was set at 28 points (with 42 contestants earning gold medals), the silver medal cutoff was 19 points (with 89 contestants earning silver medals), and the bronze medal cutoff was 15 points (with 122 contestants earning bronze medals).

In this IMO, only three contestants achieved a perfect score of 42 points, namely Zhiyu Liu from China, Iurie Boreico from Moldova, and Alexander Magazinov from Russia.

4.3 Summary

Inequalities are pervasive across various branches of mathematics, including analysis, algebra, number theory, and combinatorics, providing crucial tools for comparison, estimation, and bounding of mathematical expressions. However, more important and challenging than proving inequalities is the ability to identify and establish relationships of inequality between quantities in problems.

In the first 64 IMOs, there were a total of 33 inequality problems. These problems can be broadly categorized into three types, as depicted in Figure 4.3. The score details for these problems are presented in Table 4.2. Due to the smaller number of participating teams and missing contestant score information in early IMOs, there are several blanks in Table 4.2.

Problems 4.1–4.5 focus on "solving inequalities;" among these five problems, the one with the lowest average score is Problem 4.5 (IMO 29-4), proposed by Ireland. Problems 4.6–4.30 deal with "proving inequalities;" among these 25 problems, the one with the lowest average score is Problem 4.29 (IMO 62-2), proposed by Canada. Problems 4.31–4.33 are about "determining value ranges;" among these three problems, the one with the lowest average score is Problem 4.33 (IMO 47-3), proposed by Ireland.

These 33 problems were proposed by 18 countries, with Hungary contributing the most, totaling four problems. Ireland, South Korea, and the Netherlands each proposed three problems, while the United States, Poland, France, Germany, Russia, and Austria each contributed two problems.

From Table 4.2, it can be observed that in the first 64 IMOs, there were three inequality problems with an average score of 0–1 point; five problems

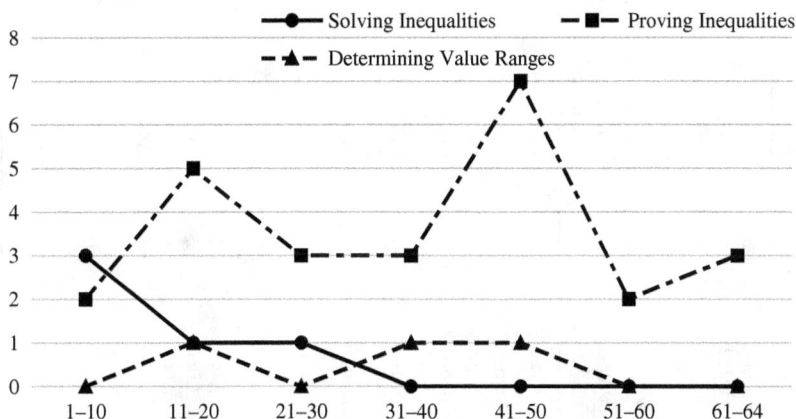

Figure 4.3 Numbers of Inequality Problems in the First 64 IMOs

with an average score of 1–2 points; 12 problems with an average score of 2–3 points; five problems with an average score of 3–4 points; eight problems with an average score above 4 points. Overall, the inequality problems were of certain difficulty, with most problems having an average score in the range of 2–3 points.

In the 24th–64th IMOs, there were a total of 21 inequality problems. Among these, three had an average score of 0–1 point; five had an average score of 1–2 points; eight had an average score of 2–3 points; one had an average score of 3–4 points; four had an average score above 4 points. Further analysis of the problem numbers of these 21 inequality problems, as shown in Table 4.3, reveals that these problems frequently appeared as the 1st/4th or 2nd/5th problem. The majority of these problems, totaling 18, were of the type proving inequalities. The other two types of inequality problems were less frequent, with only three problems appearing.

Excluding Problem 4.13 (IMO 24-6), the inequality problems from the 25th–64th IMOs are arranged in order of their average scores, from left to right, and a scatter plot of the score details is presented, as shown in Figure. 4.4.

From Table 4.2 and Figure 4.4, it is observable that in the inequality problems, the average score of the top five teams generally exceeds the average score of the problem by 3.5 points, and the average score of the 6th–15th teams typically surpasses the average score by 2 points. However, in more difficult problems, the performance of the top 15 teams is particularly outstanding, but the average score of the 16th–25th teams is very close to

Table 4.2 Score Details of Inequality Problems in the First 64 IMOs

Problem	4.1	4.2	4.3	4.4	4.5	4.6	4.7	4.8
Full points	6.000	6.000	4.000	7.000	7.000	7.000	7.000	8.000
Average score	2.600	4.529	2.738	4.909	2.332	3.111	3.294	2.438
Top five mean			3.300	5.175	6.433		4.700	3.600
6th–15th mean					3.661			
16th–25th mean					2.293			
Problem number in IMO	2-2	4-2	7-1	14-4	29-4	3-2	6-2	11-6
Proposing country	Hungary	Hungary	Yugoslavia	The Netherlands	Ireland	Poland	Hungary	The Soviet Union

Problem	4.9	4.10	4.11	4.12	4.13	4.14	4.15	4.16
Full points	5.000	6.000	6.000	6.000	7.000	7.000	7.000	7.000
Average score	2.321	4.727	3.216	5.167	2.125	4.557	2.156	2.561
Top five mean	3.354	5.725		5.775		6.833	6.233	6.133
6th–15th mean	1.836	4.797		5.224		5.767	3.600	4.133
16th–25th mean						4.167	1.200	3.467
Problem number in IMO	13-1	17-1	19-4	20-5	24-6	25-1	28-3	35-1
Proposing country	Hungary	Czechoslovakia	The United Kingdom	France	The United States	Germany	Germany	France

Problem	4.17	4.18	4.19	4.20	4.21	4.22	4.23	4.24
Full points	7.000	7.000	7.000	7.000	7.000	7.000	7.000	7.000
Average score	1.709	1.778	2.768	1.550	1.613	4.080	0.912	3.383
Top five mean	6.067	5.867	6.417	5.333	6.033	7.000	4.633	6.600
6th–15th mean	3.867	4.183	4.815	3.800	3.050	6.850	3.083	6.033
16th–25th mean	2.167	2.483	3.817	1.517	2.305	6.000	0.983	5.000
Problem number in IMO	36-2	38-3	41-2	42-2	44-5	45-4	46-3	48-1
Proposing country	Russia	Russia	The United States	South Korea	Ireland	South Korea	South Korea	New Zealand

(Continued)

Table 4.2 (*Continued*)

Problem	4.25	4.26	4.27	4.28	4.29	4.30	4.31	4.32
Full points	7.000	7.000	7.000	7.000	7.000	7.000	7.000	7.000
Average score	2.563	2.550	5.348	2.248	0.375	4.717	3.100	1.671
Top five mean	6.133	6.417	6.867	5.800	1.733	7.000	4.150	6.000
6th–15th mean	5.400	4.852	6.850	4.500	1.283	6.850	3.052	3.883
16th–25th mean	3.950	3.467	6.917	3.783	0.683	6.700		1.833
Problem number in IMO	49-2	53-2	55-1	61-2	62-2	64-4	16-5	40-2
Proposing country	Austria	Australia	Austria	Belgium	Canada	The Netherlands	The Netherlands	Poland

Problem	4.33
Full points	7.000
Average score	0.659
Top five mean	3.067
6th–15th mean	1.924
16th–25th mean	0.537
Problem number in IMO	47-3
Proposing country	Ireland

Note. Top five mean = Total score of the top five teams ÷ Total number of contestants from the top five teams,
6th–15th mean = Total score of the 6th–15th teams ÷ Total number of contestants from the 6th–15th teams,
16th–25th mean = Total score of the 16th–25th teams ÷ Total number of contestants from the 16th–25th teams.

Table 4.3   Numbers of Inequality Problems in the 24th–64th IMOs.

Inequality Problem	Problem Number			Number of Problems in the First 64 IMOs
	1, 4	2, 5	3, 6	
Solving inequalities	1	0	0	5
Proving inequalities	6	8	4	25
Determining value ranges	0	1	1	3
Total	7	9	5	33

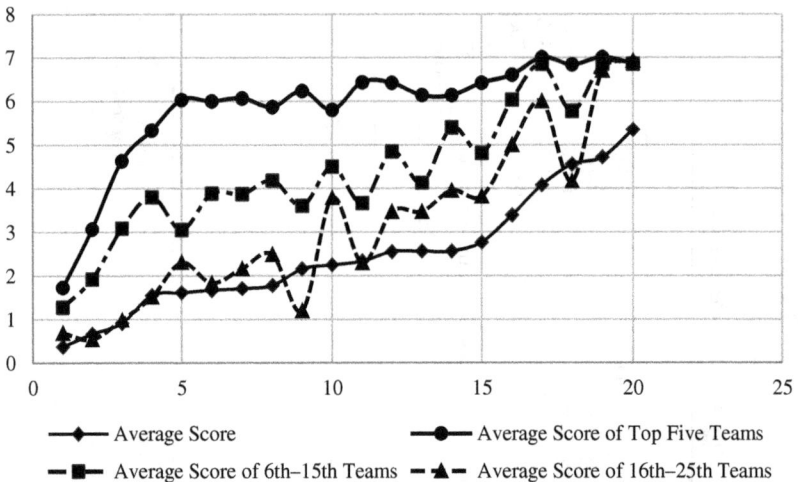

Figure 4.4   Score Details of Inequality Problems in the 25th–64th IMOs

the average score of the problem, as seen in Problem 4.33 (IMO 47-3), Problem 4.23 (IMO 46-3), and Problem 4.20 (IMO 42-2).

From Figure 4.4, it can also be observed that there are several inequality problems where the average score is higher than the average score of the 16th–25th teams, such as Problem 4.15 (IMO 28-3), Problem 4.5 (IMO 29-4), and Problem 4.14 (IMO 25-1). This phenomenon is due to the smaller number of participating teams in early IMOs. It was not until the 30th IMO in 1989 that the number of participating teams exceeded 50. Therefore, it is common to see situations where the average score is close to or even higher than the average score of the 16th–25th teams during this period.

Chapter 5

Other Algebra Problems

Algebra, deriving from the Arabic word "al-jabr," evolved in ancient times. As a plethora of problems concerning quantitative relationships arose, the need for more general methods to solve these problems led to the development of elementary algebra, with solving equations as its core focus.

The basic components of elementary algebra include numbers, expressions, and equations, which roughly align with the algebra curriculum currently established in secondary schools worldwide. However, the algebra taught in secondary schools also encompasses additional topics such as functions, sequences, inequalities, and polynomials.

In the first 64 IMOs, there had been a total of 10 other algebra problems, approximately accounting for 9.9% of all algebra problems. These problems can be primarily categorized into three types: (1) proving trigonometric identities, totaling two problems; (2) finding polynomials, totaling two problems; (3) proving properties of polynomials and sets, totaling six problems. The statistical distribution of these three types of problems in the previous IMOs is presented in Table 5.1.

It can be observed that other algebra problems were not frequent, with only 1–3 appearing in every 10 IMOs.

Specifically, there were only two problems about proving trigonometric identities. On the one hand, students begin to engage with trigonometric functions from middle school, and have developed a comprehensive understanding, so IMO problems seldom specifically focused on trigonometric concepts. Instead, these concepts were integrated into geometry, where algebraic methods can also be applied. On the other hand, since

Table 5.1 Numbers of Other Algebra Problems in the First 64 IMOs

Content	Session							Total
	1–10	11–20	21–30	31–40	41–50	51–60	61–64	
Proving trigonometric identities	2	0	0	0	0	0	0	2
Finding polynomials	0	1	0	0	1	0	0	2
Proving properties	0	1	1	1	2	0	1	6
Algebra problems	20	20	14	13	15	13	6	101
The percentage of other algebra problems among the algebra problems	10.0%	10.0%	7.1%	7.7%	20.0%	0.0%	16.7%	9.9%

both sides of an identity are necessarily equal, it merely requires a specific manipulation, making the difficulty relatively low. Therefore, it is not a primary focus.

As for polynomial problems, the divisibility and congruence of polynomials can be linked with number theory, while finding polynomials can be associated with functional equations. Consequently, there are not many polynomial problems presented in this chapter. Although polynomials are part of "Advanced Algebra" in university mathematics, many polynomial concepts and methods are quite elementary.

This chapter will be divided into three parts. The first part primarily introduces some knowledge about trigonometry and polynomials, including common trigonometric identities, common polynomials, and common techniques for polynomials.

The second part revolves around three types of problems: "proving trigonometric identities," "finding polynomials," and "proving properties of polynomials and sets." These problems are presented in chronological order, and some problems include various solutions, generalizations, and similar problems.

It is important to note that for each problem, the solutions are followed by information on the scores, including the number of contestants in each score range, the average score, and the scores of the top five teams. However, early IMOs often lacked information on contestant scores, so the number of contestants in each score range only represents the counted number of contestants, and some problems lack scores of the top five teams.

The third part provides a brief summary of this chapter.

5.1 Common Theorems, Formulas, and Methods

5.1.1 *Common trigonometric identities*

When dealing with trigonometric problems, it is generally necessary to simplify or transform expressions, highlighting the significance of formulas such as sum and difference identities. Typically, multiplication forms are more convenient than addition forms and allow for the elimination of certain elements.

(i) $\sin 3\alpha = 4 \sin \alpha \sin(60° - \alpha) \sin(60° + \alpha)$.

(ii) $\cos 3\alpha = 4 \cos \alpha \cos(60° - \alpha) \cos(60° + \alpha)$.

(iii) $\tan 3\alpha = \tan \alpha \tan(60° - \alpha) \tan(60° + \alpha)$.

(iv) $\tan(\alpha + \beta + \gamma) = \frac{\tan \alpha + \tan \beta + \tan \gamma - \tan \alpha \tan \beta \tan \gamma}{1 - \tan \alpha \tan \beta - \tan \beta \tan \gamma - \tan \gamma \tan \alpha}$.

(v) $\sin \alpha + \sin(\alpha + 2d) + \cdots + \sin(\alpha + 2nd) = \frac{\sin(n+1)d \cdot \sin(\alpha+nd)}{\sin d}$.

(vi) $\cos \alpha + \cos(\alpha + 2d) + \cdots + \cos(\alpha + 2nd) = \frac{\sin(n+1)d \cdot \cos(\alpha+nd)}{\sin d}$.

Furthermore, by utilizing the properties of unit roots, we can obtain:

(vii) $\sin \frac{\pi}{n} \sin \frac{2\pi}{n} \cdots \sin \frac{(n-1)\pi}{n} = \frac{n}{2^{n-1}}$.

(viii) $\left| \cos \frac{\pi}{n} \cos \frac{2\pi}{n} \cdots \cos \frac{(n-1)\pi}{n} \right| = \frac{1-(-1)^n}{2^n}$.

Sometimes, there exist certain quantitative relationships between angles. For example, in the triangle ABC, the following relationships can be derived:

(ix) $\sin^2 A + \sin^2 B + \sin^2 C = 2 + 2 \cos A \cos B \cos C$.

(x) $\cos^2 A + \cos^2 B + \cos^2 C = 1 - 2 \cos A \cos B \cos C$.

(xi) $\sin A + \sin B + \sin C = 4 \cos \frac{A}{2} \cos \frac{B}{2} \cos \frac{C}{2}$.

(xii) $\cos A + \cos B + \cos C = 1 + 4 \sin \frac{A}{2} \sin \frac{B}{2} \sin \frac{C}{2}$.

(xiii) $\sin \frac{A}{2} + \sin \frac{B}{2} + \sin \frac{C}{2} = 1 + 4 \sin \frac{A+B}{4} \sin \frac{B+C}{4} \sin \frac{C+A}{4}$.

(xiv) $\cos \frac{A}{2} + \cos \frac{B}{2} + \cos \frac{C}{2} = 4 \cos \frac{A+B}{4} \cos \frac{B+C}{4} \cos \frac{C+A}{4}$.

(xv) $\sin 2A + \sin 2B + \sin 2C = 4 \sin A \sin B \sin C$.

(xvi) $\cos 2A + \cos 2B + \cos 2C = -1 - 4 \cos A \cos B \cos C$.

(xvii) $\tan \frac{A}{2} \tan \frac{B}{2} + \tan \frac{B}{2} \tan \frac{C}{2} + \tan \frac{C}{2} \tan \frac{A}{2} = 1$.

Especially in a non-right-angled triangle ABC, the following relationships hold:

(xviii) $\tan A + \tan B + \tan C = \tan A \tan B \tan C$.

(xix) $\cot A \cot B + \cot B \cot C + \cot C \cot A = 1$.

(xx) $\frac{\cot A + \cot B}{\tan A + \tan B} + \frac{\cot B + \cot C}{\tan B + \tan C} + \frac{\cot C + \cot A}{\tan C + \tan A} = 1$.

Mastering basic trigonometric transformations is crucial for solving geometry problems. The technique can be employed to establish common relationships between sides and angles in triangles.

Theorem 5.1 (Projection Formula). $a = b \cos C + c \cos B$.

Theorem 5.2 (Law of Sines). $\frac{a}{\sin A} = \frac{b}{\sin B} = \frac{c}{\sin C} = 2R$.

Theorem 5.3 (Law of Cosines). $c^2 = a^2 + b^2 - 2ab \cos C$.

Theorem 5.4 (Mollweide's Formula). $\frac{a + (-1)^k b}{c} = \frac{\cos\left(\frac{k\pi}{2} - \frac{A-B}{2}\right)}{\sin\left(\frac{k\pi}{2} + \frac{C}{2}\right)}$, where $k \in \{0, 1\}$.

Theorem 5.5 (Law of Tangents). $\frac{a-b}{a+b} = \frac{\tan \frac{A-B}{2}}{\tan \frac{A+B}{2}}$.

Theorem 5.6 (Half Angle Formulas).

$$\sin \frac{A}{2} = \sqrt{\frac{(p-b)(p-c)}{bc}}, \quad \cos \frac{A}{2} = \sqrt{\frac{p(p-a)}{bc}},$$

$$\tan \frac{A}{2} = \frac{1}{p-a} \sqrt{\frac{(p-a)(p-b)(p-c)}{p}} = \frac{r}{p-a},$$

where p is the semi-perimeter and r is the inradius.

Additionally, trigonometric knowledge can be employed to characterize the area of geometric figures, which will be presented in *IMO Problems, Theorems, and Methods: Geometry*.

5.1.2   *Common methods for proving trigonometric identities*

(1)   *Transforming to the same angle*

When multiple distinct angles appear in the given conditions, one can consider expressing them using a single angle based on their quantitative relationships, thereby simplifying the problem.

Example 5.1. Prove that

$$2 \sin^4 x + \frac{3}{4} \sin^2 2x + 5 \cos^4 x - \cos 3x \cos x = 2(1 + \cos^2 x).$$

Proof. Consider transforming angles x and $3x$ into $2x$:

$$2\sin^4 x + \frac{3}{4}\sin^2 2x + 5\cos^4 x - \cos 3x \cos x$$

$$= 2\left(\frac{1-\cos 2x}{2}\right)^2 + \frac{3}{4}\sin^2 2x + 5\left(\frac{1+\cos 2x}{2}\right)^2 - \frac{1}{2}(\cos 4x + \cos 2x)$$

$$= \frac{7+6\cos 2x + 7\cos^2 2x}{4} + \frac{3}{4}\left(1-\cos^2 2x\right) - \frac{1}{2}(2\cos^2 2x - 1 + \cos 2x)$$

$$= 3 + \cos 2x$$

$$= 2(1+\cos^2 x).$$

(2) *Transforming to the same trigonometric function*

When there are multiple different trigonometric functions in the given conditions, utilize basic trigonometric identities to transform them into the same trigonometric functions.

Example 5.2. Given $\frac{\tan(A-B)}{\tan A} + \frac{\sin^2 C}{\sin^2 A} = 1$, prove that $\tan A$, $\tan C$, and $\tan B$ are in geometric progression.

Proof. The purpose is to prove that $\tan^2 C = \tan A \tan B$. Convert the sine function in the condition into the tangent function:

$$\frac{\sin^2 C}{\sin^2 A} = \frac{1-\cos^2 C}{1-\cos^2 A} = \frac{1 - \frac{1}{1+\tan^2 C}}{1 - \frac{1}{1+\tan^2 A}} = \frac{\frac{\tan^2 C}{1+\tan^2 C}}{\frac{\tan^2 A}{1+\tan^2 A}},$$

$$1 - \frac{\tan(A-B)}{\tan A} = 1 - \frac{\tan A - \tan B}{\tan A(1+\tan A \tan B)} = \frac{\tan B\left(1+\tan^2 A\right)}{(1+\tan A \tan B)\tan A}.$$

Combining these two expressions yields $\frac{\tan^2 C}{1+\tan^2 C} = \frac{\tan A \tan B}{1+\tan A \tan B}$, establishing the conclusion.

Notably, this solution also used the identity $1 = \sin^2 x + \cos^2 x = \sec^2 x - \tan^2 x$ for transformations, and such constant-containing formulas are crucial for solving extremum problems or proving inequalities.

Example 5.3. Given $\tan x_1 \tan x_2 \cdots \tan x_{2022} = 1$, find the maximum value of

$$\sin x_1 \sin x_2 \cdots \sin x_{2022}.$$

Solution. From the given condition,

$$\sin x_1 \sin x_2 \cdots \sin x_{2022} = \cos x_1 \cos x_2 \cdots \cos x_{2022}.$$

Thus, we need to find the maximum value of $\cos x_1 \cos x_2 \cdots \cos x_{2022}$. Since

$$\frac{1}{\cos^2 x_i} = 1 + \tan^2 x_i \geq 2 \left| \tan x_i \right|,$$

we have $\frac{1}{\cos^2 x_1 \cos^2 x_2 \cdots \cos^2 x_{2022}} \geq 2^{2022} \left| \tan x_1 \tan x_2 \cdots \tan x_{2022} \right| = 2^{2022}$, so

$$\sin x_1 \sin x_2 \cdots \sin x_{2022} = \cos x_1 \cos x_2 \cdots \cos x_{2022} \leq \frac{1}{2^{1011}},$$

and the equality holds when $x_1 = x_2 = \cdots = x_{2022} = \frac{\pi}{4}$.

(3) *Transforming to the same power*

When trigonometric functions of varying or higher powers appear in the conditions, consider raising or reducing their powers to transform them into functions of the same power or into simpler forms.

Example 5.4. Prove that $\cos^3 \frac{x}{3} + \cos^3 \frac{x+2\pi}{3} + \cos^3 \frac{x+4\pi}{3} = \frac{3}{4} \cos x$.

Proof. Using the triple-angle formula $\cos 3\theta = 4\cos^3 \theta - 3\cos \theta$ for cosine, we have

$$\cos^3 \frac{x}{3} = \frac{3}{4} \cos \frac{x}{3} + \frac{1}{4} \cos x,$$

$$\cos^3 \frac{x+2\pi}{3} = \frac{3}{4} \cos \frac{x+2\pi}{3} + \frac{1}{4} \cos(x+2\pi) = \frac{3}{4} \cos \frac{x+2\pi}{3} + \frac{1}{4} \cos x,$$

$$\cos^3 \frac{x+4\pi}{3} = \frac{3}{4} \cos \frac{x+4\pi}{3} + \frac{1}{4} \cos(x+4\pi) = \frac{3}{4} \cos \frac{x-2\pi}{3} + \frac{1}{4} \cos x.$$

It is evident that $\cos \frac{x}{3} + \cos \frac{x+2\pi}{3} + \cos \frac{x-2\pi}{3} = \cos \frac{x}{3} + 2\cos \frac{x}{3} \cos \frac{2\pi}{3} = 0$. Therefore, adding the both sides of the above three equalities, we have

$$\cos^3 \frac{x}{3} + \cos^3 \frac{x+2\pi}{3} + \cos^3 \frac{x+4\pi}{3} = \frac{3}{4} \cos x.$$

(4) *Elimination method*

When conditions include parameters, it is common to derive expressions for these parameters and substitute them to eliminate the parameters.

Example 5.5. Given

$$a \cos^2 \alpha + b \sin^2 \alpha = m \cos^2 \beta,$$
$$a \sin^2 \alpha + b \cos^2 \alpha = n \sin^2 \beta,$$
$$m \tan^2 \alpha = n \tan^2 \beta,$$

where $\beta \neq k\pi$, prove that $(a+b)(m+n) = 2mn$.

Proof. When $m = 0$, from $m \tan^2 \alpha = n \tan^2 \beta$, we have $n = 0$, so the conclusion holds.

When $m \neq 0$, from the given condition, $\frac{a \sin^2 \alpha + b \cos^2 \alpha}{a \cos^2 \alpha + b \sin^2 \alpha} = \frac{n}{m} \tan^2 \beta$. Also, $m \tan^2 \alpha = n \tan^2 \beta$, so $\frac{a \sin^2 \alpha + b \cos^2 \alpha}{a \cos^2 \alpha + b \sin^2 \alpha} = \tan^2 \alpha$, from which $\frac{a \tan^2 \alpha + b}{a + b \tan^2 \alpha} = \tan^2 \alpha$.

Solving this, we can get $\tan^2 \alpha = 1$. Hence $\sin^2 \alpha = \cos^2 \alpha = \frac{1}{2}$. Consequently, $\cos^2 \beta = \frac{a+b}{2m}$ and $\sin^2 \beta = \frac{a+b}{2n}$. Hence we obtain $\frac{a+b}{2m} + \frac{a+b}{2n} = 1$.

(5) *Telescoping method*

Similar to summation in sequences, when proving trigonometric identities in the summation form, the Telescoping Method can be employed for eliminations.

Example 5.6. Prove that

$$\frac{1}{\cos 0° \cos 1°} + \frac{1}{\cos 1° \cos 2°} + \cdots + \frac{1}{\cos 88° \cos 89°} = \frac{\cos 1°}{\sin^2 1°}.$$

Proof. Since $\frac{\sin 1°}{\cos k° \cos(k+1)°} = \frac{\sin((k+1)° - k°)}{\cos k° \cos(k+1)°} = \tan(k+1)° - \tan k°$,

$$\sin 1° \sum_{k=0}^{88} \frac{1}{\cos k° \cos(k+1)°} = \sum_{k=0}^{88} (\tan(k+1)° - \tan k°) = \tan 89° = \cot 1°.$$

Hence, $\frac{1}{\cos 0° \cos 1°} + \frac{1}{\cos 1° \cos 2°} + \cdots + \frac{1}{\cos 88° \cos 89°} = \frac{\cos 1°}{\sin^2 1°}.$

(6) *Mathematical induction*

For propositions related to positive integers n, mathematical induction can be considered.

Example 5.7. Prove that

$$\frac{1}{2}\tan\frac{\alpha}{2} + \frac{1}{4}\tan\frac{\alpha}{4} + \cdots + \frac{1}{2^n}\tan\frac{\alpha}{2^n} = \frac{1}{2^n}\cot\frac{\alpha}{2^n} - \cot\alpha.$$

Proof. For $n = 1$, we have $\cot\alpha = \frac{\cos\alpha}{\sin\alpha} = \frac{\cos^2\frac{\alpha}{2} - \sin^2\frac{\alpha}{2}}{2\sin\frac{\alpha}{2}\cos\frac{\alpha}{2}} = \frac{1-\tan^2\frac{\alpha}{2}}{2\tan\frac{\alpha}{2}}$. Therefore,

$$\frac{1}{2}\cot\frac{\alpha}{2} - \cot\alpha = \frac{1}{2\tan\frac{\alpha}{2}} - \frac{1-\tan^2\frac{\alpha}{2}}{2\tan\frac{\alpha}{2}} = \frac{1}{2}\tan\frac{\alpha}{2}.$$

Assume $\frac{1}{2}\tan\frac{\alpha}{2} + \frac{1}{4}\tan\frac{\alpha}{4} + \cdots + \frac{1}{2^k}\tan\frac{\alpha}{2^k} = \frac{1}{2^k}\cot\frac{\alpha}{2^k} - \cot\alpha$. For $n = k+1$,

$$\frac{1}{2^k}\cot\frac{\alpha}{2^k} + \frac{1}{2^{k+1}}\tan\frac{\alpha}{2^{k+1}} = \frac{1}{2^k}\cdot\frac{1-\tan^2\frac{\alpha}{2^{k+1}}}{2\tan\frac{\alpha}{2^{k+1}}} + \frac{1}{2^{k+1}}\tan\frac{\alpha}{2^{k+1}}$$

$$= \frac{1}{2^{k+1}}\cot\frac{\alpha}{2^{k+1}}.$$

Thus,

$$\left(\frac{1}{2}\tan\frac{\alpha}{2} + \frac{1}{4}\tan\frac{\alpha}{4} + \cdots + \frac{1}{2^k}\tan\frac{\alpha}{2^k}\right) + \frac{1}{2^{k+1}}\tan\frac{\alpha}{2^{k+1}}$$

$$= \frac{1}{2^{k+1}}\cot\frac{\alpha}{2^{k+1}} - \cot\alpha.$$

Hence, the proposition holds for any positive integer n.

5.1.3 *Common polynomials*

(1) *Univariate polynomials*

Let x be an independent variable, an algebraic expression of the form

$$f(x) = a_n x^n + a_{n-1}x^{n-1} + \cdots + a_1 x + a_0$$

is called a univariate polynomial in x, or simply a polynomial. And $a_0, a_1, \ldots, a_n$ are known as the coefficients of the polynomial. If $a_0 = a_1 = \cdots = a_n = 0$, then $f(x)$ is termed the zero polynomial.

When $a_n \neq 0$, we call n as the degree of the polynomial, denoted as $\deg f = n$. If $\deg f = 0$, then $f(x)$ is a 0th degree polynomial. However, the degree of the zero polynomial is undefined.

Note. $\mathbf{N_+}$, $\mathbf{Z}$, $\mathbf{Q}$, $\mathbf{R}$, and $\mathbf{C}$ respectively represent the sets of positive integers, integers, rational numbers, real numbers, and complex numbers. The notations $\mathbf{Z}[x]$, $\mathbf{Q}[x]$, $\mathbf{R}[x]$, and $\mathbf{C}[x]$ respectively denote the sets of univariate polynomials in x with integer, rational, real, and complex coefficients.

Proposition 5.1. *Two polynomials* $f(x) = a_n x^n + \cdots + a_1 x + a_0$ *and* $g(x) = b_m x^m + \cdots + b_1 x + b_0$ *are equal, where* $a_n b_m \neq 0$, *if and only if* $m = n$ *and* $a_i = b_i$ *for* $i = 0, 1, \cdots, n$.

Proposition 5.2. *If* $f(x)$ *and* $g(x)$ *are non-zero polynomials, then*

$$\deg(f \cdot g) = \deg f + \deg g,$$

and when $f(x) \pm g(x)$ *is a non-zero polynomial,*

$$\deg(f \pm g) \leq \max\{\deg f, \deg g\}.$$

(2) *Integer polynomials*

A polynomial with all coefficients being integers is called an integer polynomial. Similarly, one can describe a polynomial with rational, real, and complex coefficients as rational, real, and complex polynomial respectively.

Theorem 5.7 (Gauss's Theorem). If an integer polynomial $f(x)$ of degree $n(n > 0)$ can be factored into a product of nonconstant rational polynomials, then it can also be factored into a product of nonconstant integer polynomials.

Theorem 5.8 (Eisenstein's Criterion). Let $f(x) = a_n x^n + \cdots + a_1 x + a_0 (a_n \neq 0)$ be an integer polynomial. If there exists a prime number p such that:

(i) $p \nmid a_n$;
(ii) $p | a_i$ for $i = 0, 1, \ldots, n-1$;
(iii) $p^2 \nmid a_0$,

then $f(x)$ is irreducible over the rational number field.

Corollary 5.1. *Let* $f(x) = a_n x^n + \cdots + a_1 x + a_0 (a_n \neq 0)$ *be an integer polynomial. If there exist a prime number* p *and a positive number* $m \leq n$ *such that:*

(i) $p \nmid a_n$;
(ii) $p | a_i$ *for* $i = 0, 1, \ldots, m-1$;
(iii) $p^2 \nmid a_0$,

then $f(x)$ has an irreducible factor with integer coefficients, whose degree is not less than m.

Proposition 5.3. Let $f(x) = a_n x^n + \cdots + a_1 x + a_0 (a_n \neq 0)$ be an integer polynomial. If a rational number $\frac{p}{q}$ (in reduced form) is a root of $f(x)$, then

(i) $q | a_n$ and $p | a_0$;

(ii) $f(x) = \left(x - \frac{p}{q} \right) g(x)$, where $g(x) \in \mathbf{Z}[x]$.

Corollary 5.2. If the leading coefficient of an integer polynomial $f(x)$ is 1, then the rational roots of $f(x)$ can only be integers.

Corollary 5.3. The integer roots of an integer polynomial are necessarily divisors of its constant term.

Proposition 5.4. Let integer m be a divisor of the constant term of an integer polynomial $f(x)$ and $k(k \neq m)$ is any integer. Then m is not a root of $f(x)$ if $f(k)$ is not divisible by $m - k$.

Proposition 5.5. Let $f(x)$ be an irreducible rational polynomial and $g(x)$ is a rational polynomial. If $f(x)$ and $g(x)$ have a common root, then $f(x) | g(x)$.

Corollary 5.4. If a rational polynomial $f(x)$ has an irrational root $a + \sqrt{b}$, where a, b are rational and $\sqrt{b}$ is irrational, then $f(x)$ has the irrational root $a - \sqrt{b}$.

Corollary 5.5. If a rational polynomial $f(x)$ has an irrational root $a\sqrt{b} + c\sqrt{d}$, where a, b, c, d are rational, $\sqrt{b}$ and $\sqrt{d}$ are irrational and not of the same type, and $ac \neq 0$, then $f(x)$ has the irrational roots $a\sqrt{b} - c\sqrt{d}$, $-a\sqrt{b} + c\sqrt{d}$, and $-a\sqrt{b} - c\sqrt{d}$.

(3) *Integer-valued polynomials*

A polynomial $f(x)$ is called an integer-valued polynomial if, for any integer value of x, the value of $f(x)$ is an integer.

(4) *Difference Polynomials*

The polynomial $P_k(x) = \frac{1}{k!} x(x - 1) \cdots (x - k + 1)$, where k is a positive integer and $P_0(x) = 1$, is referred to as the difference polynomial of degree k and denoted as C_x^k.

If x is a positive integer n greater than or equal to k, then $P_k(n) = C_n^k$ is an integer. If x is a negative integer $-m$, then $P_k(-m) = (-1)^k C_{m+k-1}^k$ is also an integer. If $x \in \{0, 1, \ldots, k-1\}$, then $P_k(x) = 0$. Therefore, difference polynomials are integer-valued polynomials.

Similar to the properties of binomial coefficients, $P_k(x+1) - P_k(x) = P_{k-1}(x)$. Furthermore, we can define the first order difference of a polynomial $f(x)$ as

$$\Delta f(x) = f(x+1) - f(x),$$

and $(r+1)$th order difference of $f(x)$ as

$$\Delta^{r+1} f(x) = \Delta(\Delta^r f(x)) = \Delta^r f(x+1) - \Delta^r f(x).$$

Proposition 5.6. $\Delta^n f(x) = \sum_{i=0}^{n}(-1)^{n-i} C_n^i f(x+i) = \sum_{i=0}^{n}(-1)^i C_n^i f(x+n-i).$

Proposition 5.7. *If $f(x)$ is a polynomial of degree n with the leading coefficient a_n, then $\Delta^n f(x) = n!a_n$.*

Proposition 5.8. *If $\Delta^{n+1} f(x) = 0$ for $n \in \mathbf{N}$, then $f(x)$ is a polynomial in x of degree not greater than n.*

Proposition 5.9. *Any polynomial $f(x)$ of degree n can be uniquely expressed as*

$$f(x) = b_n C_x^n + b_{n-1} C_x^{n-1} + \cdots + b_1 C_x^1 + b_0,$$

where $b_k (k = 0, 1, \ldots, n)$ are constants.

Obviously, when $b_k (k = 0, 1, \ldots, n)$ are all integers, $f(x)$ is an integer-valued polynomial. Conversely, if $f(x)$ is an integer-valued polynomial, then $b_k (k = 0, 1, \ldots, n)$ are all integers.

Proposition 5.10. *Let $f(x)$ be a polynomial of degree n. Then*

$$\sum_{k=0}^{n} f(k) = \sum_{i=0}^{n} \Delta^i f(0) \cdot C_{n+1}^{i+1}.$$

Proposition 5.11. *A polynomial $f(x)$ of degree n is an integer-valued polynomial if and only if, when x takes $n+1$ consecutive integer values, the values of $f(x)$ are all integers.*

Proposition 5.12. *For integer-valued polynomials*

$$f_k(x) = L(k) \cdot C_x^k \quad (k = 1, 2, \ldots)$$

and any two distinct integers a and b,

$$(a - b)|(f_k(a) - f_k(b)),$$

where $L(k)$ is the least common multiple of $1, 2, \ldots, k$.

Proposition 5.13. *Let $f(x) = \sum_{k=0}^{d} a_k C_x^k$ be an integer-valued polynomial of degree d. If for integers i, j satisfying $0 \leq i \neq j \leq d$,*

$$(i - j)|(f(i) - f(j)),$$

then $L(k)|a_k(0 \leq k \leq d)$, and for all distinct integers a and b,

$$(a - b)|(f(a) - f(b)),$$

where $L(k)$ is the least common multiple of $1, 2, \ldots, k$.

(5) *Multivariate polynomials*

A polynomial of the form

$$f(x_1, x_2, \ldots, x_n) = \sum a_{k_1 k_2 \cdots k_n} x_1^{k_1} x_2^{k_2} \cdots x_n^{k_n}$$

is termed an n-variable polynomial of degree m, where $x_1, x_2, \ldots, x_n$ are variables, and $m = \max\{k_1 + k_2 + \cdots + k_n\}$ with $k_i \in \mathbf{N}(i = 1, 2, \ldots, n)$.

If every monomial of $f(x_1, x_2, \ldots, x_n)$ is of degree m, then $f(x_1, x_2, \ldots, x_n)$ is called a homogeneous polynomial of degree m, and in this case,

$$f(tx_1, tx_2, \ldots, tx_n) = t^m f(x_1, x_2, \ldots, x_n).$$

If substituting x_1 with x_2, x_2 with $x_3, \ldots, x_{n-1}$ with x_n, and x_n with x_1 in $f(x_1, x_2, \ldots, x_{n-1}, x_n)$ always results in

$$f(x_1, x_2, \ldots, x_{n-1}, x_n) = f(x_2, x_3, \ldots, x_n, x_1),$$

then $f(x_1, x_2, \ldots, x_n)$ is called a cyclic polynomial.

If for any permutation $i_1, i_2, \ldots, i_n$ of $1, 2, \ldots, n$,

$$f(x_1, x_2, \ldots, x_n) = f(x_{i_1}, x_{i_2}, \ldots, x_{i_n}),$$

then $f(x_1, x_2, \ldots, x_n)$ is called a symmetric polynomial. Clearly, every symmetric polynomial is a cyclic polynomial, but not every cyclic polynomial is symmetric.

For n-variable symmetric polynomials, we introduce the following notations:

$$\sigma_1 = x_1 + x_2 + \cdots + x_n,$$

$$\sigma_2 = \sum_{1 \leq i_1 < i_2 \leq n} x_{i_1} x_{i_2},$$

$$\cdots\cdots\cdots\cdots\cdots\cdots$$

$$\sigma_k = \sum_{1 \leq i_1 < i_2 < \cdots < i_k \leq n} x_{i_1} x_{i_2} \cdots x_{i_k},$$

$$\cdots\cdots\cdots\cdots\cdots\cdots$$

$$\sigma_n = x_1 x_2 \cdots x_n,$$

where $\sigma_1, \sigma_2, \ldots, \sigma_n$ are called the elementary symmetric polynomials in $x_1, x_2, \ldots, x_n$.

Theorem 5.9 (Fundamental Theorem of Symmetric Polynomials).
Any symmetric polynomial $f(x_1, x_2, \ldots, x_n)$ can be uniquely expressed as a polynomial in $\sigma_1, \sigma_2, \ldots, \sigma_n$, i.e.,

$$f(x_1, x_2, \ldots, x_n) = \varphi(\sigma_1, \sigma_2, \ldots, \sigma_n).$$

Theorem 5.10 (Newton's Identities). Let $S_k = x_1^k + x_2^k + \cdots + x_n^k (k \in \mathbf{N}_+)$, where $x_1, x_2, \ldots, x_n$ are roots of the polynomial

$$f(x) = x^n + a_1 x^{n-1} + \cdots + a_{n-1} x + a_n.$$

Then,
  (i) for $k \leq n$, there is $S_k + a_1 S_{k-1} + \cdots + a_{k-1} S_1 + k a_k = 0$;
  (ii) for $k > n$, there is $S_k + a_1 S_{k-1} + \cdots + a_n S_{k-n} = 0$.

(6) *Cyclotomic polynomials*

Let n be a positive integer, and $\varepsilon_n = e^{\frac{2\pi i}{n}}$ be one of the nth roots of unity. The polynomial $\Phi_n(x) = \prod_{1 \leq k < n, (k,n)=1} \left(x - \varepsilon_n^k\right)$ is termed the nth cyclotomic polynomial.

Proposition 5.14. *For a positive integer n, $x^n - 1 = \prod_{d|n} \Phi_d(x)$.*

Proposition 5.15. *For a positive integer n, $\Phi_n(x)$ is an irreducible integer polynomial.*

Proposition 5.16. $\Phi_n(x) = x^{\varphi(n)} \Phi_n\left(\frac{1}{x}\right)$, *where $\varphi(n)$ is the Euler's totient function.*

Proposition 5.17. *Let p be a prime number.*

(i) *If $p \nmid n$, then $\Phi_{np}(x) = \frac{\Phi_n(x^p)}{\Phi_n(x)}$.*

(ii) *If $p \mid n$, then $\Phi_{np}(x) = \Phi_n(x^p)$.*

Proposition 5.18. $\Phi_n(x) = \prod_{d \mid n}(x^d - 1)^{\mu\left(\frac{n}{d}\right)}$, *where $\mu(n)$ is the Möbius function.*

(7) Chebyshev polynomials

Let $x = \cos\theta$. Then $\cos n\theta$ can be expressed as a polynomial of degree n in x, known as the Chebyshev polynomial, denoted as $T_n(x) = \cos(n \arccos x)$, where $|x| \leq 1$. Alternatively, $T_n(x)$ can be represented as

$$T_n(x) = \sum_{k=0}^{\lfloor \frac{n}{2} \rfloor}(-1)^k C_n^{2k} x^{n-2k}(1 - x^2)^k,$$

where $\lfloor x \rfloor$ represents the greatest integer less than or equal to x.

Chebyshev polynomials satisfy the recurrence relation

$$T_{n+1}(x) = 2x T_n(x) - T_{n-1}(x).$$

Proposition 5.19. *The polynomial $T_n(x)$ is of degree n, with the leading coefficient 2^{n-1}.*

Proposition 5.20. *For all $x \in \mathbf{C}$ and $n \in \mathbf{N}$, $T_n(-x) = (-1)^n T_n(x)$, i.e., $T_n(x)$ is an odd (even) function when n is odd (even).*

Proposition 5.21. *If $|x| \leq 1$ with $x \in \mathbf{R}$, then $|T_n(x)| \leq 1$.*

Proposition 5.22. *For all $m \in \mathbf{N}$, $T_{2m+1}(0) = 0$ and $T_{2m}(0) = (-1)^m$.*

Proposition 5.23. *The polynomial $T_n(x)$ has n distinct real roots in the interval $[-1, 1]$, given by*

$$x_k = \cos\frac{(2k - 1)\pi}{2n} (k = 1, 2, \ldots, n).$$

Proposition 5.24. *The polynomial $T_n(x)$ has $n + 1$ points in the interval $[-1, 1]$, $x_k' = \cos\frac{k\pi}{n} (k = 0, 1, 2, \ldots, n)$, where it alternates between the maximum value 1 and the minimum value -1.*

Proposition 5.25. *Another recurrence relation for $T_n(x)$ is $T_n\left(\frac{y + y^{-1}}{2}\right) = \frac{y^n + y^{-n}}{2}$, for $y \in \mathbf{C}$ with $y \neq 0$.*

Proposition 5.26. *The generating function for the sequence* $\{T_n(x)\}$ *is*

$$\sum_{n \geq 0} T_n(x)t^n = \frac{1 - xt}{1 - 2xt + t^2}, \quad \text{for } t \in \mathbf{R} \quad \text{with } |t| \leq 1.$$

5.1.4 *Common techniques for polynomials*

(1) *Factorization*

A polynomial can be completely or partially factorized into a product of factors or decomposed into a sum of several parts to reveal its properties. Especially when it is known that a polynomial $f(x)$ has a root α, it can be factored as

$$f(x) = (x - \alpha)q(x).$$

Example 5.8. Let $f(x)$ be an integer polynomial. Prove that if the absolute values of $f(x)$ are all 1 when x takes three distinct integers, then $f(x)$ has no integer roots.

Proof. Suppose $f(x)$ has an integer root x_0. Then $f(x)$ can be factored as $f(x) = (x - x_0)q(x)$, where $q(x)$ is an integer polynomial. Let a, b, c be three distinct integers such that $|f(a)| = |f(b)| = |f(c)| = 1$. Then

$$|a - x_0| = |b - x_0| = |c - x_0| = 1,$$

implying that a, b, c must be $x_0 \pm 1$. Consequently, at least two of these values must be equal, which is a contradiction.

(2) *Parity analysis*

When dealing with problems related to integer polynomials, parity analysis is often employed to analyze the parity of coefficients, integer roots, and the values of the polynomial, so as to deduce relationships among these quantities.

Example 5.9. Prove that if an integer coefficient equation $ax^2 + bx + c = 0 (a \neq 0)$ has a rational root, then at least one of a, b, c must be even.

Proof. The equation has a rational root if and only if its discriminant is a perfect square. If a, b, c are all odd, assume $b = 2m + 1$, then $\Delta = 4(m(m+1) - ac) + 1$. Clearly, $m(m+1) - ac$ is odd, denoted as $2n + 1$, so $\Delta = 8n + 5$, implying Δ is an odd number and leaves a remainder of 5 when divided by 8.

However, the square of any odd number $2k+1$ is $4k(k+1)+1$, which is an integer that leaves a remainder of 1 when divided by 8. Thus, Δ cannot be a perfect square, indicating that at least one of a, b, c must be even.

(3) Factor analysis

When dealing with problems involving integer polynomials, if the polynomial $f(x)$ can be factored as $f(x) = g(x)h(x)$, then taking a specific integer value n for x implies that $g(n)|f(n)$. A subsequent analysis can then be conducted on the values of $f(n)$ and $g(n)$. This is typically applicable in situations where $f(x)$ equals a given integer for multiple integer values of x.

Example 5.10. Let a, b, c, d be distinct integers, and the equation

$$(x-a)(x-b)(x-c)(x-d) = 25$$

has an integer root. Prove that $4|(a+b+c+d)$.

Proof. Assume x_0 is an integer root. Then 25 is the product of four distinct factors $x_0 - a, x_0 - b, x_0 - c, x_0 - d$, implying that these four numbers must be ± 1 and ± 5. Therefore,

$$(x_0 - a) + (x_0 - b) + (x_0 - c) + (x_0 - a) = 1 + 5 + (-1) + (-5) = 0,$$

which implies that $a + b + c + d = 4x_0$.

(4) Root analysis

By utilizing properties related to the roots of a polynomial, such as the number of roots not exceeding the degree of the polynomial, an analysis can be conducted on the number of roots and their multiplicities, thereby deducing special properties of the polynomial or relationships between two polynomials.

Example 5.11. Find the polynomial $f(x)$ such that $f(x^2 + 1) = f^2(x) + 1$ with $f(0) = 0$.

Solution. From the given conditions,

$$f(1) = f(0^2 + 1) = f^2(0) + 1 = 1,$$
$$f(2) = f(1^2 + 1) = f^2(1) + 1 = 2,$$
$$f(5) = f(2^2 + 1) = f^2(2) + 1 = 5,$$
$$\dotsb\dotsb\dotsb\dotsb\dotsb\dotsb\dotsb\dotsb\dotsb$$

Construct a sequence $\{x_n\}$ with $x_0 = 0$ and $x_n = x_{n-1}^2 + 1$ for $n \in \{1, 2, \ldots\}$. By mathematical induction, we can prove $f(x_n) = x_n$ for all $n \in \{1, 2, \ldots\}$, so the polynomial $f(x) - x$ has infinitely many roots, which implies $f(x) - x = 0$, i.e., $f(x) = x$.

There is a similar problem:

• **(Polish Mathematical Olympiad Finals 2000, Problem 6).** Suppose that $P(x)$ is a polynomial of an odd degree satisfying

$$P(x^2 - 1) = (P(x))^2 - 1 \quad \text{for all } x.$$

Prove that $P(x) = x$ for all x.

(5) *Mathematical Induction*

When proving properties of polynomials or solving polynomial equations, the characteristics of some polynomials are often given, allowing the use of mathematical induction based on the degree of the polynomial, the number of iterations of the polynomial, the number of roots, etc.

Example 5.12. Prove that for any positive integer n, the polynomial $(x + 1)^{2n+1} + x^{n+2}$ is divisible by the polynomial $x^2 + x + 1$.

Proof. To prove this, use mathematical induction. When $n = 1$, the conclusion is obviously true. Assume the conclusion holds for $n \in \{1, 2, \ldots, k - 1\}$.

When $n = k$,

$$(x + 1)^{2k+1} + x^{k+2} = (x + 1)^2 (x + 1)^{2k-1} + x \cdot x^{k+1}$$
$$= (x^2 + 2x + 1)(x + 1)^{2k-1} + x \cdot x^{k+1}$$
$$= (x^2 + x + 1)(x + 1)^{2k-1} + x((x + 1)^{2k-1} + x^{k+1}),$$

and thus, the conclusion is valid.

(6) *Inequality analysis*

If relationships among the coefficients, roots, and values of a polynomial are known, an inequality analysis can be employed to deduce the polynomial's properties.

Example 5.13. Prove that if a real coefficient equation $x^3 + ax - 2 = 0$ has three real roots, then at least one real root must be greater than or equal to 2.

Proof. Let the three real roots be x_1, x_2, x_3. Then $x_1 + x_2 + x_3 = 0$ and $x_1 x_2 x_3 = 2$, implying that the roots consist of two negatives and one positive.

Assume $x_1 > 0$ and $x_2, x_3 < 0$. Then $x_1 = (-x_2) + (-x_3) \geq 2\sqrt{x_2 x_3}$. Since $2 = x_1 x_2 x_3 \leq x_1 \left(\frac{1}{2} x_1\right)^2$, it follows that $x_1 \geq 2$.

(7) *Polynomial construction*

Based on the given conditions, construct an appropriate polynomial to transform the original problem into a polynomial problem, and then use properties related to polynomials for analysis.

Example 5.14. Let a, b, c be real numbers with absolute values less than 1. Prove that $ab + bc + ca + 1 > 0$.

Proof. Construct a polynomial $f(x) = (b + c)x + bc + 1$ for $|x| < 1$. Its graph is a line segment excluding the two endpoints $(-1, f(-1))$ and $(1, f(1))$. If it can be proven that the function values at these two endpoints, $f(-1)$ and $f(1)$, are both greater than 0, then for every point x in its domain, $f(x)$ is always greater than 0, thereby proving the proposition. Since

$$f(-1) = -(b + c) + bc + 1 = (1 - b)(1 - c) > 0,$$

$$f(1) = (b + c) + bc + 1 = (1 + b)(1 + c) > 0,$$

it follows that $f(a) = a(b + c) + bc + 1 > 0$.

(8) *Difference method*

This method is typically suitable for dealing with situations where the variable x in a polynomial $f(x)$ takes several consecutive integer values. By using the difference formula, $\Delta^n f(x)$ can be represented by $f(x + 1), f(x + 2), \ldots, f(x + n)$, and then analyzed and solved via relevant properties.

Example 5.15. Let $p(x)$ be a polynomial of degree $2n$ satisfying

$$p(0) = p(2) = \cdots = p(2n) = 0,$$

$$p(1) = p(3) = \cdots = p(2n - 1) = 2,$$

$$p(2n + 1) = -30.$$

Find n.

Solution. From $\Delta^{2n+1}p\,(0) = \sum_{k=0}^{2n+1} (-1)^k \, C_{2n+1}^k p\,(2n+1-k) = 0$,

$$(-30) + \sum_{j=1}^{n} C_{2n+1}^{2j} \cdot 2 = 0,$$

which simplifies to $(-30) + 2\left(\frac{1}{2} \cdot 2^{2n+1} - 1\right) = 0$. Thus, $n = 2$.

5.1.5 *Other important theorems*

Theorem 5.11 (Division Theorem). For polynomials $f(x)$ and $g(x)$ with $g(x) \neq 0$, there exist unique polynomials $q(x)$ and $r(x)$ such that

$$f(x) = g(x)q(x) + r(x),$$

where $r(x) = 0$ or $\deg r(x) < \deg g(x)$. The polynomial $q(x)$ is called the quotient of dividing $f(x)$ by $g(x)$, and $r(x)$ is the remainder.

Theorem 5.12 (Remainder Theorem). The remainder of dividing a polynomial $f(x)$ by $x - a$ is $f(a)$.

Theorem 5.13 (Factor Theorem). A polynomial $f(x)$ has a factor $x-a$ if and only if $f(a) = 0$.

Theorem 5.14 (Bezout's Identity). If the greatest common divisor of polynomials $f(x)$ and $g(x)$ is $d(x)$, then there exist polynomials $u(x)$ and $v(x)$ such that $f(x)u(x) + g(x)v(x) = d(x)$.

Corollary 5.6. *Two polynomials $f(x)$ and $g(x)$ are coprime if and only if there exist polynomials $u(x)$ and $v(x)$ such that $f(x)u(x) + g(x)v(x) = 1$.*

Theorem 5.15 (Unique Factorization Theorem). A non-constant polynomial $f(x)$ over a field $\mathbf{F}$ can be uniquely factored into the following form

$$f(x) = a p_1^{k_1}(x) p_2^{k_2}(x) \cdots p_t^{k_t}(x),$$

where a is the leading coefficient of $f(x)$ and $p_1(x), p_2(x), \ldots, p_t(x)$ are distinct irreducible polynomials with leading coefficients of 1. Here, $p_i(x)$ $(i = 1, 2, \ldots, t)$ are the k_i-fold factors of $f(x)$.

Theorem 5.16 (Fundamental Theorem of Algebra). Any complex polynomial of degree $n(n > 0)$ has, counted with multiplicity, exactly n roots in the complex field.

Corollary 5.7. *Let $f(x)$ be a complex polynomial of degree $n(n > 0)$. Then it can be uniquely factored in the complex field as*

$$f(x) = a(x - \alpha_1)^{m_1} (x - \alpha_2)^{m_2} \cdots (x - \alpha_t)^{m_t},$$

where α_i $(i = 1, 2, \ldots, t)$ are distinct complex numbers and m_i-fold roots of $f(x)$, with $\sum_{i=1}^{t} m_i = n$, and a is the leading coefficient of $f(x)$.

Theorem 5.17 (Complex Conjugate Root Theorem). If a real polynomial $f(x)$ has a non-real complex root α, then its conjugate $\bar{\alpha}$ is also a root of $f(x)$, and α and $\bar{\alpha}$ have the same multiplicity. In other words, the non-real roots of a real polynomial appear in conjugate pairs.

Corollary 5.8. *Any real polynomial of odd degree has at least one real root.*

Theorem 5.18 (Lagrange Interpolating Polynomial). Let $f(x)$ be a single-valued function with $x_0 < x_1 < \cdots < x_n$, and $f(x_i) = y_i$ for $i = 0, 1, \ldots, n$. There exists a unique polynomial $p(x)$ with $\deg p(x) \leq n$ such that $p(x_i) = y_i$, where

$$p(x) = \sum_{i=0}^{n} f(x_i) r_i(x) \quad \text{and} \quad r_i(x) = \prod_{\substack{j=0 \\ j \neq i}}^{n} \frac{x - x_j}{x_i - x_j}.$$

Corollary 5.9. *If $f(x)$ is a polynomial of degree n, then for any integer k,*

$$n! a_n = \sum_{i=0}^{n} (-1)^{n-i} C_n^i f(k + i),$$

where a_n is the leading coefficient of $f(x)$.

5.2 Problems and Solutions

5.2.1 *Proving trigonometric identities*

Problem 5.1 (IMO 5-5, proposed by the German Democratic. Republic). Prove that $\cos \frac{\pi}{7} - \cos \frac{2\pi}{7} + \cos \frac{3\pi}{7} = \frac{1}{2}$.

Proof 1. It is evident that

$$\cos \frac{\pi}{7} - \cos \frac{2\pi}{7} + \cos \frac{3\pi}{7} = \frac{2 \cos \frac{\pi}{14} \left(\cos \frac{\pi}{7} - \cos \frac{2\pi}{7} + \cos \frac{3\pi}{7} \right)}{2 \cos \frac{\pi}{14}}.$$

Using the product-to-sum formula for cosine, we have

$$2\cos\frac{\pi}{14}\cos\frac{\pi}{7} = \cos\frac{3\pi}{14} + \cos\frac{\pi}{14},$$

$$2\cos\frac{\pi}{14}\cos\frac{2\pi}{7} = \cos\frac{5\pi}{14} + \cos\frac{3\pi}{14},$$

$$2\cos\frac{\pi}{14}\cos\frac{3\pi}{7} = \cos\frac{7\pi}{14} + \cos\frac{5\pi}{14}.$$

Thus,

$$\cos\frac{\pi}{7} - \cos\frac{2\pi}{7} + \cos\frac{3\pi}{7} = \frac{\cos\frac{7\pi}{14} + \cos\frac{\pi}{14}}{2\cos\frac{\pi}{14}} = \frac{1}{2}.$$

Proof 2. As illustrated in Figure 5.1, construct $\angle MON = \frac{\pi}{7}$.

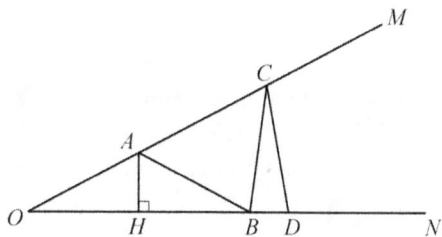

Figure 5.1 $\angle MON$.

On the line segment OM, choose a point A such that $OA = 1$. On the line segment ON, distinct from O, choose a point B such that $AB = 1$. On OM, distinct from A, choose a point C such that $BC = 1$. On ON, distinct from B, choose a point D such that $CD = 1$. Then,

$$\angle CAB = \angle ACB = \frac{2\pi}{7} \quad\text{and}\quad \angle CBD = \angle CDB = \frac{3\pi}{7}.$$

Thus, $\angle OCD = \frac{3\pi}{7}$ and $OC = OD$.

Construct $AH \perp OB$. Then $OB = 2OH = 2\cos\frac{\pi}{7}$. Similarly, $AC = 2\cos\frac{2\pi}{7}$ and $BD = 2\cos\frac{3\pi}{7}$. From $OA + AC = OC = OD = OB + BD$,

$$1 + 2\cos\frac{2\pi}{7} = 2\cos\frac{\pi}{7} + 2\cos\frac{3\pi}{7},$$

which implies $\cos\frac{\pi}{7} - \cos\frac{2\pi}{7} + \cos\frac{3\pi}{7} = \frac{1}{2}$.

Proof 3. Consider the roots of unity from $z^7 = 1$:

$$z_k = \cos \frac{2(k-1)\pi}{7} + i \sin \frac{2(k-1)\pi}{7} \quad \text{for } k = 1, 2, \ldots, 7.$$

It is evident that $z_1 + z_2 + \cdots + z_7 = 0$. Thus

$$0 = 1 + \cos \frac{2\pi}{7} + \cos \frac{4\pi}{7} + \cos \frac{6\pi}{7} + \cos \frac{8\pi}{7} + \cos \frac{10\pi}{7} + \cos \frac{12\pi}{7}$$

$$= 1 + 2 \left(\cos \frac{2\pi}{7} + \cos \frac{4\pi}{7} + \cos \frac{6\pi}{7} \right).$$

Since $\cos \frac{6\pi}{7} = -\cos \frac{\pi}{7}$ and $\cos \frac{4\pi}{7} = -\cos \frac{3\pi}{7}$,

$$\cos \frac{\pi}{7} - \cos \frac{2\pi}{7} + \cos \frac{3\pi}{7} = \frac{1}{2}.$$

Note. When $z^n = 1$, each nth root of 1 is called an nth root of unity. There are n such nth roots of unity, denoted as $\varepsilon_0, \varepsilon_1, \ldots, \varepsilon_{n-1}$, where $\varepsilon_k = \cos \frac{2k\pi}{n} + i \sin \frac{2k\pi}{n}$ for $k = 0, 1, 2, \ldots, n-1$. This indicates that the product of two nth roots of unity is still an nth root of unity, and the integer power of an nth root of unity is also an nth root of unity.

(i) Let m be an integer and n be a positive integer greater than 1. Then

$$1 + \varepsilon_1^m + \varepsilon_2^m + \cdots + \varepsilon_{n-1}^m = \begin{cases} n, & \text{if } n \mid m, \\ 0, & \text{if } n \nmid m. \end{cases}$$

In particular, $1 + \varepsilon_1 + \varepsilon_2 + \cdots + \varepsilon_{n-1} = 0$.

By (i), we can prove the following proposition.

(ii) Let n be a positive integer. Then

$$\cos^k \frac{\pi}{2n+1} + \cos^k \frac{3\pi}{2n+1} + \cdots + \cos^k \frac{(2n-1)\pi}{2n+1}$$

$$= \frac{1}{2} \quad \text{for } k = 1, 3, \ldots, 2n-1.$$

(iii) $\sin \frac{\pi}{n} \sin \frac{2\pi}{n} \cdots \sin \frac{(n-1)\pi}{n} = \frac{n}{2^{n-1}}$.

(iv) $\left| \cos \frac{\pi}{n} \cos \frac{2\pi}{n} \cdots \cos \frac{(n-1)\pi}{n} \right| = \frac{1-(-1)^n}{2^n}$.

(v) $\sin \frac{\pi}{2n} \sin \frac{2\pi}{2n} \cdots \sin \frac{(n-1)\pi}{2n} = \frac{\sqrt{n}}{2^{n-1}}$.

(vi) $\cos \frac{\pi}{2n+1} \cos \frac{2\pi}{2n+1} \cdots \cos \frac{n\pi}{2n+1} = \frac{1}{2^n}$.

(vii) $\sin \frac{\pi}{2n+1} \sin \frac{2\pi}{2n+1} \cdots \sin \frac{n\pi}{2n+1} = \frac{\sqrt{2n+1}}{2^n}$.

【Score Situation】This particular problem saw the following distribution of scores among contestants: 3 contestants scored 6 points, 3 contestants scored 5 points, no contestant scored 4 points, 1 contestant scored 3 points, no contestant scored 2 points, 2 contestants scored 1 point, and 7 contestants scored 0 point. The average score of this problem is 2.375, indicating that it had a certain level of difficulty.

Among the top five teams in the team scores, the Soviet Union team achieved a total score of 271 points, the Hungary team achieved a total score of 234 points, the Romania team achieved a total score of 191 points, the Yugoslavia team achieved a total score of 162 points, and the Czechoslovakia team achieved a total score of 151 points.

The gold medal cutoff for this IMO was set at 35 points (with 7 contestants earning gold medals), the silver medal cutoff was 28 points (with 11 contestants earning silver medals), and the bronze medal cutoff was 21 points (with 17 contestants earning bronze medals).

In this IMO, no contestant achieved a perfect score of 40 points.

Problem 5.2 (IMO 8-4, proposed by Yugoslavia). Prove that for every positive integer n and real number $x \neq \frac{k\pi}{2^t}$ ($t = 0, 1, \ldots, n$; k any integer),

$$\frac{1}{\sin 2x} + \frac{1}{\sin 4x} + \cdots + \frac{1}{\sin 2^n x} = \cot x - \cot 2^n x.$$

Proof. Use mathematical induction.

For $x \neq \frac{k\pi}{2^t}$, the number $\cot 2^t x$ is well defined. When $n = 1$,

$$
\begin{aligned}
\cot x - \cot 2x &= \frac{\cos x}{\sin x} - \frac{\cos 2x}{\sin 2x} \\
&= \frac{\sin 2x \cos x - \cos 2x \sin x}{\sin x \sin 2x} \\
&= \frac{\sin x}{\sin x \sin 2x} \\
&= \frac{1}{\sin 2x}.
\end{aligned}
$$

Therefore, the conclusion holds for $n = 1$.

Assume the conclusion holds for $n = i$. When $n = i + 1$,

$$\frac{1}{\sin 2x} + \frac{1}{\sin 4x} + \cdots + \frac{1}{\sin 2^i x} + \frac{1}{\sin 2^{i+1} x}$$

$$= \cot x - \cot 2^i x + \frac{1}{\sin 2^{i+1} x}.$$

From the case for $n = 1$, it is evident that

$$\frac{1}{\sin 2^{i+1} x} = \cot 2^i x - \cot 2^{i+1} x.$$

Thus, the conclusion holds for any positive integer n.

Note. (i) $\tan \alpha + 2 \tan 2\alpha + \cdots + 2^n \tan 2^n \alpha = \cot \alpha - 2^{n+1} \cot 2^{n+1} \alpha$.

(ii) $\tan \alpha \tan 2\alpha + \tan 2\alpha \tan 3\alpha + \cdots + \tan(n-1)\alpha \tan n\alpha = \frac{\tan n\alpha}{\tan \alpha} - n$.

Furthermore, there is a similar problem:

- **(William Lowell Putnam Mathematical Competition 1973, B6).** On the domain $0 \leq \theta \leq 2\pi$:

  (a) Prove that $\sin^2 \theta \cdot \sin 2\theta$ takes its maximum at $\frac{\pi}{3}$ and $\frac{4\pi}{3}$. (Hence, its minimum at $\frac{2\pi}{3}$ and $\frac{5\pi}{3}$.)

  (b) Show that

  $$\left| \sin^2 \theta \cdot \sin^3 2\theta \cdot \sin^3 4\theta \cdots \sin^3 2^{n-1}\theta \cdot \sin 2^n \theta \right|$$

  takes its maximum at $\theta = \frac{\pi}{3}$. (The maximum may also be attained at other points.)

  (c) Derive the inequality:

  $$\sin^2 \theta \cdot \sin^2 2\theta \cdot \sin^2 4\theta \cdots \sin^2 2^n \theta \leq \left(\frac{3}{4}\right)^n.$$

【Score Situation】This particular problem saw the following distribution of scores among contestants: 23 contestant scored 5 points, 3 contestants scored 4 points, no contestant scored 3 points, no contestants scored 2 points, no contestants scored 1 point, and 1 contestant scored 0 point. The average score of this problem is 4.704, indicating that it was simple.

Among the top five teams in the team scores, the Soviet Union team achieved a total score of 293 points, the Hungary team achieved a total score of 281 points, the German Democratic Republic team achieved a total score of 280 points, the Poland team achieved a total score of 269 points, and the Romania team achieved a total score of 257 points.

The gold medal cutoff for this IMO was set at 39 points (with 13 contestants earning gold medals), the silver medal cutoff was 34 points (with 15 contestants earning silver medals), and the bronze medal cutoff was 31 points (with 11 contestants earning bronze medals).

In this IMO, a total of 11 contestants achieved a perfect score of 40 points.

5.2.2 *Finding polynomials*

Problem 5.3 (IMO 17-6, proposed by the United Kingdom). Find all polynomials P in two variables with the following properties:

(i) for a positive integer n and all real t, x, and y,

$$P(tx, ty) = t^n P(x, y) \text{ (that is, } P \text{ is homogeneous of degree } n\text{);}$$

(ii) for all real a, b, and c,

$$P(b + c, a) + P(c + a, b) + P(a + b, c) = 0;$$

(iii) $P(1, 0) = 1$.

Solution 1. From conditions (ii) and (iii), $P(x, y)$ is not a constant. Setting $a = b$ and $c = -2a$ in (ii), we obtain

$$0 = P(2a, -2a) + P(-a, a) + P(-a, a)$$
$$= ((-2)^n + 2) P(-a, a).$$

For $n > 1$, we have $P(-a, a) = 0$, implying $P(x, y) = 0$ for any $y = -x$. It follows that $P(x, y)$ is divisible by $x + y$. Therefore, we can assume

$$P(x, y) = (x + y) P_1(x, y).$$

It is evident that $P_1(x, y)$ is a homogeneous polynomial of degree $n - 1$, satisfying conditions (i), (ii), and (iii). Consequently,

$$P(x, y) = (x + y) P_1(x, y)$$
$$= (x + y)^2 P_2(x, y)$$
$$\cdots\cdots\cdots\cdots\cdots$$
$$= (x + y)^{n-1} P_{n-1}(x, y).$$

Here, $P_{n-1}(x, y) = Ax + By$ is a homogeneous polynomial of degree 1, still meeting conditions (i), (ii), and (iii).

Setting $a = b = c = x$, we get $3P_{n-1}(2x, x) = 0$, which implies $2Ax + Bx = 0$. The arbitrariness of x indicates $2A + B = 0$. Further, from $P_{n-1}(1, 0) = 1$, we deduce $A = 1$. Thus $B = -2$ and $P_{n-1}(x, y) = x - 2y$.

Hence, $P(x, y) = (x + y)^{n-1}(x - 2y)$, which includes the case $n = 1$.

Solution 2. Setting $b = 1 - a$ and $c = 0$ in (ii), we have

$$P(1 - a, a) + P(a, 1 - a) + P(1, 0) = 0.$$

Since $P(1, 0) = 1$,

$$P(1 - a, a) = -1 - P(a, 1 - a). \tag{1}$$

Setting $c = 1 - a - b$ in (ii), we get

$$P(1 - a, a) + P(1 - b, b) + P(a + b, 1 - a - b) = 0,$$

and by (1), this implies that

$$-2 - P(a, 1 - a) - P(b, 1 - b) + P(a + b, 1 - a - b) = 0. \tag{2}$$

Therefore, with $f(x) = P(x, 1 - x) + 2$, we see that (2) can take the form

$$f(a + b) = f(a) + f(b).$$

Since $P(x, y)$ is continuous, so is $f(x)$. Obviously, $f(x)$ is known as Cauchy's functional equation (which was presented in Chapter 2). Hence,

$$f(x) = f(1) \cdot x = (P(1, 0) + 2)x = 3x.$$

By definition $f(x) = P(x, 1 - x) + 2$, so $P(x, 1 - x) = 3x - 2$. If $a + b \neq 0$, then

$$P(a, b) = (a + b)^n P\left(\frac{a}{a + b}, \frac{b}{a + b}\right) = (a + b)^n \left(\frac{3a}{a + b} - 2\right)$$

$$= (a + b)^{n-1}(a - 2b).$$

Since $P(x, y)$ is continuous, it follows that $P(x, y) = (x + y)^{n-1}(x - 2y)$ even when $x + y = 0$. It is easily verified that the polynomial

$$P(x, y) = (x + y)^{n-1}(x - 2y)$$

meets (i), (ii), and (iii).

【Score Situation】 This particular problem saw the following distribution of scores among contestants: 19 contestants scored 8 points, 1 contestant scored 7 points, 5 contestants scored 6 points, 5 contestants scored 5 points, 8 contestants scored 4 points, 13 contestants scored 3 points, 27 contestants scored 2 points, 19 contestants scored 1 point, and 31 contestants scored 0 point. The average score for this problem is 2.797, indicating that it had a certain level of difficulty.

Among the top five teams in the team scores, the scores of this problem are as follows: the Hungary team scored 33 points (with a total team score of 258 points), the German Democratic Republic team scored 33 points (with a total team score of 249 points), the United States team scored 39 points (with a total team score of 247 points), the Soviet Union team scored 28 points (with a total team score of 246 points), and the United Kingdom team scored 44 points (with a total team score of 241 points).

The gold medal cutoff for this IMO was set at 38 points (with 8 contestants earning gold medals), the silver medal cutoff was 32 points (with 25 contestants earning silver medals), and the bronze medal cutoff was 23 points (with 36 contestants earning bronze medals).

In this IMO, a total of six contestants achieved a perfect score of 40 points.

Problem 5.4 (IMO 45-2, proposed by South Korea). Find all polynomials f with real coefficients such that for all reals a, b, c satisfying $ab + bc + ca = 0$, we have the following relation:

$$f(a - b) + f(b - c) + f(c - a) = 2f(a + b + c).$$

Solution. For any given $a, b, c \in \mathbf{R}$ with $ab + bc + ca = 0$,

$$f(a - b) + f(b - c) + f(c - a) = 2f(a + b + c). \tag{1}$$

Setting $a = b = c = 0$ in (1), we obtain $f(0) = 0$. Setting $b = c = 0$ in (1), we find $f(-a) = f(a)$ for any real number a. Consequently, all coefficients of the odd-degree terms of $f(x)$ are 0. Assume

$$f(x) = a_n x^{2n} + a_{n-1} x^{2(n-1)} + \cdots + a_1 x^2 (a_n \neq 0).$$

Setting $b = 2a$ and $c = -\frac{2}{3}a$ in (1), we have

$$f(-a) + f\left(\frac{8}{3}a\right) + f\left(-\frac{5}{3}a\right) = 2f\left(\frac{7}{3}a\right),$$

which implies $\sum_{i=1}^{n} a_i \left(1 + \left(\frac{8}{3}\right)^{2i} + \left(\frac{5}{3}\right)^{2i} - 2\left(\frac{7}{3}\right)^{2i}\right) a^{2i} = 0$ for all $a \in \mathbf{R}$.

Since $a_n \neq 0$ for $n \geq 3$, from $8^6 = 262144 > 235298 = 2 \times 7^6$, it follows that $\left(\frac{8}{7}\right)^{2n} \geq \left(\frac{8}{7}\right)^6 > 2$, and thus $1 + \left(\frac{8}{3}\right)^{2n} + \left(\frac{5}{3}\right)^{2n} - 2\left(\frac{7}{3}\right)^{2n} > 0$. This contradicts

$$\sum_{i=1}^{n} a_i \left(1 + \left(\frac{8}{3}\right)^{2i} + \left(\frac{5}{3}\right)^{2i} - 2\left(\frac{7}{3}\right)^{2i}\right) a^{2i} = 0$$

for all $a \in \mathbf{R}$.

Therefore, $n \leq 2$. Let $f(x) = \alpha x^4 + \beta x^2$, where $\alpha, \beta \in \mathbf{R}$. It can be easily verified that $f(x) = \alpha x^4 + \beta x^2$ satisfies the given conditions.

Note. We can also set $a = (1-\sqrt{3})x$, $b = x$, and $c = (1+\sqrt{3})x$. It follows that

$$f(-\sqrt{3}x) + f(-\sqrt{3}x) + f(2\sqrt{3}x) = 2f(3x).$$

For the coefficient of x^n to be nonzero, we must have $2(-\sqrt{3})^n + (2\sqrt{3})^n = 2 \cdot 3^n$, which only holds for $n = 2, 4$.

There is a similar problem:

- **(United States of America Mathematical Olympiad 2019, Problem 6).** Find all polynomials P with real coefficients such that

$$\frac{P(x)}{yz} + \frac{P(y)}{zx} + \frac{P(z)}{xy} = P(x-y) + P(y-z) + P(z-x)$$

for all nonzero real numbers x, y, z satisfying $2xyz = x + y + z$.

【Score Situation】 This particular problem saw the following distribution of scores among contestants: 83 contestants scored 7 points, 31 contestants scored 6 points, 23 contestants scored 5 points, 23 contestants scored 4 points, 32 contestants scored 3 points, 57 contestants scored 2 points, 158 contestants scored 1 point, and 79 contestants scored 0 point. The average score for this problem is 2.761, indicating that it had a certain level of difficulty.

Among the top five teams in the team scores, the scores of this problem are as follows: the China team scored 41 points (with a total team score of 220 points), the United States team scored 41 points (with a total team score of 212 points), the Russia team scored 34 points (with a total team score of 205 points), the Vietnam team scored 28 points (with a total team score of 196 points), and the Bulgaria team scored 36 points (with a total team score of 194 points).

The gold medal cutoff for this IMO was set at 32 points (with 45 contestants earning gold medals), the silver medal cutoff was 24 points (with 78 contestants earning silver medals), and the bronze medal cutoff was 16 points (with 120 contestants earning bronze medals).

In this IMO, a total of four contestants achieved a perfect score of 42 points.

5.2.3 *Proving properties*

Problem 5.5 (IMO 16-6, proposed by Sweden). Let P be a non-constant polynomial with integer coefficients. If $n(P)$ is the number of distinct integers k such that $(P(k))^2 = 1$, prove that $n(P) - \deg(P) \leq 2$, where $\deg(P)$ denotes the degree of the polynomial P.

Proof 1. Since $(P(x))^2 - 1 = (P(x) - 1)(P(x) + 1)$, the integer roots of $(P(x))^2 = 1$ are all integer roots of either $P(x) = 1$ or $P(x) = -1$. We will prove that one of the equations $P(x) = 1$ or $P(x) = -1$ must have at most two integer roots, and by the Fundamental Theorem of Algebra, the other equation has at most $\deg(P)$ integer roots, thus proving the conclusion.

We use a proof by contradiction. Assume that both equations $P(x) + 1 = 0$ and $P(x) - 1 = 0$ have at least three distinct integer roots, and these roots are different from each other. Among these six distinct integers, let the smallest be a.

Without loss of generality, assume a is a root of $P(x) + 1 = 0$. Then $P(x) + 1 = (x - a)Q(x)$, where $Q(x)$ is an integer polynomial.

Let b, c, and d be three distinct integer roots of $P(x) - 1 = 0$, all greater than a. Since $P(x) - 1 = (x - a)\,Q(x) - 2$ and $P(b) - 1 = P(c) - 1 = P(d) - 1 = 0$, it follows that

$$2 = (b - a)Q(b) = (c - a)Q(c) = (d - a)Q(d),$$

where $b - a$, $c - a$, and $d - a$ are three distinct positive integers. Clearly, at least one of them must be greater than 2 and cannot be a divisor of 2. This leads to a contradiction.

Proof 2. First, we prove a lemma.

Lemma. *If m is an integer root of the integer polynomial $F(x)$. Then the integer roots of the polynomial $F(x) + p$ or $F(x) - p$ can only be $m - p$, $m - 1$, $m + 1$, or $m + p$, where p is a prime number.*

Proof of Lemma. Suppose $F(x) = (x - m)G(x)$. Then

$$F(x) \pm p = (x - m)\,G(x) \pm p,$$

where $G(x)$ is an integer polynomial. If $F(x_0) \pm p = 0$ for an integer x_0, then $(x_0 - m)G(x_0) = \mp p$. Therefore, $x_0 - m$ divides p, and since p is a prime number, $x_0 - m = \pm 1$ or $x_0 - m = \pm p$. The lemma is proved.

Returning to the original problem, assume that $(P(x))^2 = 1$ has integer roots, with m being the smallest one, and without loss of generality, let it be a root of $P(x) - 1 = 0$.

Since $P(x) + 1 = P(x) - 1 + 2$, which is the case with $p = 2$ in the lemma, the integer roots of $P(x) + 1$ can only be at most two, $m + 1$ or $m + 2$. The rest of the proof is similar to Proof 1.

【Score Situation】This particular problem saw the following distribution of scores among contestants: 35 contestants scored 8 points, 8 contestants scored 7 points, 3 contestants scored

6 points, 5 contestants scored 5 points, 1 contestant scored 4 points, 3 contestants scored 3 points, 7 contestants scored 2 points, 28 contestants scored 1 point, and 50 contestants scored 0 point. The average score of this problem is 3.100, indicating that it was relatively straightforward.

Among the top five teams in the team scores, the scores of this problem are as follows: the Soviet Union team scored 40 points (with a total team score of 256 points), the United States team scored 58 points (with a total team score of 243 points), the Hungary team scored 54 points (with a total team score of 237 points), the German Democratic Republic team scored 42 points (with a total team score of 236 points), and the Yugoslavia team scored 36 points (with a total team score of 216 points).

The gold medal cutoff for this IMO was set at 38 points (with 10 contestants earning gold medals), the silver medal cutoff was 30 points (with 24 contestants earning silver medals), and the bronze medal cutoff was 23 points (with 37 contestants earning bronze medals).

In this IMO, a total of six contestants achieved a perfect score of 40 points.

Problem 5.6 (IMO 26-3, proposed by the Netherlands). For any polynomial $P(x) = a_0 + a_1x + \cdots + a_kx^k$ with integer coefficients, the number of coefficients which are odd is denoted by $w(P)$. For $i = 0, 1, \ldots$, let $Q_i(x) = (1+x)^i$. Prove that if $i_1, i_2, \ldots, i_n$ are integers such that $0 \leq i_1 < i_2 < \cdots < i_n$, then

$$w(Q_{i_1} + Q_{i_2} + \cdots + Q_{i_n}) \geq w(Q_{i_1}).$$

Proof. Let R and S be two polynomials with integer coefficients, and let m be a positive integer. There are the following two lemmas:

Lemma 1. *If $k = 2^m$, then $(1+x)^k \equiv 1 + x^k \pmod 2$.*

Lemma 2. *If the degree of R is less than k, then*

$$w(R + x^k S) = w(R) + w(S).$$

These two lemmas are evidently true.

Define $Q = Q_{i_1} + Q_{i_2} + \cdots + Q_{i_n}$. When $i_n \in \{0, 1\}$, the conclusion $w(Q) \geq w(Q_{i_1})$ is obviously true. By mathematical induction, assume the conclusion $w(Q) \geq w(Q_{i_1})$ holds for $i_n < 2^m (m \geq 1)$. When $k = 2^m \leq i_n < 2^{m+1}$, consider two cases:

Case 1: $i_1 < k.$

Suppose $i_r < k \leq i_{r+1}$. Then $Q = R + (1+x)^k S$, where

$$R = Q_{i_1} + Q_{i_2} + \cdots + Q_{i_r}, \quad S = (1+x)^{-k}(Q_{i_{r+1}} + Q_{i_{r+2}} + \cdots + Q_{i_n}).$$

The degrees of R and S are both less than k. By Lemma 1, in $(1+x)^k$, the terms that affect the number of odd coefficients are only 1 and x^k. Thus,

$$w(Q) = w(R + (1+x)^k S) = w(R + S + x^k S)$$
$$= w(R + S) + w(S) \geq w(R).$$

The last step follows from the triangle inequality. R has $w(R)$ odd coefficients, and after adding S, there are t of them that become even, implying that S must have at least t odd coefficients, i.e.,

$$w(R + S) \geq w(R) - t \quad \text{and} \quad w(S) \geq t.$$

Adding these two inequalities yields $w(R + S) + w(S) \geq w(R)$.

By the induction hypothesis, $w(R) \geq w(Q_{i_1})$, so $w(Q) \geq w(Q_{i_1})$.

Case 2: $i_1 \geq k.$

Suppose $Q_{i_1} = (1+x)^k R$ and $Q = (1+x)^k S$, where $R = (1+x)^{-k} Q_{i_1} = (1+x)^{i_1-k}$ and $S = (1+x)^{-k} Q = (1+x)^{i_1-k} + (1+x)^{i_2-k} + \cdots + (1+x)^{i_n-k}$.

Then, the degrees of R and S are both less than k. By Lemma 1 and Lemma 2,

$$w(Q) = w((1+x)^k S) = w((1+x^k)S) = 2w(S),$$
$$w(Q_{i_1}) = w((1+x)^k R) = w((1+x^k)R) = 2w(R).$$

Furthermore, by the induction hypothesis, $w(S) \geq w(R)$, so $w(Q) \geq w(Q_{i_1})$.

【Score Situation】 This particular problem saw the following distribution of scores among contestants: 12 contestants scored 7 points, 3 contestants scored 6 points, no contestant scored 5 points, 1 contestant scored 4 points, 5 contestants scored 3 points, 8 contestants scored 2 points, 27 contestants scored 1 point, and 153 contestants scored 0 point. The average score for this problem is 0.785, indicating that it was extremely difficult.

Among the top five teams in the team scores, the scores of this problem are as follows: the Romania team scored 16 points (with a total team score of 201 points), the United States team scored 34 points (with a total team score of 180 points), the Hungary team scored 17 points (with a total team score of 168 points), the Bulgaria team scored 10 points (with a total team score of 165 points), and the Vietnam team scored 2 points (with a total team score of 144 points).

The gold medal cutoff for this IMO was set at 34 points (with 14 contestants earning gold medals), the silver medal cutoff was 22 points (with 35 contestants earning silver medals), and the bronze medal cutoff was 15 points (with 52 contestants earning bronze medals).

In this IMO, only two contestants achieved a perfect score of 42 points, namely Géza Kós from Hungary and Daniel Tătaru from Romania.

Problem 5.7 (IMO 34-1, proposed by Ireland). Let $f(x) = x^n + 5x^{n-1} + 3$, where $n > 1$ is an integer. Prove that $f(x)$ cannot be expressed as the product of two nonconstant polynomials with integer coefficients.

Proof 1. Suppose $f(x)$ can be factored into the product of two integer polynomials, i.e., $f(x) = g(x)h(x)$, where

$$g(x) = x^p + a_{p-1}x^{p-1} + \cdots + a_1 x + a_0,$$
$$h(x) = x^q + b_{q-1}x^{q-1} + \cdots + b_1 x + b_0,$$

with $a_p = b_q = 1$, a_i and b_j are integers, and p and q are positive integers.

First, prove that both p and q are greater than or equal to 2. Otherwise, assume $p = 1$. Then $f(x) = (x + a_0)h(x)$, and from $a_0 b_0 = 3$, we have $a_0 = \pm 1$ or $a_0 = \pm 3$, i.e., $f(x)$ has roots ± 1 or ± 3. However, when x is odd, $f(x) \equiv 1 \pmod 2$, so $f(x)$ cannot be zero, implying p and q are both greater than or equal to 2.

Let $2 \le p \le q \le n - 2$, and from $a_0 b_0 = 3$, if $a_0 = \pm 3$, since a_p is not a multiple of 3, then there must be a k such that $3|a_i$ for $0 \le i \le k - 1$ and a_k is not a multiple of 3.

Considering the coefficients of x^k in the expansions of $f(x)$ and $g(x)h(x)$, we have $0 = a_k b_0 + a_{k-1} b_1 + \cdots + a_0 b_k$, i.e., $a_k b_0 = -(a_{k-1} b_1 + \cdots + a_0 b_k)$. This implies $3|a_k$, a contradiction.

Similarly, if $b_0 = \pm 3$, then the same conclusion can be drawn. Therefore, $f(x)$ cannot be expressed as the product of two nonconstant polynomials with integer coefficients.

Proof 2. Similar to Proof 1, suppose $f(x)$ can be factored into the product of two polynomials with integer coefficients, both having leading coefficients of 1, i.e., $f(x) = g(x)h(x)$.

Since $g(0)h(0) = f(0) = 3$, either $|g(0)| = 1$ or $|h(0)| = 1$. Without loss of generality, assume $|g(0)| = 1$, and let $g(x)$ have k roots

$a_1, a_2, \ldots, a_k (k < n)$. Then, by Vieta's formulas,

$$|a_1 a_2 \cdots a_k| = 1. \tag{1}$$

Since a_i is a root of $f(x)$, we see that $a_i^n + 5a_i^{n-1} + 3 = 0$ for $i \in \{1, 2, \ldots, k\}$. Therefore,

$$|(a_i + 5)a_i^{n-1}| = 3. \tag{2}$$

Thus,

$$\prod_{i=1}^{k} |(a_i + 5)a_i^{n-1}| = 3^k = \prod_{i=1}^{k} |a_i + 5| \cdot \prod_{i=1}^{k} |a_i|^{n-1} = \prod_{i=1}^{k} |a_i + 5|.$$

Assume without loss of generality that $|a_k + 5| = \min_{1 \leq i \leq k} |a_i + 5|$. Then $|a_k + 5| \leq 3$. Also, since $|-a_k| + |a_k + 5| \geq 5$, we have $|a_k| \geq 5 - |a_k + 5| \geq 2$. From (1), we know that $k \geq 2$.

Assume without loss of generality that $|a_1| = \min_{1 \leq i \leq k-1} |a_i|$. Then

$$|a_1|^{k-1} \leq |a_1||a_2| \cdots |a_{k-1}| = \frac{1}{|a_k|} \leq \frac{1}{2}.$$

Hence, $|a_1| \leq \frac{1}{\sqrt[k-1]{2}}$, and $|a_1 + 5| \leq 5 + \frac{1}{\sqrt[k-1]{2}} < 6$ and $|a_1|^{n-1} \leq \frac{1}{2^{\frac{n-1}{k-1}}} < \frac{1}{2}$. Therefore, $|a_1 + 5| \cdot |a_1|^{n-1} < 3$, which contradicts (2).

In conclusion, $f(x)$ cannot be factored into the product of two nonconstant polynomials with integer coefficients.

Note. A famous irreducibility criterion that requires no information on the canonical decomposition of the coefficients of an integer polynomial is the following result of Perron:

Let $f(x) = x^n + a_{n-1}x^{n-1} + \cdots + a_1 x + a_0 \in \mathbf{Z}[x]$ with $a_0 \neq 0$. If

$$|a_{n-1}| > 1 + |a_{n-2}| + \cdots + |a_1| + |a_0|,$$

then f is irreducible in $\mathbf{Z}[x]$.

Furthermore, there are several similar problems:

- **(Asian Pacific Mathematics Olympiad 2018, Problem 5).** Find all polynomials $P(x)$ with integer coefficients such that for all real numbers s and t, if $P(s)$ and $P(t)$ are both integers, then $P(st)$ is also an integer.
- **(Romania Team Selection Test 2001, Problem 2).** (a) Let $f, g : \mathbf{Z} \to \mathbf{Z}$ be injective maps. Show that the function $h : \mathbf{Z} \to \mathbf{Z}$, defined by $h(x) = f(x)g(x)$ for all $x \in \mathbf{Z}$ cannot be surjective.
  (b) Let $f : \mathbf{Z} \to \mathbf{Z}$ be a surjective map. Show that there exist surjective functions $g, h : \mathbf{Z} \to \mathbf{Z}$ such that $f(x) = g(x)h(x)$ for all $x \in \mathbf{Z}$.

- **(Estonia Team Selection Test 1999, Problem 2).** A polynomial $a_n x^n + a_{n-1} x^{n-1} + \cdots + a_1 x + a_0$ is called *alternating* when $n \geq 1$ and the coefficients a_i and a_{i-1} are nonzero real numbers with different signs for $i = 1, 2, \ldots, n$.

  Let $P(x)$ and $Q(x)$ be arbitrary alternating polynomials. Prove that the polynomial $R(x) = P(x)Q(x)$ is alternating.

- **(Japan Mathematical Olympiad 1999, Final Round, Problem 4).** Prove that the polynomial

$$f(x) = \left(x^2 + 1\right)\left(x^2 + 2^2\right) \cdots \left(x^2 + n^2\right) + 1$$

  cannot be expressed as a product of two polynomials with integer coefficients with degree greater than 1.

- **(Korean Mathematical Olympiad 1995, 2nd Round, Problem 5).** Let a, b be integers and p be a prime number such that:

  (i) p is the greatest common divisor of a and b;
  (ii) p^2 divides a.

  Prove that the polynomial $x^{n+2} + ax^{n+1} + bx^n + a + b$ cannot be decomposed into the product of two polynomials with integer coefficients and degree greater than 1.

【Score Situation】 This particular problem saw the following distribution of scores among contestants: 92 contestants scored 7 points, 13 contestants scored 6 points, 1 contestant scored 5 points, 1 contestant scored 4 points, 1 contestant scored 3 points, 6 contestants scored 2 points, 91 contestants scored 1 point, and 208 contestants scored 0 point. The average score for this problem is 2,027, indicating that it had a certain level of difficulty.

Among the top five teams in the team scores, the scores of this problem are as follows: the China team scored 35 points (with a total team score of 215 points), the Germany team scored 29 points (with a total team score of 189 points), the Bulgaria team scored 24 points (with a total team score of 178 points), the Russia team scored 28 points (with a total team score of 177 points), and the Chinese Taiwan team scored 42 points (with a total team score of 162 points).

The gold medal cutoff for this IMO was set at 30 points (with 35 contestants earning gold medals), the silver medal cutoff was 20 points (with 66 contestants earning silver medals), and the bronze medal cutoff was 11 points (with 97 contestants earning bronze medals).

In this IMO, only two contestants achieved a perfect score of 42 points, namely Hong Zhou from China and Hung-Wu Wu from Chinese Taiwan.

Problem 5.8 (IMO 47-5, proposed by Romania). Let $P(x)$ be a polynomial of degree $n > 1$ with integer coefficients and let k be a positive integer. Consider the polynomial $Q(x) = P(P(\cdots P(P(x))\cdots))$, where P occurs k times. Prove that there are at most n integers t such that $Q(t) = t$.

Proof. If every integer fixed point of Q is also a fixed point of P and the polynomial $P(x) - x$ has at most n roots, then the conclusion holds.

If not, then there exists an integer x_0 such that $Q(x_0) = x_0$ but $P(x_0) \neq x_0$. Define $x_{i+1} = P(x_i)$ for $i = 0, 1, 2, \ldots$. Then $x_k = x_0$ and $x_{i+1} \neq x_i$ for $i = 1, 2, \ldots, k$. Clearly, $(u - v) | (P(u) - P(v))$ for distinct integers u and v. Thus, for the following (non-zero) differences, the previous term divides the subsequent one:

$$x_0 - x_1, x_1 - x_2, \ldots, x_{k-1} - x_k, x_k - x_{k+1}.$$

Since $x_k - x_{k+1} = x_0 - x_1$, the absolute values of the above differences are equal.

Consider $x_m = \min(x_1, x_2, \ldots, x_k)$. Then $x_{m-1} - x_m = -(x_m - x_{m+1})$, so $x_{m-1} = x_{m+1} \neq x_m$, implying that successive differences have opposite signs, and $x_0, x_1, \ldots$ take two distinct values. Thus, x_0 is a fixed point of the polynomial $P(P(x))$.

(It is evident that the integer fixed points of P are fixed points of the polynomial $P(P(x))$. Thus the integer fixed points of Q are all fixed points of the polynomial $P(P(x))$.)

Assume $a = x_0$, and let $b = P(a) \neq a$. Then $a = P(b)$. If $P(P(x))$ has only two fixed points a and b, since $n \geq 2$, then the conclusion holds.

If not, for any integer fixed point α of $P(P(x))$, let $\beta = P(\alpha)$. Then $\alpha = P(\beta)$, where α and β can be the same, but distinct from a and b. From the previous proof, we can know that, for the four pairs $(\alpha, a), (\beta, b), (\alpha, b), (\beta, a)$, the numbers $\alpha - a$ and $\beta - b$ divide each other, and $\alpha - b$ and $\beta - a$ also divide each other, yielding

$$\alpha - b = \pm(\beta - a), \qquad \alpha - a = \pm(\beta - b).$$

If the plus sign is taken in both equations, then $\alpha - b = \beta - a$ and $\alpha - a = \beta - b$, implying $a - b = b - a$, which contradicts $a \neq b$. Thus, at least one of the equations takes the minus sign, resulting in $\alpha + \beta = a + b$, i.e., $P(\alpha) + \alpha - (a + b) = 0$.

Let C represent $a + b$. We have shown that every fixed point of Q, not equal to a or b, is a root of the polynomial $F(x) = P(x) + x - C$, and a and b are also the roots of $F(x)$.

Since the polynomial $F(x)$ has the same degree as $P(x)$, both being polynomials of degree n, there are at most n distinct integer roots.

Note. There are also several similar problems:

- **(All-Russian Mathematical Olympiad 2021, Grade 10, Problem 6).** Given a real-coefficient polynomial $P(x)$ of degree $n > 1$, it is known that the equation $P(P(P(x))) = P(x)$ has exactly n^3 distinct real roots. Prove that these n^3 roots can be divided into two groups such that the arithmetic means of the roots in each group are equal.

- **(Japan Team Selection Test 2019, Problem 9).** Let $P(x)$ be a rational coefficient polynomial and suppose $P(P(x))$ and $P(P(P(x)))$ are both integer coefficient polynomials. Prove that $P(x)$ is an integer coefficient polynomial.

- **(All-Russian Mathematical Olympiad 2018, Grade 11, Problem 1).** Given a polynomial $P(x)$ such that $P(P(x))$ and $P(P(P(x)))$ are strictly monotone on the whole real axis, prove that $P(x)$ is also strictly monotone on the whole real axis.

- **(All-Russian Mathematical Olympiad 2002, 4th Round, Grade 11, Problem 5).** Let $P(x)$ be a polynomial of an odd degree. Prove that the equation $P(P(x)) = 0$ has at least as many different real roots as the equation $P(x) = 0$ does.

- **(Romania Team Selection Test 2000, Problem 11).** Suppose P, Q are monic complex polynomials such that $P(P(x)) = Q(Q(x))$. Prove that $P = Q$.

- **(William Lowell Putnam Mathematical Competition 2000, A6).** Let $f(x)$ be a polynomial with integer coefficients. Define a sequence $a_0, a_1, \ldots$ of integers such that $a_0 = 0$ and $a_{n+1} = f(a_n)$ for all $n \geq 0$. Prove that if there exists a positive integer m for which $a_m = 0$, then either $a_1 = 0$ or $a_2 = 0$.

- **(Turkey Mathematical Olympiad 2000, 2nd Round, Problem 6).** Find all continuous functions $f : [0, 1] \to [0, 1]$ for which there exists a positive integer n such that $f^n(x) = x$ for $x \in [0, 1]$, where $f^0(x) = x$ and $f^{k+1}(x) = f(f^k(x))$ for every positive integer k.

- **(United States of America Mathematical Olympiad 1974, Problem 1).** Let a, b, c denote three distinct integers and P denote a polynomial with integer coefficients. Show that it is impossible that $P(a) = b$, $P(b) = c$, and $P(c) = a$.

【Score Situation】 This particular problem saw the following distribution of scores among contestants: 48 contestants scored 7 points, 2 contestants scored 6 points, 5 contestants scored 5 points, 6 contestants scored 4 points, 25 contestants scored 3 points, 8 contestants scored 2 points, 101 contestants scored 1 point, and 303 contestants scored 0 point. The average score for this problem is 1.183, indicating that it was relatively challenging.

Among the top five teams in the team scores, the scores of this problem are as follows: the China team scored 38 points (with a total team score of 214 points), the Russia team scored 28 points (with a total team score of 174 points), the South Korea team scored 29 points (with a total team score of 170 points), the Germany team scored 32 points (with a total team score of 157 points), and the United States team scored 20 points (with a total team score of 154 points).

The gold medal cutoff for this IMO was set at 28 points (with 42 contestants earning gold medals), the silver medal cutoff was 19 points (with 89 contestants earning silver medals), and the bronze medal cutoff was 15 points (with 122 contestants earning bronze medals).

In this IMO, only three contestants achieved a perfect score of 42 points, namely Zhiyu Liu from China, Iurie Boreico from Moldova, and Alexander Magazinov from Russia.

Problem 5.9 (IMO 48-6, proposed by the Netherlands). Let n be a positive integer. Consider

$$S = \{(x, y, z) : x, y, z \in \{0, 1, \ldots, n\}, x + y + z > 0\}$$

as a set of $(n+1)^3 - 1$ points in a three-dimensional space. Determine the smallest possible number of planes, the union of which contains S but does not include $(0, 0, 0)$.

Solution. It is easy to see that $3n$ planes satisfy the conditions. Take the planes $x = i$, $y = i$, and $z = i$ for $i = 1, 2, \ldots, n$.

Suppose there are m planes $a_i x + b_i y + c_i z - d_i = 0$ satisfying the conditions, where $1 \le i \le m$ and $d_i \ne 0$. Then the polynomial of degree m

$$f(x, y, z) = \prod_{i=1}^{m} (a_i x + b_i y + c_i z - d_i)$$

is zero at every point in S but not at the origin, i.e., $f(0, 0, 0) \ne 0$. Define the difference operator $\Delta \in \{\Delta_x, \Delta_y, \Delta_z\}$ as follows:

$$\Delta_x f(x, y, z) = f(x + 1, y, z) - f(x, y, z),$$
$$\Delta_y f(x, y, z) = f(x, y + 1, z) - f(x, y, z),$$
$$\Delta_z f(x, y, z) = f(x, y, z + 1) - f(x, y, z).$$

For every positive integer k, let Δ^k represent the mth order difference of $f(x, y, z)$.

If $m < 3n$, then $\Delta_x^n \Delta_y^n \Delta_z^n f(x,y,z) \equiv 0$. Using the difference formula for univariate polynomials, we have:

$$\Delta^n P(x) = \sum_{i=0}^{n} (-1)^{n-i} C_n^i P(x+i),$$

so $\Delta_x^n \Delta_y^n \Delta_z^n f(x,y,z) = \sum_{(i,j,k) \in (S \bigcup (0,0,0))} (-1)^{3n-i-j-k} C_n^i C_n^j C_n^k f(x + i, y+j, z+k) \equiv 0$.

Setting $x = y = z = 0$, and knowing that $f(i,j,k) = 0$ for $i,j,k \in \{0,1,\dots,n\}$ with $i+j+k > 0$, we get

$$f(0,0,0) = \sum_{(i,j,k) \in S} (-1)^{i+j+k+1} C_n^i C_n^j C_n^k f(i,j,k) = 0,$$

which contradicts $f(0,0,0) \neq 0$. Therefore, $m \geq 3n$.

Note. For a non-zero polynomial $F(x_1, x_2, \dots, x_k)$ with $F(0,0,\dots,0) \neq 0$, if all points $(x_1, x_2, \dots, x_k)$ satisfying $x_1 + x_2 + \cdots + x_k > 0$ are zeros of $F(x_1, x_2, \dots, x_k)$, where $x_1, x_2, \dots, x_k \in \{0,1,\dots,n\}$, then $\deg F \geq kn$.

Furthermore, this problem is related to Alon's papers from 1993 and 1999. Theorem 4 in the 1993 paper "Covering the Cube by Affine Hyperplanes" is:

Let $\mathbf{F}$ be an arbitrary field, let $S_1, S_2, \dots, S_n$ be non-empty subsets of $\mathbf{F}$, $|S_i| = s_i$, and let B be the set $S_1 \times S_2 \times \cdots \times S_n$. If m hyperplanes do not cover B completely, then they miss at least

$$M\left(s, \left(\sum_{i=1}^{n} s_i\right) - m\right)$$

points of B.

For a sequence of positive integers $s = (s_1, s_2, \dots, s_n)$, let $M(s,l)$ denote the minimum of the product of n positive integers $y_i \leq s_i$ the sum of which is at least l. For $l \leq n$, we have $M = 1$, and for $l \geq \sum_{i=1}^{n} s_i$ we define $M = \prod_{i=1}^{n} s_i$.

Theorem 1.2 in the 1999 paper "Combinatorial Nullstellensatz" is:

Let $\mathbf{F}$ be an arbitrary field, and let $f = f(x_1, x_2, \dots, x_n)$ be a polynomial in $\mathbf{F}[x_1, x_2, \dots, x_n]$. Suppose the degree $\deg(f)$ of f is $\sum_{i=1}^{n} t_i$, where each t_i is a non-negative integer, and suppose the coefficient of $\prod_{i=1}^{n} x_i^{t_i}$ in f is nonzero. If $S_1, S_2, \dots, S_n$ are subsets of $\mathbf{F}$ with $|S_i| > t_i$, then there are $s_1 \in S_1, s_2 \in S_2, \dots, s_n \in S_n$ such that $f(s_1, s_2, \dots, s_n) \neq 0$.

There is a similar problem:

- (**All-Russian Mathematical Olympiad 2010, Grade 11, Problem 4**). Given an integer $n \geq 3$, what is the smallest possible value of k if the following statements are true?
  For any n points $A_i(x_i, y_i)(1 \leq i \leq n)$ on a plane, where no three points are colinear, and for any n real numbers $c_i(1 \leq i \leq n)$, there exists a polynomial $P(x, y)$ of degree no higher than k such that
  $$P(x_i, y_i) = c_i \quad \text{for } i = 1, 2, \ldots, n.$$

Note. A function in the following form is called a bivariate polynomial:
$$P(x, y) = a_{0,0} + a_{1,0}x + a_{0,1}y + a_{2,0}x^2 + a_{1,1}xy + a_{0,2}y^2 + \cdots$$
$$+ a_{k,0}x^k + a_{k-1,1}x^{k-1}y + \cdots + a_{0,k}y^k,$$
where the degree of non-zero monomial $a_{i,j}x^iy^j$ is $i+j$, and the degree of polynomial $P(x, y)$ is the greatest degree of the degrees of its monomials.

【Score Situation】 This particular problem saw the following distribution of scores among contestants: 5 contestants scored 7 points, no contestant scored 6 points, no contestant scored 5 points, no contestant scored 4 points, no contestant scored 3 points, 2 contestants scored 2 points, 40 contestants scored 1 point, and 473 contestants scored 0 point. The average score for this problem is 0.152, indicating that it was extremely difficult.

Among the top five teams in the team scores, the scores of this problem are as follows: the Russia team scored 9 points (with a total team score of 184 points), the China team scored 3 points (with a total team score of 181 points), the South Korea team scored 2 points (with a total team score of 168 points), the Vietnam team scored 0 point (with a total team score of 168 points), and the United States team scored 4 points (with a total team score of 155 points).

The gold medal cutoff for this IMO was set at 29 points (with 39 contestants earning gold medals), the silver medal cutoff was 21 points (with 83 contestants earning silver medals), and the bronze medal cutoff was 14 points (with 131 contestants earning bronze medals).

In this IMO, no contestant achieved a perfect score of 42 points.

Problem 5.10 (IMO 62-6, proposed by Austria). Let $m \geq 2$ be an integer, A be a finite set of (not necessarily positive) integers, and $B_1, B_2, B_3, \ldots, B_m$ be subsets of A. Assume that for each $k = 1, 2, \ldots, m$ the sum of the elements of B_k is m^k. Prove that A contains at least $\frac{m}{2}$ elements.

Proof. Denote $A = \{a_1, \ldots, a_n\}$. By contradiction, assume $n = |A| < \frac{m}{2}$. Note that the following m^m sums are distinct:

$$f(c_1, \ldots, c_m) := c_1 m + c_2 m^2 + \cdots + c_m m^m,$$

where $c_j \in \{0, 1, \ldots, m-1\}$ for $j = 1, 2, \ldots, m$. (Similar to the expansion in base m.)

However, by replacing each m^j ($1 \leq j \leq m$) with the sum of elements in B_j, we can rewrite $f(c_1, \ldots, c_m)$ as the following form of the sum of elements:

$$\alpha_1 a_1 + \alpha_2 a_2 \cdots + \alpha_n a_n,$$

where each $\alpha_i \in \{0, 1, \ldots, m(m-1)\}$. There are only

$$(m(m-1)+1)^n < m^{2n} < m^m$$

such distinct expressions, which contradicts the earlier statement that all $f(c_1, \ldots, c_m)$ are distinct. Therefore, $n \geq \frac{m}{2}$.

Note. The problem statement also holds if A is a set of real numbers.

【Score Situation】This particular problem saw the following distribution of scores among contestants: 37 contestants scored 7 points, no contestant scored 6 points, 2 contestants scored 5 points, 1 contestant scored 4 points, 3 contestants scored 3 points, 2 contestants scored 2 points, 12 contestants scored 1 point, and 562 contestants scored 0 point. The average score for this problem is 0.481, indicating that it was extremely difficult.

Among the top five teams in the team scores, the scores of this problem are as follows: the China team scored 42 points (with a total team score of 208 points), the Russia team scored 28 points (with a total team score of 183 points), the South Korea team scored 36 points (with a total team score of 172 points), the United States team scored 29 points (with a total team score of 165 points), and the Canada team scored 14 points (with a total team score of 151 points).

The gold medal cutoff for this IMO was set at 24 points (with 52 contestants earning gold medals), the silver medal cutoff was 19 points (with 103 contestants earning silver medals), and the bronze medal cutoff was 12 points (with 148 contestants earning bronze medals).

In this IMO, only one contestant achieved a perfect score of 42 points, namely Yichuan Wang from China.

5.3 Summary

Trigonometry originated from the needs of navigation, astronomy, and similar fields, and later evolved into an independent branch of mathematics. Polynomial, on the other hand, is a classical subject in algebra. Therefore, this chapter also introduces some knowledge of trigonometric identities and polynomials.

In the first 64 IMOs, there were a total of 10 other algebra problems. These problems can be broadly categorized into three types, as depicted in Figure 5.2. The score details for these problems are presented in Table 5.2. Due to the smaller number of participating teams and missing contestant score information in early IMOs, there are several blanks in Table 5.2.

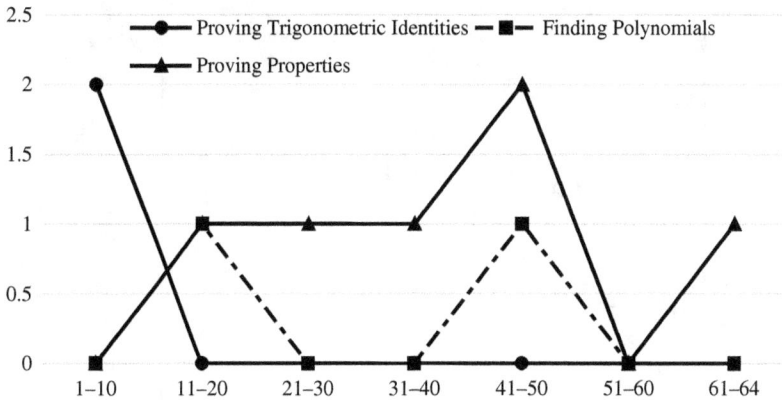

Figure 5.2 Numbers of Other Algebra Problems in the First 64 IMOs.

Problems 5.1–5.2 focus on "proving trigonometric identities;" among these two problems, the one with the lowest average score is Problem 5.1 (IMO 5-5), proposed by the German Democratic Republic. Problems 5.3–5.4 deal with "finding polynomials;" among these two problems, the one with the lowest average score is Problem 5.4 (IMO 45-2), proposed by South Korea. Problems 5.5–5.10 are about "proving properties of polynomials and sets;" among these six problems, the one with the lowest average score is Problem 5.9 (IMO 48-6), proposed by the Netherlands.

These 10 problems were proposed by nine countries, and the Netherlands contributed two problems.

Table 5.2 Score Details of Other Algebra Problems in the First 64 IMOs

Problem	5.1	5.2	5.3	5.4	5.5	5.6	5.7	5.8
Full points	6.000	5.000	8.000	7.000	8.000	7.000	7.000	7.000
Average score	2.375	4.704	2.797	2.761	3.100	0.785	2.027	1.183
Top five mean			4.425	6.000	5.750	2.633	5.267	4.900
6th–15th mean			2.329	5.350	2.416	1.150	4.636	2.682
16th–25th mean				3.667		0.167	2.463	2.296
Problem number in IMO	5-5	8-4	17-6	45-2	16-6	26-3	34-1	47-5
Proposing country	The German Democratic Republic	Yugoslavia	The United Kingdom	South Korea	Sweden	The Netherlands	Ireland	Romania

Problem	5.9	5.10
Full points	7.000	7.000
Average score	0.152	0.481
Top five mean	0.600	4.967
6th–15th mean	0.467	1.317
16th–25th mean	0.367	0.600
Problem number in IMO	48-6	62-6
Proposing country	The Netherlands	Austria

Note. Top five mean = Total score of the top five teams ÷ Total number of contestants from the top five teams,

    6th–15th mean = Total score of the 6th–15th teams ÷ Total number of contestants from the 6th–15th teams,

    16th–25th mean = Total score of the 16th–25th teams ÷ Total number of contestants from the 16th–25th teams.

From Table 5.2, it can be observed that in the first 64 IMOs, there were three other algebra problems with an average score of 0–1 point; one problem with an average score of 1–2 points; four problems with an average score of 2–3 points; one problem with an average score of 3–4 points; one problem with an average score above 4 points. Overall, the other algebra problems were relatively difficult.

In the 24th–64th IMOs, there were a total of six other algebra problems. Among these, three had an average score of 0–1 point; one had an average score of 1–2 points; two had an average score of 2–3 points; no problem had an average score of 3–4 points; no problem had an average score above 4 points. Further analysis of the problem numbers of these six other algebra problems, as shown in Table 5.3, reveals that these problems frequently appeared as the 3rd/6th problem. The majority of these problems, totaling five, were of the type proving properties of polynomials and sets.

Table 5.3 Numbers of Other Algebra Problems in the 24th–64th IMOs

Other Algebra Problems	Problem Number			Number of Problems in the First 64 IMOs
	1, 4	2, 5	3, 6	
Proving trigonometric identities	0	0	0	2
Finding polynomials	0	1	0	2
Proving properties	1	1	3	6
Total	1	2	3	10

From Table 5.2, it can be observed that in other algebra problems, the average score of the top five teams is generally 2.5 points higher than the average score of the problem. Meanwhile, the average scores of the 6th–15th teams and the 16th–25th teams are both within 1 point of the average score of the problem.

Appendix A

IMO General Information

Session	Year	Host	Number of Participating Teams	Number of Contestants	Gold Silver Bronze (Cutoffs/Numbers of Medalists)		
IMO 1	1959	Romania	7	52	37/3	36/3	33/5
IMO 2	1960	Romania	5	39	40/4	37/4	33/4
IMO 3	1961	Hungary	6	48	37/3	34/4	30/4
IMO 4	1962	Czechoslovakia	7	56	41/4	34/12	29/15
IMO 5	1963	Poland	8	64	35/7	28/11	21/17
IMO 6	1964	The Union of Soviet Socialist Republics	9	72	38/7	31/9	27/19
IMO 7	1965	The German Democratic Republic	10	80	38/8	30/12	20/17
IMO 8	1966	Bulgaria	9	72	39/13	34/15	31/11
IMO 9	1967	Yugoslavia	13	99	38/11	30/14	22/26
IMO 10	1968	The Union of Soviet Socialist Republics	12	96	39/22	33/22	26/20
IMO 11	1969	Romania	14	112	40/3	30/20	24/21
IMO 12	1970	Hungary	14	112	37/7	30/11	19/40

(*Continued*)

(*Continued*)

Session	Year	Host	Number of Participating Teams	Number of Contestants	Gold (Cutoffs/Numbers of Medalists)	Silver	Bronze
IMO 13	1971	Czechoslovakia	15	115	35/7	23/12	11/29
IMO 14	1972	Poland	14	107	40/8	30/16	19/30
IMO 15	1973	The Union of Soviet Socialist Republics	16	125	35/5	27/15	17/48
IMO 16	1974	The German Democratic Republic	18	140	38/10	30/24	23/37
IMO 17	1975	Bulgaria	17	135	38/8	32/25	23/36
IMO 18	1976	Austria	18	139	34/9	23/28	15/45
IMO 19	1977	Yugoslavia	21	155	34/13	24/29	17/35
IMO 20	1978	Romania	17	132	35/5	27/20	22/38
IMO 21	1979	The United Kingdom	23	166	37/8	29/32	20/42
IMO 22	1981	The United States of America	27	185	41/36	34/37	26/30
IMO 23	1982	Hungary	30	119	37/10	30/20	21/31
IMO 24	1983	France	32	186	38/9	26/27	15/57
IMO 25	1984	Czechoslovakia	34	192	40/14	26/35	17/49
IMO 26	1985	Finland	38	209	34/14	22/35	15/52
IMO 27	1986	Poland	37	210	34/18	26/41	17/48
IMO 28	1987	Cuba	42	237	42/22	32/42	18/56
IMO 29	1988	Australia	49	268	32/17	23/48	14/65
IMO 30	1989	Germany	50	291	38/20	30/55	18/72
IMO 31	1990	The People's Republic of China	54	308	34/23	23/56	16/76
IMO 32	1991	Sweden	56	318	39/20	31/51	19/84
IMO 33	1992	The Russian Federation	56	322	32/26	24/55	14/74

(*Continued*)

Session	Year	Host	Number of Participating Teams	Number of Contestants	Gold	Silver	Bronze
					(Cutoffs/Numbers of Medalists)		
IMO 34	1993	Turkey	73	413	30/35	20/66	11/97
IMO 35	1994	Chinese Hong Kong	69	385	40/30	30/64	19/98
IMO 36	1995	Canada	73	412	37/30	29/71	19/100
IMO 37	1996	India	75	424	28/35	20/66	12/99
IMO 38	1997	Argentina	82	460	35/39	25/70	15/122
IMO 39	1998	Chinese Taiwan	76	419	31/37	24/66	14/102
IMO 40	1999	Romania	81	450	28/38	19/70	12/118
IMO 41	2000	Republic of Korea	82	461	30/39	21/71	11/119
IMO 42	2001	The United States of America	83	473	30/39	20/81	11/122
IMO 43	2002	The United Kingdom	84	479	29/39	23/73	14/120
IMO 44	2003	Japan	82	457	29/37	19/69	13/104
IMO 45	2004	Greece	85	486	32/45	24/78	16/120
IMO 46	2005	Mexico	91	513	35/42	23/79	12/128
IMO 47	2006	Slovenia	90	498	28/42	19/89	15/122
IMO 48	2007	Vietnam	93	520	29/39	21/83	14/131
IMO 49	2008	Spain	97	535	31/47	22/100	15/120
IMO 50	2009	Germany	104	565	32/49	24/98	14/135
IMO 51	2010	Kazakhstan	95	522	27/47	21/103	15/115
IMO 52	2011	The Netherlands	101	563	28/54	22/90	16/137
IMO 53	2012	Argentina	100	547	28/51	21/88	14/137
IMO 54	2013	Colombia	97	527	31/45	24/92	15/141
IMO 55	2014	South Africa	101	560	29/49	22/113	16/133
IMO 56	2015	Thailand	104	577	26/39	19/100	14/143
IMO 57	2016	Chinese Hong Kong	109	602	29/44	22/101	16/135

(*Continued*)

(*Continued*)

Session	Year	Host	Number of Participating Teams	Number of Contestants	Gold Silver Bronze (Cutoffs/Numbers of Medalists)		
IMO 58	2017	Brazil	111	615	25/48	19/90	16/153
IMO 59	2018	Romania	107	594	31/48	25/98	16/143
IMO 60	2019	The United Kingdom	112	621	31/52	24/94	17/156
IMO 61	2020	The Russian Federation	105	616	31/49	24/112	15/155
IMO 62	2021	The Russian Federation	107	619	24/52	19/103	12/148
IMO 63	2022	Norway	104	589	34/44	29/101	23/140
IMO 64	2023	Japan	112	618	32/54	25/90	18/170

Appendix B

IMO Algebra Problem Index

Problem Number in the IMO	Proposing Country	Category	Problem Number in the Book	Page Number
IMO 1-2	Romania	Finding Solutions, Equations	Problem 1.1	31
IMO 1-3	Hungary	Investigating Conditions, Equations	Problem 1.12	53
IMO 2-2	Hungary	Solving Inequalities, Inequalities	Problem 4.1	237
IMO 3-1	Hungary	Investigating Conditions, Equations	Problem 1.13	54
IMO 3-2	Poland	Proving Inequalities, Inequalities	Problem 4.6	244
IMO 3-3	Bulgaria	Finding Solutions, Equations	Problem 1.4	38
IMO 4-2	Hungary	Solving Inequalities, Inequalities	Problem 4.2	238
IMO 4-4	Romania	Finding Solutions, Equations	Problem 1.2	32
IMO 5-1	Czechoslovakia	Finding Solutions, Equations	Problem 1.5	39
IMO 5-4	The Soviet Union	Finding Solutions, Equations	Problem 1.6	40

(Continued)

(Continued)

Problem Number in the IMO	Proposing Country	Category	Problem Number in the Book	Page Number
IMO 5-5	The German Democratic Republic	Proving Trigonometric Identities, Others	Problem 5.1	336
IMO 6-2	Hungary	Proving Inequalities, Inequalities	Problem 4.7	248
IMO 7-1	Yugoslavia	Solving Inequalities, Inequalities	Problem 4.3	239
IMO 7-2	Poland	Proving Relationships, Equations	Problem 1.8	43
IMO 7-4	The Soviet Union	Finding Solutions, Equations	Problem 1.3	33
IMO 8-4	Yugoslavia	Proving Trigonometric Identities, Others	Problem 5.2	339
IMO 8-5	Czechoslovakia	Finding Solutions, Equations	Problem 1.7	42
IMO 9-5	The Soviet Union	Determining Values, Sequences	Problem 3.1	171
IMO 10-3	Bulgaria	Proving Relationships, Equations	Problem 1.9	45
IMO 10-5	The German Democratic Republic	Proving Properties, Functions	Problem 2.1	92
IMO 11-2	Hungary	Proving Properties, Functions	Problem 2.2	94
IMO 11-6	The Soviet Union	Proving Inequalities, Inequalities	Problem 4.8	250
IMO 12-3	Sweden	Existence Problems, Sequences	Problem 3.5	180
IMO 13-1	Hungary	Proving Inequalities, Inequalities	Problem 4.9	253
IMO 14-4	The Netherlands	Solving Inequalities, Inequalities	Problem 4.4	240

(Continued)

Problem Number in the IMO	Proposing Country	Category	Problem Number in the Book	Page Number
IMO 14-5	Bulgaria	Proving Properties, Functions	Problem 2.3	96
IMO 15-3	Sweden	Investigating Conditions, Equations	Problem 1.14	57
IMO 15-5	Poland	Proving Properties, Functions	Problem 2.4	98
IMO 15-6	Sweden	Existence Problems, Sequences	Problem 3.6	182
IMO 16-5	The Netherlands	Determining Value Ranges, Inequalities	Problem 4.31	304
IMO 16-6	Sweden	Proving Properties, Others	Problem 5.5	344
IMO 17-1	Czechoslovakia	Proving Inequalities, Inequalities	Problem 4.10	256
IMO 17-6	The United Kingdom	Finding Polynomials, Others	Problem 5.3	341
IMO 18-2	Finland	Proving Relationships, Equations	Problem 1.10	48
IMO 18-5	The Netherlands	Proving Relationships, Equations	Problem 1.11	51
IMO 18-6	The United Kingdom	Proving Quantitative Relationships, Sequences	Problem 3.10	194
IMO 19-4	The United Kingdom	Proving Inequalities, Inequalities	Problem 4.11	257
IMO 19-6	Bulgaria	Proving Properties, Functions	Problem 2.5	100
IMO 20-3	The United Kingdom	Determining Values, Functions	Problem 2.10	111
IMO 20-5	France	Proving Inequalities, Inequalities	Problem 4.12	259

(Continued)

(*Continued*)

Problem Number in the IMO	Proposing Country	Category	Problem Number in the Book	Page Number
IMO 21-5	Israel	Investigating Conditions, Equations	Problem 1.15	59
IMO 22-6	Finland	Determining Values, Functions	Problem 2.11	114
IMO 23-1	The United Kingdom	Determining Values, Functions	Problem 2.12	116
IMO 23-3	The Soviet Union	Existence Problems, Sequences	Problem 3.7	184
IMO 24-1	The United Kingdom	Deriving Expressions, Functions	Problem 2.14	120
IMO 24-6	The United States	Proving Inequalities, Inequalities	Problem 4.13	261
IMO 25-1	Germany	Proving Inequalities, Inequalities	Problem 4.14	262
IMO 26-3	The Netherlands	Proving Properties, Others	Problem 5.6	346
IMO 26-6	Sweden	Proving Quantitative Relationships, Sequences	Problem 3.11	196
IMO 27-5	The United Kingdom	Deriving Expressions, Functions	Problem 2.15	122
IMO 28-3	Germany	Proving Inequalities, Inequalities	Problem 4.15	265
IMO 28-4	Vietnam	Proving Properties, Functions	Problem 2.6	101
IMO 29-3	The United Kingdom	Determining Values, Functions	Problem 2.13	118
IMO 29-4	Ireland	Solving Inequalities, Inequalities	Problem 4.5	242

(*Continued*)

(*Continued*)

Problem Number in the IMO	Proposing Country	Category	Problem Number in the Book	Page Number
IMO 31-4	Turkey	Deriving Expressions, Functions	Problem 2.16	123
IMO 32-6	The Netherlands	Existence Problems, Sequences	Problem 3.8	187
IMO 33-2	India	Deriving Expressions, Functions	Problem 2.17	125
IMO 34-1	Ireland	Proving Properties, Others	Problem 5.7	348
IMO 34-5	Germany	Proving Properties, Functions	Problem 2.7	104
IMO 35-1	France	Proving Inequalities, Inequalities	Problem 4.16	267
IMO 35-5	The United Kingdom	Deriving Expressions, Functions	Problem 2.18	128
IMO 36-2	Russia	Proving Inequalities, Inequalities	Problem 4.17	268
IMO 36-4	Poland	Determining Values, Sequences	Problem 3.2	173
IMO 37-6	France	Proving Quantitative Relationships, Sequences	Problem 3.12	199
IMO 38-3	Russia	Proving Inequalities, Inequalities	Problem 4.18	270
IMO 40-2	Poland	Determining Value Ranges, Inequalities	Problem 4.32	307
IMO 40-6	Japan	Deriving Expressions, Functions	Problem 2.19	129
IMO 41-2	The United States	Proving Inequalities, Inequalities	Problem 4.19	272

(*Continued*)

(*Continued*)

Problem Number in the IMO	Proposing Country	Category	Problem Number in the Book	Page Number
IMO 42-2	South Korea	Proving Inequalities, Inequalities	Problem 4.20	274
IMO 43-5	India	Deriving Expressions, Functions	Problem 2.20	130
IMO 44-5	Ireland	Proving Inequalities, Inequalities	Problem 4.21	279
IMO 45-2	South Korea	Finding Polynomials, Others	Problem 5.4	343
IMO 45-4	South Korea	Proving Inequalities, Inequalities	Problem 4.22	281
IMO 46-3	South Korea	Proving Inequalities, Inequalities	Problem 4.23	285
IMO 47-3	Ireland	Determining Value Ranges, Inequalities	Problem 4.33	309
IMO 47-5	Romania	Proving Properties, Others	Problem 5.8	351
IMO 48-1	New Zealand	Proving Inequalities, Inequalities	Problem 4.24	288
IMO 48-6	The Netherlands	Proving Properties, Others	Problem 5.9	353
IMO 49-2	Austria	Proving Inequalities, Inequalities	Problem 4.25	290
IMO 49-4	South Korea	Deriving Expressions, Functions	Problem 2.21	133
IMO 50-3	The United States	Proving Quantitative Relationships, Sequences	Problem 3.13	200
IMO 50-5	France	Deriving Expressions, Functions	Problem 2.22	134

(*Continued*)

(*Continued*)

Problem Number in the IMO	Proposing Country	Category	Problem Number in the Book	Page Number
IMO 51-1	France	Deriving Expressions, Functions	Problem 2.23	137
IMO 51-6	Iran	Proving Quantitative Relationships, Sequences	Problem 3.14	202
IMO 52-3	Belarus	Proving Properties, Functions	Problem 2.8	107
IMO 53-2	Australia	Proving Inequalities, Inequalities	Problem 4.26	294
IMO 53-4	South Africa	Deriving Expressions, Functions	Problem 2.24	138
IMO 54-5	Bulgaria	Proving Properties, Functions	Problem 2.9	109
IMO 55-1	Austria	Proving Inequalities, Inequalities	Problem 4.27	296
IMO 56-5	Albania	Deriving Expressions, Functions	Problem 2.25	140
IMO 57-5	Russia	Investigating Conditions, Equations	Problem 1.16	61
IMO 58-1	South Africa	Determining Values, Sequences	Problem 3.3	174
IMO 58-2	Albania	Deriving Expressions, Functions	Problem 2.26	142
IMO 59-2	Slovakia	Determining Values, Sequences	Problem 3.4	177
IMO 60-1	South Africa	Deriving Expressions, Functions	Problem 2.27	146
IMO 61-2	Belgium	Proving Inequalities, Inequalities	Problem 4.28	298

(*Continued*)

(Continued)

Problem Number in the IMO	Proposing Country	Category	Problem Number in the Book	Page Number
IMO 62-2	Canada	Proving Inequalities, Inequalities	Problem 4.29	300
IMO 62-6	Austria	Proving Properties, Others	Problem 5.10	355
IMO 63-2	The Netherlands	Deriving Expressions, Functions	Problem 2.28	147
IMO 64-3	Malaysia	Existence Problems, Sequences	Problem 3.9	191
IMO 64-4	The Netherlands	Proving Inequalities, Inequalities	Problem 4.30	303

www.ingramcontent.com/pod-product-compliance
Lightning Source LLC
Chambersburg PA
CBHW061233220326
41599CB00028B/5410